INSTRUCTOR'S SOLUTIONS MANUAL

CALCULUS
from Graphical, Numerical, and Symbolic Points of View

Volume 2

Arnold Ostebee
Paul Zorn
St. Olaf College

Saunders College Publishing

Harcourt Brace College Publishers

Fort Worth Philadelphia San Diego New York Orlando Austin
San Antonio Toronto Montreal London Sydney Tokyo

Copyright © 1997 by Harcourt Brace & Company

All rights reserved. No part of this publication may be reproduced or transmitted in any form or by any means, electronic or mechanical, including photocopy, recording, or any information storage and retrieval system, without permission in writing from the publisher.

Requests for permission to make copies of any part of the work should be mailed to: Permissions Department, Harcourt Brace & Company, 6277 Sea Harbor Drive, Orlando, Florida 32887-6777.

Portions of this work were published in previous editions.

Printed in the United States of America.

Ostebee & Zorn: Instructor's Solutions Manual to accompany *Calculus from Graphical, Numerical, and Symbolic Points of View, Vol 2,2E*.

ISBN 0-03-017413-9

9012 021 987654

5.1 Areas and Integrals

1. The area of the entire rectangle shown is only 250, so $\int_1^2 g(x)\,dx < 250$. Furthermore, since $g(x) > 0$ for all x, $\int_1^2 g(x)\,dx > 0$. Finally, it is clear from the picture that $\int_{1.75}^2 g(x)\,dx > 12.5$ (the area of one dotted rectangle). Thus, the only possible choice is $\int_1^2 g(x)\,dx \approx 45$.

2. (a) -65
   (b) -1.85

3. (a) Since $2 < f(x) < 5$ on $[1, 6]$, $10 < \int_1^6 f(x)\,dx < 25$. Thus, the best estimate for the value of the integral is 20. [NOTE: Be sure to take into account that the interval of integration is $[1, 6]$ and not $[0, 8]$!]
   (b) $A = 12$ and $B = 20$ since $3 \leq f(x) \leq 5$ if $3 \leq x \leq 7$. (Many other answers are possible.)
   (c) This approximation underestimates the exact value of the integral. (Count squares.)
   (d) $\frac{1}{2}\int_0^2 f(x)\,dx \approx \frac{11}{2}$
   (e) too large

4. (a) $\int_0^2 f(x)\,dx = 6$, $\int_1^4 f(x)\,dx = 45/2$, $\int_{-5}^{-1} f(x)\,dx = -36$, $\int_{-2}^3 f(x)\,dx = 15/2$
   (b) $\int_0^2 f(x)\,dx = 14$, $\int_1^4 f(x)\,dx = 30$, $\int_{-5}^{-1} f(x)\,dx = -4$, $\int_{-2}^3 f(x)\,dx = 30$
   (c) $\int_0^2 f(x)\,dx = 6$, $\int_1^4 f(x)\,dx = 0$, $\int_{-5}^{-1} f(x)\,dx = 44$, $\int_{-2}^3 f(x)\,dx = 20$

5. (a) $\int_0^2 f(x)\,dx = 4$ since this definite integral is the area of a trapezoid with base 2 and heights 1 and 3.
   (b) $\int_2^5 f(x)\,dx = 9\pi/4$ since this definite integral is the area of a quarter-circle with radius 3.
   (c) $\int_0^5 f(x)\,dx = \int_0^2 f(x)\,dx + \int_2^5 f(x)\,dx = 4 + 9\pi/4$
   (d) $\int_5^9 f(x)\,dx = -4\pi$ since this definite integral is the area of a quarter-circle with radius 4 that lies below the x-axis.
   (e) $\int_5^5 f(x)\,dx = 0$
   (f) $\int_0^{15} f(x)\,dx = \int_0^2 f(x)\,dx + \int_2^5 f(x)\,dx + \int_5^9 f(x)\,dx + \int_9^{15} f(x)\,dx$
   $= 4 + 9\pi/4 - 4\pi - 12 \approx -13.5$
   (g) $\int_0^{15} |f(x)|\,dx = \int_0^2 f(x)\,dx + \int_2^5 f(x)\,dx - \int_5^9 f(x)\,dx - \int_9^{15} f(x)\,dx$
   $= 4 + 9\pi/4 + 4\pi + 12 \approx 35.6$
   since $|f(x)| = -f(x)$ if $f(x) < 0$.
   (h) $\int_{15}^9 f(x)\,dx = -\int_9^{15} f(x)\,dx = 12$ since the second definite integral is the area of a triangle with base 6 and height 4 that lies below the x-axis.
   (i) $\int_{12}^{15} f(x)\,dx = -3$
   (j) $\int_9^{12} f(x)\,dx = -9$ since this definite integral is the area of a trapezoid with base 3 and heights 4 and 2 that lies below the x-axis.

6. (a) $\int_{-2}^5 (f(x) + g(x))\,dx = 23$
   (b) $\int_{-2}^5 (f(x) + h(x))\,dx = 7$
   (c) $\int_{-2}^5 (f(x) + g(x) - h(x))\,dx = 34$
   (d) $\int_{-2}^5 3f(x)\,dx = 54$
   (e) $\int_{-2}^5 -4g(x)\,dx = -20$

(f) $\int_5^{-2} f(x)\,dx = -18$

(g) $\int_{-2}^5 (h(x)+1)\,dx = -4$

(h) $\int_0^7 g(x-2)\,dx = \int_{-2}^5 g(x)\,dx = 5$

7. Jack's answer is too big — the area of the entire rectangle shown is only $\pi/2$. Since $\cos^8 x \geq 0$ on the entire interval $[0, \pi/2]$, the value of the integral must be positive. This rules out Ed's answer. Furthermore, the value of the integral is approximately the area of the triangle with vertices at $(0,0)$, $(0,1)$, and $(1,0)$. Since this triangle has area $1/2$, Lesley's answer is too big. Therefore, Joan's answer must be the correct answer.

   [NOTE: Since $\cos^8 x \leq 1$ if $0 \leq x \leq 1/2$ and $\cos^8 x \leq 2/5$ if $1/2 \leq x \leq 1$, $\int_0^1 \cos^8 x\,dx = \int_0^{1/2} \cos^8 x\,dx + \int_{1/2}^1 \cos^8 x\,dx \leq 1/2 + 1/5 = 7/10.$]

8. $\int_6^{10} h(x)\,dx < h'(5) < \frac{1}{10}(h(10)-h(0)) < \frac{1}{10}\int_0^{10} h(x)\,dx < \int_0^{10} h(x)\,dx < \int_0^5 h(x)\,dx$ (i.e., (vi) < (i) < (iii) < (ii) < (iv) < (v))

9. (a) $h'(5) \approx -2.2$

   (b) $\frac{1}{10}\int_0^{10} h(x)\,dx \approx 17/10 = 1.7$

   (c) $\frac{1}{10}(h(10)-h(0)) = -4/10 = -0.4$

   (d) $\int_0^{10} h(x)\,dx \approx 17$

   (e) $\int_0^5 h(x)\,dx \approx 20$

   (f) $\int_6^{10} h(x)\,dx \approx -4.7$

10. (a) After 10 seconds, the object was approximately 17 meters east of its starting point.

    (b) The object's average velocity over the time interval $[0, 10]$ was approximately 1.7 meters/second.

    (c) The object's average acceleration over the time interval $[0, 10]$ was approximately -0.4 meters/second^2.

    (d) The object's speed at time $t = 8$ seconds was approximately 1.5 meters/second.

    (e) Over the time interval $[0, 10]$, the object traveled a distance of approximately 26 meters.

    (f) The object's average speed over the time interval $[0, 10]$ was approximately 2.6 meters/second.

11. (a) $\int_0^4 v(t)\,dt = 80$. At time $t = 4$ the car is 80 miles east of its starting point.

    (b) The car's average (eastward) velocity is 20 mph $= \frac{1}{4}\int_0^4 v(t)\,dt$.

    (c) Since $s(t) = |v(t)|$, $\int_0^4 s(t)\,dt = 100$. Over the 4 hours, the car travels a total distance of 100 miles.

    (d) 100 miles/4 hours = 25 mph

12. (a) $y = -2$ and $y = -4x$ are two possibilities

    (b) $y = 8 - 20x$ is one possibility

13. (a) $\int_0^2 f(x)\,dx = 2$, $\int_1^4 f(x)\,dx = 3$, $\int_{-5}^{-1} f(x)\,dx = 4$, $\int_{-2}^3 f(x)\,dx = 5$

    (b) $\int_0^2 f(x)\,dx = 2a$, $\int_1^4 f(x)\,dx = 3a$, $\int_{-5}^{-1} f(x)\,dx = 4a$, $\int_{-2}^3 f(x)\,dx = 5a$

    (c) $\int_0^2 f(x)\,dx = -2a$, $\int_1^4 f(x)\,dx = -3a$, $\int_{-5}^{-1} f(x)\,dx = -4a$, $\int_{-2}^3 f(x)\,dx = -5a$

14. (a) The trapezoid with vertices $(2, 0)$, $(2, 3)$, $(4, 2)$, and $(4, 0)$ has area 5 and lies below the graph of f over the interval $[2, 4]$.

    (b) The trapezoid with verticles $(6, 0)$, $(6, -2)$, $(9, -4)$, and $(9, 0)$ has area 9.

    (c) Yes, since $\int_5^6 f(x)\,dx < -1.25$ (draw a triangle with vertices at $(5, 0)$, $(6, 0)$, and $(6, -2.5)$) and $\int_0^5 f(x)\,dx = 4 + 9\pi/4 \approx 11.069 < 11.25$.

Section 5.1 Areas and Integrals

(d) $\int_2^7 f(x)\,dx > 0$ because $\int_2^5 f(x)\,dx = 9\pi/4 > 7$ and $\int_5^7 f(x)\,dx > -6$ (draw a trapezoid with vertices at $(5, 0)$, $(5, -2)$, $(7, -4)$, and $(7, 0)$).

15. (a) $\int_{-3}^3 (x+2)\,dx = 12$

    (b) $\int_{-3}^3 |x+2|\,dx = 13$

    (c) $\int_{-3}^3 (|x|+2)\,dx = 21$

    (d) $\int_{-3}^3 (2-|x|)\,dx = 3$

16. $\int_0^2 f(x)\,dx = \int_0^1 f(x)\,dx + \int_1^2 f(x)\,dx = \frac{3}{2} + \frac{1}{2} = 2$

17. $\int_0^1 \sqrt{1-(x-1)^2}\,dx = \pi/4$
    [NOTE: The integral is the area of a quarter-circle with radius 1 and center at $(1, 0)$.]

18. $\int_1^3 \left(6 - \sqrt{4-(x-3)^2}\,dx\right) dx = 12 - \pi$
    [NOTE: $y = -\sqrt{4-(x-3)^2}$ is the bottom half of a circle with radius 2 and center at $(3, 6)$.]

19. $\int_0^3 f(x)\,dx = 3 + 3\pi/4$

20. The regions whose areas are represented by the three integrals are horizontal translates of each other.

21. (a) $\int_1^4 f(x)\,dx = \int_1^2 f(x)\,dx + \int_2^4 f(x)\,dx = -1 + 7 = 6$

    (b) $\int_0^4 3f(x)\,dx = 3\left(\int_0^2 f(x)\,dx + \int_2^4 f(x)\,dx\right) = 3 \cdot 9 = 27.$

    (c) $\int_0^1 f(x)\,dx = \int_0^2 f(x)\,dx - \int_1^2 f(x)\,dx = 2 - (-1) = 3.$

    (d) $\int_0^1 f(x+1)\,dx = \int_1^2 f(x)\,dx = -1$

    (e) $\int_0^1 (f(x)+1)\,dx = \int_0^1 f(x)\,dx + \int_0^1 dx = 3 + 1 = 4.$

    (f) $\int_2^4 f(x-2)\,dx = \int_0^2 f(x)\,dx = 2.$

    (g) $\int_2^4 (f(x)-2)\,dx = \int_2^4 f(x)\,dx - 2\int_2^4 dx = 7 - 2 \cdot 2 = 3.$

    (h) $-1 = \int_1^2 f(x)\,dx < 0$ implies that $f(x) < 0$ over some (or all) of the interval $[1, 2]$.

    (i) $6 = \int_2^4 3\,dx < \int_2^4 f(x)\,dx = 7$ implies that $f(x) > 3$ over some (or all) of the interval $[0, 2]$.

    (j) One possibility: $f(x) = \begin{cases} -8(x-1)-1 & 0 \le x \le 1 \\ -1 & 1 < x \le 2 \\ 9(x-2)-1 & x > 2 \end{cases}$

22. (a) If f is an *even* function, $f(x) = f(-x)$ so the signed area of the region enclosed by the graph of f between $x = -3$ and $x = 0$ is the same as the signed area of the region enclosed by the graph of f between $x = 0$ and $x = 3$. Thus, $\int_{-3}^3 f(x)\,dx = \int_{-3}^0 f(x)\,dx + \int_0^3 f(x)\,dx = 2\int_0^3 f(x)\,dx = 2 \cdot -1 = -2.$

    (b) If f is an *odd* function, $f(x) = -f(-x)$ so the signed area of the region enclosed by the graph of f between $x = -3$ and $x = 0$ has the same magnitude but the opposite sign as the signed area of the region enclosed by the graph of f between $x = 0$ and $x = 3$. Thus, $\int_{-3}^3 f(x)\,dx = \int_{-3}^0 f(x)\,dx + \int_0^3 f(x)\,dx = 0.$

23. (a) $\int_0^\pi \cos x\,dx = 0$

    (b) $\int_{\pi/2}^{3\pi/2} \sin x\,dx = 0$

    (c) $\int_{-2}^2 (7x^5 + 3)\,dx = 12$

    (d) $\int_{-1}^1 (4x^3 - 2x)\,dx = 0$

24. (a) $\int_0^{2\pi} \sin x \, dx = 0$

(b) $\int_0^{2\pi} |\sin x| \, dx = 4$

(c) $\int_0^{\pi} (1 + \sin x) \, dx = 2 + \pi$

(d) $\int_0^{\pi/2} \sin x \, dx = 1$

(e) $\int_{-\pi/2}^{\pi/2} \cos x \, dx = \int_{-\pi/2}^{\pi/2} \sin(x + \pi/2) \, dx = \int_0^{\pi} \sin x \, dx = 2$

(f) $\int_0^{\pi/2} (x + \cos x) \, dx = \int_0^{\pi/2} x \, dx + \int_0^{\pi/2} \cos x \, dx = 1 + \pi^2/4$

(g) $\int_0^{100\pi} \sin x \, dx = 50 \int_0^{2\pi} \sin x = 0$

(h) $\int_0^{100\pi} |\sin x| \, dx = 100 \int_0^{\pi} |\sin x| = 200$

(i) $\int_0^{100\pi} \cos x \, dx = 50 \int_0^{2\pi} \cos x \, dx = 0$

25. $\int_1^3 \frac{1-x}{x^2} dx = \int_1^2 \frac{1-x}{x^2} dx + \int_2^3 \frac{1-x}{x^2} dx$. Since the integrand $((1-x)/x^2)$ is negative over the interval $[2, 3]$, $\int_2^3 \frac{1-x}{x^2} dx < 0$. This implies that $\int_1^3 \frac{1-x}{x^2} dx < \int_1^2 \frac{1-x}{x^2} dx$.

26. No. Since $2x^3 - x^2 < 3x^2 + x$ if $0 < x \leq 2$, $\int_0^2 (2x^3 - x^2) \, dx < \int_0^2 (3x^2 + x) \, dx$.

27. (a) $1 + \cos x \geq 0$ for all x.

(b) Since $\int_0^{\pi} \cos x \, dx = \int_{\pi}^{2\pi} \cos x \, dx = 0$, $\int_0^{2\pi} (1 + \cos x) \, dx = \int_0^{2\pi} dx = 2\pi > 0$.

28. $0.442 < \frac{5\pi}{12} - \frac{\sqrt{3}}{2} \leq f(x) \leq \frac{\pi}{12} + \frac{\sqrt{3}}{2} < 1.128$ if $0 \leq x \leq 3$. Therefore,
$1.3 < 1.326 = 3 \cdot 0.442 \leq \int_0^3 f(x) \, dx \leq 3 \cdot 1.128 = 3.384 < 3.5$.

29. Let $f(x) = e^x \sin x$. Then, $2.25 < e \sin 1 \leq f(x) \leq e^{3\pi/4} \frac{\sqrt{2}}{2} < 7.5$ if $1 \leq x \leq 3$. Therefore,
$4.5 = 2 \cdot 2.25 \leq \int_1^3 f(x) \, dx \leq 2 \cdot 7.5 = 15$.

30. $0 \leq x \sin x \leq x$ if $0 \leq x \leq \pi$, so $0 \leq \int_0^{\pi} x \sin x \, dx \leq \int_0^{\pi} x \, dx = \pi^2/2$

31. Since $1/2 \leq \sin x \leq 1$ if $\pi/6 \leq x \leq \pi/2$, $\pi/6 = \frac{1}{2} \cdot (\pi/2 - \pi/6) \leq \int_{\pi/6}^{\pi/2} \sin x \, dx \leq 1 \cdot (\pi/2 - \pi/6) = \pi/3$.

32. Since $-1 \leq \cos x \leq -1/2$ if $2\pi/3 \leq x \leq \pi$, $-(\pi - 2\pi/3) \leq \int_{2\pi/3}^{\pi} \cos x \, dx \leq -\frac{1}{2}(\pi - 2\pi/3) = -\pi/6$

33. (a) If $0 \leq x \leq \pi$, then $0 \leq \sin x \leq 1 \implies 1 \leq e^{\sin x} \leq e$. Thus, $\pi \leq \int_0^{\pi} e^{\sin x} \, dx \leq \pi e$.
If $\pi \leq x \leq 2\pi$, $-1 \leq \sin x \leq 0 \implies 1/e \leq e^{\sin x} \leq 1$. Thus, $\pi/e \leq \int_{\pi}^{2\pi} e^{\sin x} \, dx \leq \pi$.
Since $\int_0^{2\pi} e^{\sin x} \, dx = \int_0^{\pi} e^{\sin x} \, dx + \int_{\pi}^{2\pi} e^{\sin x} \, dx$, the inequalities above lead to
$\pi(1 + 1/e) \leq \int_0^{2\pi} e^{\sin x} \, dx \leq \pi(1 + e)$.

(b) Since the integrand is periodic with period 2π, $25\pi(1 + 1/e) \leq \int_0^{50\pi} e^{\sin x} \, dx \leq 25\pi(1 + e)$.

34. (a) Since $1/2 \leq \cos x \leq 1$ if $0 \leq x \leq \pi/3$, $\pi/6 \leq \int_0^{\pi/3} \cos x \, dx \leq \pi/3$. Similarly, since $0 \leq \cos x \leq 1$ if $\pi/3 \leq x \leq \pi/2$, $0 \leq \int_{\pi/3}^{\pi/2} \cos x \, dx \leq \pi/12$. Thus, $\pi/6 \leq \int_0^{\pi/2} \cos x \, dx \leq \pi/3 + \pi/12 = 5\pi/12$.

Section 5.1 Areas and Integrals

(b) $\int_0^{\sqrt{\pi/2}} \cos x^2\, dx = \int_0^{\sqrt{\pi/3}} \cos x^2\, dx + \int_{\sqrt{\pi/3}}^{\sqrt{\pi/2}} \cos x^2\, dx$ so

$$\frac{1}{2}\sqrt{\frac{\pi}{3}} \le \int_0^{\sqrt{\pi/2}} \cos x^2\, dx \le \sqrt{\frac{\pi}{3}} + \frac{1}{2}\left(\sqrt{\frac{\pi}{2}} - \sqrt{\frac{\pi}{3}}\right) = \frac{1}{2}\left(\sqrt{\frac{\pi}{2}} + \sqrt{\frac{\pi}{3}}\right)$$

35. Let $f(x) = \sin(e^x)$. Then, $0.4 < \sin(e) \le f(x) \le 1$ if $0 \le x \le 1$. Therefore, $0.4 \le \int_0^1 f(x)\, dx \le 1$.

36. One function with the desired property is $f(x) = x - 2$. f has this property because it is continuous and takes on both positive and negative values over the interval of integration.

37. If $0 \le x \le \pi$, $\cos 1 \le \cos(\sin x) \le 1$. Therefore,
$$\frac{\pi}{2} < \cos 1 \cdot (\pi - 0) \le \int_0^\pi \cos(\sin x)\, dx \le 1 \cdot (\pi - 0) = \pi.$$

38. (a) Since the graph of the arctangent function is concave down, the secant line through the points $(0, \arctan 0) = (0, 0)$ and $(1, \arctan 1) = (1, \pi/4)$ lies below the curve $y = \arctan x$.

    (b) Let $f(x) = \arctan x$. Then $f(0) = 0$, $f'(x) = 1/(1 + x^2)$, and so $f'(0) = 1$. Therefore, the equation of the tangent line is $y = x$.

    (c) Let g be secant line from part (a) [$g(x) = \pi x/4$] and $h(x) = x$ be the tangent line from part (b). Then
    $g(x) \le \arctan x \le h(x)$ if $0 \le x \le 1$. Therefore, $\dfrac{\pi}{8} = \int_0^1 g(x)\, dx \le \int_0^1 \arctan x\, dx \le \int_0^1 h(x)\, dx = \dfrac{1}{2}$.

39. (a) Let $f(x) = \sin x - 2x/\pi$. Then $f'(x) = \cos x - 2/\pi$, so f has a single stationary point (a local maximum) in the interior of the interval $[0, 1]$. Since $f(0) = 0$ and $f(1) > 0$, $2x/\pi \le \sin x$ if $0 \le x \le 1$.
    Let $g(x) = x - \sin x$. Then $g'(x) = 1 - \cos x \ge 0$ if $0 \le x \le 1$. Since $g(0) = 0$, $\sin x \le x$ if $0 \le x \le 1$.

    (b) Since $\int_0^{\pi/4} x\, dx = \pi^2/32$, the results in part (a) lead directly to the desired inequalities.

40. (a) Let $f(x) = e^x - (1 + x)$ and $g(x) = 1 + 3x - e^x$. Then $f'(x) = e^x - 1 \ge 0$ and $g'(x) = 3 - 3^x \ge 0$ if $0 \le x \le 1$. Since $f(0) = g(0) = 0$, the inequalities $1 + x \le e^x \le 1 + 3x$ are valid if $0 \le x \le 1$.

    (b) Since $\int_0^1 x\, dx = 1/2$, the inequalities in part (a) imply that $3/2 \le \int_0^1 e^x\, dx \le 5/2$.

41. (a) Since $0 \le x^k f(x) \le f(x)$ if $0 \le x \le 1$ and $k \ge 0$ is an integer, $\int_0^1 x^k f(x)\, dx \le \int_0^1 f(x)\, dx$.

    (b) No. For example, let $f(x) = -1$. Then $-1 = \int_0^1 f(x)\, dx < \int_0^1 x f(x)\, dx = -1/2$.

42. $\int_0^{\pi/2} \cos x\, dx > \int_0^{\pi/4} \cos(2x)\, dx$. Since $\int_{\pi/4}^{\pi/2} \cos x\, dx > 0$ and $\cos(2x) \le \cos x$ if $0 \le x \le \pi/4$,
    $\int_0^{\pi/2} \cos x\, dx = \int_0^{\pi/4} \cos x\, dx + \int_{\pi/4}^{\pi/2} \cos x\, dx > \int_0^{\pi/4} \cos x\, dx \ge \int_0^{\pi/4} \cos(2x)\, dx$.

43. (a) $\int_0^{\pi/2} \cos^2 x\, dx = \int_0^{\pi/2} \sin^2(x - \pi/2)\, dx = \int_{-\pi/2}^0 \sin^2 x\, dx = \int_0^{\pi/2} \sin^2 x\, dx$

    (b) $\int_0^{\pi/2} \sin^2 x\, dx + \int_0^{\pi/2} \cos^2 x\, dx = \int_0^{\pi/2}\left(\sin^2 x + \cos^2 x\right) dx = \int_0^{\pi/2} dx = \dfrac{\pi}{2}$

    (c) $\int_0^{\pi/2} \sin^2 x\, dx + \int_0^{\pi/2} \cos^2 x\, dx = 2\int_0^{\pi/2} \sin^2 x\, dx = \dfrac{\pi}{2}$ so $\int_0^{\pi/2} \sin^2 x\, dx = \dfrac{\pi}{4}$.

44. $\sqrt{1 + \cos(2x)} = \sqrt{2\cos^2 x} = \sqrt{2}\cos x$ if $0 \le x \le \pi/2$.

45. Since f is an odd function, $\int_{-a}^{0} f(x)\,dx = -\int_{0}^{a} f(x)\,dx$. Thus, $\int_{-a}^{a} f(x)\,dx = 0$.

46. Since f is an even function, $\int_{-a}^{a} f(x)\,dx = 2\int_{0}^{a} f(x)\,dx$. Thus, the average value of f over the interval $[-a, a]$ is $\dfrac{\int_{-a}^{a} f(x)\,dx}{2a} = \dfrac{2\int_{0}^{a} f(x)\,dx}{2a} = \dfrac{\int_{0}^{a} f(x)\,dx}{a}$.

47. The bounds on f imply that $-4 \le \int_{1}^{3} f(x)\,dx \le 10$. Thus, $-2 \le \dfrac{\int_{1}^{3} f(x)\,dx}{2} \le 5$.

48. The information implies that $\int_{0}^{1} f(x)\,dx = 1 \cdot 2 = 2$ and $\int_{1}^{3} f(x)\,dx = 2 \cdot 4 = 8$. Thus, $\dfrac{\int_{0}^{3} f(x)\,dx}{3} = \dfrac{10}{3}$.

49. $\int_{-3}^{1} f(x)\,dx = 2 \cdot 4 = 8$ and $\int_{-3}^{7} f(x)\,dx = 10 \cdot 5 = 50$, so $\int_{1}^{7} f(x)\,dx = 42$. Therefore, the average value of f over the interval $[1, 7]$ is $42/6 = 7$.

50. Let $V(c) = \int_{a}^{b} (f(x) - c)^2\,dx = \int_{a}^{b} (f(x))^2\,dx - 2c\int_{a}^{b} f(x)\,dx + c^2 \int_{a}^{b} dx$. Now $V'(c) = -2\int_{a}^{b} f(x)\,dx + 2c(b-a)$ is zero if $c = \left(\int_{a}^{b} f(x)\,dx\right)/(b-a)$ (i.e., if c is the average value of f on the interval $[a, b]$). Since $V''(c) = 2(b-a) > 0$, this value of c corresponds to the minimum value of V.

51. Both integrals measure the area of the same region.

52. $-\int_{a}^{b} |f(x)|\,dx \le \int_{a}^{b} f(x)\,dx \le \int_{a}^{b} |f(x)|\,dx \implies \left|\int_{a}^{b} f(x)\,dx\right| \le \int_{a}^{b} |f(x)|\,dx$.

53. (a) No. Let $f(x) = 0$ and $g(x) = x$. Then $\int_{-1}^{1} f(x)\,dx = \int_{-1}^{1} g(x)\,dx = 0$ but $f(x) \ge g(x)$ if $-1 \le x \le 0$.

    (b) Yes. If $f(x) > g(x)$ for *every* x such that $a \le x \le b$, then $\int_{a}^{b} f(x)\,dx > \int_{a}^{b} g(x)\,dx$ would be true — a contradiction.

54. (a) $\int_{-a}^{a} x^2\,dx$ is the area of the region bounded by the lines $x = -a$, $x = a$, and $y = 0$, and the curve $y = x^2$. The area of this region is is the area of the rectangle bounded by the lines $x = -a$, $x = a$, and $y = 0$, and $y = a^2$ minus the area of the parabolic arch enclosed by the lines $x = -a$ and $x = a$, and the curve $y = x^2$: $2a^3 - 4a^3/3 = 2a^3/3$.

    (b) $\dfrac{2a^3}{3} = \int_{-a}^{a} x^2\,dx = \int_{-a}^{0} x^2\,dx + \int_{0}^{a} x^2\,dx = 2\int_{0}^{a} x^2\,dx$, so $\int_{0}^{a} x^2\,dx = \dfrac{a^3}{3}$.

    (c) $\int_{a}^{b} x^2\,dx = \int_{0}^{b} x^2\,dx - \int_{0}^{a} x^2\,dx = \dfrac{b^3}{3} - \dfrac{a^3}{3} = \dfrac{b^3 - a^3}{3}$

55. (b) $\int_{1}^{2} f(x)\,dx = 6$

    $\int_{1}^{3} f(x)\,dx = 14$

    (c) $x = 2f^{-1}(x) + 3 \implies f^{-1}(x) = \tfrac{1}{2}(x - 3)$

    (d) $\int_{5}^{9} f^{-1}(x)\,dx = 8$

    $\int_{5}^{7} f^{-1}(x)\,dx = 3$

SECTION 5.1 AREAS AND INTEGRALS

(e) $2f(2) - f(1) - \int_{f(1)}^{f(2)} f^{-1}(x)\, dx = 2 \cdot 7 - 1 \cdot 5 - \int_{5}^{7} f^{-1}(x)\, dx = 14 - 5 - 3 = 6 = \int_{1}^{2} f(x)\, dx$

$3f(3) - f(1) - \int_{f(1)}^{f(3)} f^{-1}(x)\, dx = 3 \cdot 9 - 1 \cdot 5 - \int_{5}^{9} f^{-1}(x)\, dx = 27 - 5 - 8 = 14 = \int_{1}^{3} f(x)\, dx$

56. (b) Let $f(x) = \ln x$. Then $f^{-1}(x) = e^x$ so

$$\int_{1}^{e} \ln x\, dx = e \ln e - \ln 1 - \int_{0}^{1} e^x\, dx = e - (e - 1) = 1.$$

(c) Let $f(x) = \sqrt{x}$. Then $f^{-1}(x) = x^2$ so

$$\int_{a}^{b} \sqrt{x}\, dx = b\sqrt{b} - a\sqrt{a} - \int_{\sqrt{a}}^{\sqrt{b}} x^2\, dx$$
$$= b^{3/2} - a^{3/2} - \int_{\sqrt{a}}^{\sqrt{b}} x^2\, dx$$
$$= b^{3/2} - a^{3/2} - \tfrac{1}{3}b^{3/2} + \tfrac{1}{3}a^{3/2}$$
$$= \tfrac{2}{3}b^{3/2} - \tfrac{2}{3}a^{3/2}.$$

57. $\int_{0}^{a} \sqrt{x}\, dx = a^{3/2} - \int_{0}^{\sqrt{a}} x^2\, dx = a^{3/2} - \tfrac{1}{3}a^{3/2} = \tfrac{2}{3}a^{3/2}.$

5.2 The Area Function

1. Suppose that $x < 0$. Then, $\int_x^0 t\, dt = -x^2/2$ and, therefore $\int_0^x t\, dt = x^2/2$.

2. (a) $F(x) = x$; $G(x) = x - 1$; $H(x) = x + 2$. Yes.
   (b) $F(x) = 3x^2/2$; $G(x) = 3(x^2 - 1)/2$; $H(x) = 3(x^2 - 4)/2$. Yes.
   (c) $F(x) = 3x^2/2 + x$; $G(x) = 3(x^2 - 1)/2 + (x - 1) = 3x^2/2 + x - 5/2$; $H(x) = 3(x^2 - 4)/2 + (x+2) = 3x^2/2 + x - 4$. Yes.

3. (a) $F(x) = ax$; $G(x) = a(x - 2)$; $H(x) = a(x + 1)$. Yes.
   (b) $F(x) = bx^2/2$; $G(x) = b(x^2 - 4)/2$; $H(x) = b(x^2 - 1)/2$. Yes.
   (c) $F(x) = bx^2/2 + ax$; $G(x) = b(x^2 - 4)/2 + a(x - 2)$; $H(x) = b(x^2 - 1)/2 + a(x + 1)$. Yes.

4. (a) $A_f(a) = 0$
   (b) $A_f(x) = 2x - x^2/2 - (2a - a^2/2) = 2(x - a) - (x^2 - a^2)/2$
   (c) $\dfrac{d}{dx} A_f(x) = \dfrac{d}{dx}(2x - x^2/2) = 2 - x$

5. $G(x) = \int_b^x f(t)\, dt = \int_a^x f(t)\, dt - \int_a^b f(t)\, dt = F(x) + C$ where $C = -\int_a^b f(t)\, dt$.

6. (a) f is non-negative on $[a, b]$
   (b) F is decreasing on $[a, b]$
   (c) F is concave up on $[a, b]$
   (d) $f'(x) \leq 0$ when $a \leq x \leq b$
   (e) $G(x) = F(x) + C$ where C is a constant

7. (a) $A_f(\pi) = 2$, $A_f(3\pi/2) = 1$, $A_f(2\pi) = 0$, $A_f(-\pi/2) = 1$, $A_f(-\pi) = 2$, $A_f(-3\pi/2) = 1$, $A_f(-2\pi) = 0$

   (b) Since f is 2π-periodic, $A_f(x) = \int_0^x f(t)\, dt = \int_{2\pi}^{x+2\pi} f(t)\, dt$. Now, $A_f(2\pi) = 0$, so $\int_{2\pi}^{x+2\pi} f(x)\, dx = \int_0^{x+2\pi} f(x)\, dx = A_f(x + 2\pi)$. Thus, $A_f(x) = A_f(x + 2\pi)$ which implies that A_f is 2π-periodic.

   (c) Since f is positive on the interval $[0, \pi]$, A_f is an increasing function on this interval. Thus, $A_f(0) = 0 \leq A_f(x) \leq A_f(\pi) = 2$ when $0 \leq x \leq \pi$. Since f is negative on the interval $[\pi, 2\pi]$, A_f is decreasing on this interval. Thus, $A_f(\pi) = 2 \geq A_f(x) \geq A_f(2\pi) = 0$ when $0 \leq x \leq \pi$. Finally, since A_f is 2π-periodic, we may conclude that $0 \leq A_f(x) \leq 2$ for all x.

   (d) $A_f(x) = 1 - \cos x$

8. Let $A_f(x) = \int_0^x \cos t\, dt$. Then A_f is an increasing function on the intervals $[0, \pi/2]$ and $[3\pi/2, 2\pi]$; it is a decreasing function on the interval $[\pi/2, 3\pi/2]$. Arguments similar to those used in parts (b) and (c) of the previous exercise can be used to establish the desired inequalities.

9. (a) $\int_{\sqrt{\pi/2}}^x f(t)\, dt = \int_0^x f(t)\, dt - \int_0^{\sqrt{\pi/2}} f(t)\, dt = \sin x^2 - \sin(\pi/2) = \sin x^2 - 1$

   (b) $\int_{-\sqrt{3\pi/2}}^x f(t)\, dt = -\int_0^{-\sqrt{3\pi/2}} f(t)\, dt + \int_0^x f(t)\, dt = -\sin(3\pi/2) + \sin x^2 = 1 + \sin x^2$

10. (a) $A_f(x) = \begin{cases} 2x - x^2/2, & x \geq 0 \\ 2x + x^2/2, & x < 0 \end{cases}$

    (c) $[-2, 2]$; f is non-negative on this interval

Section 5.2 The Area Function

(d) $(-\infty, -2]$ and $[2, \infty)$; f is non-positive on these intervals

(e) at $x = -2$ and at $x = 2$; f has a root at each of these locations

(f) $(-\infty, 0)$; f is increasing on this interval

(g) $(0, \infty)$; f is decreasing on this interval

(h) Yes, at $x = 0$. This point corresponds to a local extremum of f.

(i) No changes.

11. (a) $A_f(5)$ is larger because f is positive on the interval $[1, 5]$

    (b) $A_f(7)$ is larger because f is negative on the interval $[7, 10]$

    (c) $A_f(-2) < A_f(-1) < 0$

    (d) A_f is increasing on the interval $(-2, 6)$

    (e) It is a local maximum because f changes from positive to negative (i.e., A_f changes from increasing to decreasing).

    (f) $A_f(x) = \int_0^x f(t)\,dt = \int_{-2}^x f(t)\,dt - \int_{-2}^0 f(t)\,dt = F(x) + C$ where $C = -\int_{-2}^0 f(t)\,dt$. $C < 0$ because $f(x) > 0$ when $-2 \leq x \leq 0$.

    (g) $0 < A_f(-1) - A_f(-2) < A_f(0) - A_f(-1) < A_f(1) - A_f(0) < A_f(2) - A_f(1)$

    (h) These values suggest that A_f is concave down on the interval $[3, 8]$—the slopes of the secant lines are decreasing.

12. (a) Let y and z be numbers such that $a \leq y < z \leq b$. Then,
    $$A_f(z) - A_f(y) = \int_c^z f(t)\,dt - \int_c^y f(t)\,dt = \int_y^z f(t)\,dt > 0$$
    so A_f is an increasing function on the interval $[a, b]$.

    (b) If f is negative on $[a, b]$, then $A_f(z) - A_f(y) = \int_y^z f(t)\,dt < 0$.

    (c) Parts (a) and (b) imply that A_f changes from increasing to decreasing (or from decreasing to increasing) wherever f changes sign.

13. (a) $G(3)$

    (b) $-G(0)$

    (c) $G(2) - G(-2)$

14. (a) $\int_{-1}^4 \sqrt{1 + u^4}\,du = F(4)$

    (b) $\int_0^{-1} \sqrt{1 + z^4}\,dz = -F(0)$

    (c) $\int_{-2}^3 \sqrt{1 + t^4}\,dt = F(3) - F(-2)$

15. (b) $g(1) = \int_1^1 f(t)\,dt = 0$

    (c) $g(x) = h(x) + \int_1^3 f(t)\,dt$ and $\int_1^3 f(t)\,dt > 0$ since f is positive on the interval $[1, 3]$ ($f(0) = 0$ and $f'(x) \geq 0$ on the interval).

(d) $h(0) < h(4) < g(3) < g(7)$

$h(0) = \int_3^0 f(t)\,dt = -\int_0^3 f(t)\,dt < 0$ since $f(x) > 0$ on the interval $[0, 3)$

$6 < g(3) = \int_1^3 f(t)\,dt < 12$ since $f(1) = 3$, $f(3) = 6$, and $f'(x) \geq 0$ if $1 \leq x \leq 3$

$4 < h(4) = \int_3^4 f(t)\,dt < 6$ since $f(3) = 6$, $f(4) = 4$, and $f'(x) \leq 0$ if $3 \leq x \leq 4$

$g(7) = g(3) + h(4) + \int_4^7 f(t)\,dt$. Now $f(x) \geq 0$ if $4 \leq x \leq 5$ and $f(x) \geq -2$ if $5 \leq x \leq 7$ imply that $-4 \leq \int_4^7 f(t)\,dt$. Thus, $h(4) + \int_4^7 f(t)\,dt > 0$ so $g(3) < g(7)$.

(e) one — $g(1) = 0$. The information given implies that $g'(x) = f(x) \geq 0$ if $0 \leq x \leq 5$ and $g'(x) = f(x) \leq 0$ if $5 \leq x \leq 7$. Thus, g has only one root in the interval $[0.7]$.

(f)

16. (a) Suppose that $a \leq u \leq x \leq v \leq b$. Then

$$\frac{\int_u^v f(t)\,dt}{v-u} - \frac{\int_u^x f(t)\,dt}{x-u} = \frac{\int_u^v f(t)\,dt}{v-u} - \frac{v-u}{x-u}\frac{\int_u^x f(t)\,dt}{v-u}$$

$$= \frac{\int_u^v f(t)\,dt}{v-u} - \frac{(v-x)+(x-u)}{x-u}\frac{\int_u^x f(t)\,dt}{v-u}$$

$$= \frac{1}{v-u}\left(\int_u^v f(t)\,dt - \int_u^x f(t)\,dt\right) - \frac{v-x}{(x-u)(v-u)}\int_u^x f(t)\,dt$$

$$= \frac{1}{v-u}\int_x^v f(t)\,dt - \frac{v-x}{(x-u)(v-u)}\int_u^x f(t)\,dt$$

Now, f is increasing on the interval $[u, v]$ and $u \leq x \leq v$, so $\int_x^v f(t)\,dt > f(x)(v-x)$ and $\int_u^x f(t)\,dt < f(x)(x-u)$. Therefore,

$$\frac{1}{v-u}\int_x^v f(t)\,dt - \frac{v-x}{(x-u)(v-u)}\int_u^x f(t)\,dt > \frac{v-x}{v-u}f(x) - \frac{v-x}{v-u}f(x) = 0$$

This implies that A_f is concave up on the interval $[a, b]$.

(b) Since f is decreasing on the interval $[a, b]$, the function $-f$ is increasing on this interval. By part (a), $A_{-f} = -A_f$ is concave up and, therefore, A_f is concave down.

(c) Since f changes from increasing to decreasing at a, the concavity of A_f changes from concave up to concave down at a (i.e., A_f has an inflection point at a).

17. (a) $G(x) = H(x) + \int_{-3}^2 f(t)\,dt = H(x) - 2\pi + 1/2$. The values of G and H differ by the signed area of the region bounded by f over the interval $[-3, 2]$.

(b) $(-5, -3)$ and $(1, 5)$

(c) $G'(x) = f(x)$ changes sign (from negative to positive) at $x = 1$.

(d) G achieves its minimum value at $x = -5$; $G(-5) = -3$. G achieves its maximum value at $x = -3$; $G(-3) = 0$.

(e) H achieves its minimum value at $x = 1$; $H(1) = -1/2$. H achieves its maximum value at $x = -3$; $H(-3) = 2\pi - 1/2$.

(f) G is concave down on $(-5, -1)$ and $(3, 5)$

(g) H has points of inflection at $x = -1$ and at $x = 3$.

18. $G(x) - F(x) = \int_a^x g(t)\,dt - \int_a^x f(t)\,dt = \int_a^x \big(g(t) - f(t)\big)\,dt \geq 0$ since $g(t) - f(t) \geq 0$ for all $t \geq a$.

SECTION 5.2 THE AREA FUNCTION

19. $G(x) - F(x) = \int_b^x g(t)\,dt - \int_a^x f(t)\,dt = \int_b^c g(t)\,dt + \int_c^x g(t)\,dt - \int_a^c f(t)\,dt - \int_c^x f(t)\,dt$

    $= G(c) - F(c) + \int_c^x \big(g(t) - f(t)\big)\,dt \geq 0$ when $x \geq c$.

20. (a) When $x \geq 0$, $\int_0^x \sqrt{1-t^2}\,dt$ can be evaluated by adding together the area of the circular sector with angle θ and the area of the right triangle shown in the diagram. Since $\theta = \arcsin x$, the areas are $\tfrac{1}{2}\arcsin x$ and $\tfrac{1}{2}x\sqrt{1-x^2}$, respectively.
    A similar justification can be given for the $x < 0$ case.

    (b) $\left(\tfrac{1}{2}x\sqrt{1-x^2} + \tfrac{1}{2}\arcsin x\right)' = \sqrt{1-x^2}$

21. (a) $A_f(x) = \int_{-1/2}^0 f(t)\,dt + \int_0^x f(t)\,dt = \dfrac{\sqrt{3}}{8} + \dfrac{\pi}{12} + \dfrac{1}{2}x\sqrt{1-x^2} + \dfrac{1}{2}\arcsin x$

    (b) Yes.

22. $\int_0^x \sqrt{a^2 - t^2}\,dt = \dfrac{1}{2}x\sqrt{a^2 - x^2} + \dfrac{a^2}{2}\arcsin(x/a)$

23. (a) $A_f(x) = x^3/3$

    (b) Yes.

24. (a) $A_f(x) = x^3/3 + 9$

    (b) Yes.

5.3 The Fundamental Theorem of Calculus

1. (a) Graph A is the winner. The point is that $F' = f$. Graph C is wrong because it goes down in the middle. Graph B has the wrong direction of concavity.

   (b) The g-graph is just like the F-graph, but *raised* two units vertically (i.e., $g(x) = F(x) + 2$). As always in this section, the point is that F and g have the same derivative, f.

2. (a) $F(-1) = -1.5$, $F(0) = 0$, $F(2) = 4$.

   (b) x corresponds to a critical point of F if $F'(x)$ equals zero or is undefined, or if x is an endpoint of the interval. Since $F'(x) = f(x)$, The critical points of F are at $x = -5$, $x = -3.5$, $x = -2$, $x = 2$, and $x = 5$. The second of these corresponds to a local minimum in the graph of F and fourth to a local maximum; the first, third, and fifth critical points are neither local maxima nor local minima.

   (c) x corresponds to an inflection point in the graph of F if F'' changes sign at x. Since $F''(x) = f'(x)$, it follows that F has inflection points at $x = -3$, $x = -2$, $x = 1$, and $x = 3$.

   (d) Reasoning geometrically, $\int_{-5}^{-3} f(x)\,dx = -2$, $\int_{-3}^{-1} f(x)\,dx = 2 - \pi/2$, and $\int_{-1}^{5} f(x)\,dx = 0$. Thus, $\int_{-5}^{5} f(x)\,dx = -\pi/2$ so the average value of f over the interval $[-5, 5]$ is $-\pi/20$.

   (e) $G''(x) = F'(x) = f(x)$. Thus, G is concave up on intervals where f is positive: $(-3.5, 2)$.

3. (a) $\displaystyle\int_1^4 \left(x + x^{3/2}\right) dx = 199/10$.

   (b) $\displaystyle\int_0^\pi \cos x\, dx = 0$

   (c) $\displaystyle\int_{-2}^5 \frac{dx}{x+3} = \ln 8$

   (d) $\displaystyle\int_0^b x^2\, dx = b^3/3$

   (e) $\displaystyle\int_1^b x^n\, dx = \frac{b^{n+1}}{n+1} - \frac{1}{n+1} \quad [n \neq -1]$

   (f) $\displaystyle\int_2^{2.001} \frac{x^5}{1000}\, dx = \frac{2.001^6 - 2^6}{6000} \approx 0.00003204$

   (g) $\displaystyle\int_0^{0.001} \frac{\cos x}{1000}\, dx = \frac{\sin 0.001}{1000} \approx 0.000001$

   (h) $\displaystyle\int_0^{\sqrt{\pi}} x \sin\left(x^2\right) dx = -\frac{1}{2}\cos(x^2)\bigg]_0^{\sqrt{\pi}} = 1$

4. (a) $F'(x) = f(x)$. Therefore, $F'(x) > 0$ when $0 < x < 1$, $3 < x < 5$, and $7 < x < 9$; $F'(x) < 0$ when $1 < x < 3$, $5 < x < 7$, and $9 < x \leq 10$. Thus, $F(x)$ has local maxima at $x = 1$, $x = 5$, and $x = 9$.

   (b) $F(x)$ attains its minimum at $x = 3$.

   (c) $f'(x) = F''(x) > 0$ when $2 < x < 4$, and $6 < x < 8$. Therefore, the graph of F is concave up on these intervals.

5. By the FTC, $\displaystyle\frac{d}{dx}\left[\int_a^x f(t)\, dt\right] = f(x)$. The FTC also implies that

$$\int_a^x \left[\frac{d}{dt} f(t)\right] dt = \int_a^x f'(t)\, dt = f(x) - f(a).$$

Thus if $f(a) \neq 0$, $\displaystyle\frac{d}{dx}\left[\int_a^x f(t)\, dt\right] \neq \int_a^x \left[\frac{d}{dt} f(t)\right] dt$.

SECTION 5.3 THE FUNDAMENTAL THEOREM OF CALCULUS

6. (a) Here is a function with the desired properties:

$$f(x) = \begin{cases} 2x & \text{if } 0 \le x < 2 \\ x^4 - 10x^3 + 36x^2 - 54x + 32 & \text{if } 2 \le x < 3 \\ -\frac{3}{80}x^5 + \frac{21}{20}x^4 - \frac{427}{40}x^3 + \frac{99}{2}x^2 - \frac{8559}{80}x + \frac{371}{4} & \text{if } 3 \le x < 5 \\ -\frac{39}{40}x^6 + \frac{703}{20}x^5 - \frac{2628}{5}x^4 + \frac{83,447}{20}x^3 - \frac{148,337}{8}x^2 + \frac{437,277}{10}x - \frac{85,477}{2} & \text{if } 5 \le x \le 7 \end{cases}$$

(b) $g'(4) = f(4) = 4$

(c) g is concave down where $g'' = f' < 0$. That is, on the interval $[3, 6]$.

(d) $g(x)$ achieves its maximum value at $x = 5$.

(e) The minimum value of $g(x)$ occurs at $x = 0$. To see this, note that the minimum value occurs at either $x = 0$ or $x = 7$. Now $-15 \le h(0) = \int_3^0 f(t)\,dt \le -15/2$ since $\frac{5}{3}t \le f(t) \le 5$ if $0 \le t \le 3$. (The first inequality follows from the fact that f is concave down on the interval $[0, 3]$.) However, $1 \le h(7) = \int_3^7 f(t)\,dt \le 10$ since $h(7) = \int_3^7 f(t)\,dt = \int_3^5 f(t)\,dt + \int_5^7 f(t)\,dt$, $5 \le \int_3^5 f(t)\,dt \le 10$, and $-4 \le \int_5^7 f(t)\,dt \le 0$.

(f) The average value of f' over the interval $[2, 5]$ is $\dfrac{\int_2^5 f'(x)\,dx}{5-2} = \dfrac{f(5) - f(2)}{3} = -\dfrac{4}{3}$.

7. (a) $g'(4) = f(4) = 2$

(b) g is concave up on $[1, 4]$ (i.e., where f is increasing)

(c) Two — $g(1) = 0$; g also has a root in the interval $[2, 3]$ since $g(2) < 0$ and $g(3) > 0$

(d) $-1 < g(2) < 0 < g(0) < 1 < g(4)$

(e) The average value of g' over the interval $[0, 3]$ is $\dfrac{\int_0^3 g'(x)\,dx}{3-0} = \dfrac{g(3) - g(0)}{3}$. Now, $0 < g(0) < 1$ and $0 < g(3) < 1$, so $-1 = 0 - 1 < g(3) - g(0) < 1 - 0 = 1$. Therefore, the average value of g' over the interval $[0, 3]$ is less than 1.

ALTERNATE SOLUTION: The average value of g' over the interval $[0, 3]$ is $\frac{1}{3}\int_0^3 g'(x)\,dx = \frac{1}{3}\int_0^3 f(x)\,dx$. Since $\int_0^3 f(x)\,dx < 3$, the average value of g' over the interval $[0, 3]$ is less than 1.

8. (a) The graph of G is a vertical translate of the graph of g since $G(x) = g(x) + C$ where $C = \int_0^5 g'(x)\,dx - 1$.

(b) Two. Begin by observing that $g(x) = 1 + \int_5^x g'(t)\,dt = G(x) - \int_0^5 g'(x)\,dx + 1$. It is obvious from the graph of g' that $\int_0^5 g'(x)\,dx > 1$ so $g(0) < 0$. Since g is a continuous, increasing function on the interval $(0, 5)$ and $g(5) = 1 > 0$, g must have exactly one root in this interval. Similarly, $\int_5^{10} g'(x)\,dx < -1$ so $g(10) < 0$. This implies that g has at least one root in the interval $(5, 10)$. It has exactly one root in this interval because $g(5) > 0$, $g(10) < 0$, and g has exactly one local maximum and one local minimum in the interval $[5, 10]$.

(c) None. $\displaystyle\int_0^{5.5} g'(t)\,dt > -\int_{5.5}^9 g'(t)\,dt$.

9. Let $F(x) = \int_0^x f(t)\,dt = 3x^2 + e^x - \cos x$. Since $F'(x) = f(x) = 6x + e^x + \sin x$, $f(2) = 12 + e^2 + \sin 2$.

10. By the fundamental theorem, $F'(x) = \sqrt[3]{x^2 + 7}$, so $F'(1) = \sqrt[3]{8} = 2$. Thus the slope of the tangent line is 2; since the graph of F passes through the point $(1, 0)$, $y = 2x - 2$ is an equation for the tangent line.

11. (a) Let $g(x) = \displaystyle\int_a^x \sqrt[3]{1 + t^2}\,dt$. By the FTC, $g'(x) = \sqrt[3]{1 + x^2} = f'(x)$. Therefore, since f and g have the same derivative function, $f(x) = g(x) + C$ for some number C. Since $f(1) = 0 = g(1)$, it follows that $C = 0$ (i.e., $f = g$).

(b) $\int_0^3 \sqrt[3]{1+x^2}\,dx = f(3) - f(0)$

12. Using the chain rule, $\dfrac{d}{dx}\left[F(x^2)\right] = F'(x^2) \cdot 2x = 2x\sqrt[3]{1+x^4}$ since $F'(x) = \sqrt[3]{1+x^2}$.

13. The answer is $f(x) = 4x^3 - 3x^2 + 10$. To find it, observe:

    (i) $f'(x) = ax^2 + bx \implies f(x) = ax^3/3 + bx^2/2 + c$

    (ii) $f''(x) = 2ax + b$.

    (iii) $f'(1) = 6 \implies a + b = 6$

    (iv) $f''(1) = 18 \implies 2a + b = 18$

    (v) $\int_1^2 f(x)\,dx = 18 = ax^3/3 + bx^2/2 + c\Big]_1^2 = \dfrac{5a}{4} + c + \dfrac{7b}{6}$.

    The last three observations provide three equations that must be satisfied by a, b, and c. The solution of this system of three equations in three unknowns is $a = 12$, $b = -6$, and $c = 10$.

14. (a) $\int_{-1}^1 f(x)\,dx = \int_{-1}^0 f(x)\,dx + \int_0^1 f(x)\,dx = -2 + 1 = -1$

    (b) $F(x) = \int_{-2}^x f(t)\,dt$. So when $x < 0$, $F(x) = -(2x + 4)$ and when $x \geq 0$, $F(x) = x - 4$. Here's the picture:

    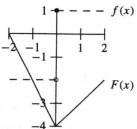

    (c) $F'(1) = f(1) = 1$

    (d) $F'(-1) = f(-1) = -2$

    (e) $F'(0)$ does not exist because $F'(x) = 1$ for all $x > 0$ and $F'(x) = -2$ for all $x < 0$. At $x = 0$, F has a sharp bend. This result doesn't contradict the FTC because f is not continuous at $x = 0$.

15. (a) The demand at time t is $D(t) = 800 - 10t$ where t measures time in months. The production at time t is $P(t) = 900$.

    (b) $D(t)$ represents the demand for the product in units/month. Since $D(t)$ is a rate function, $\int_0^t D(s)\,ds$ is the accumulated demand over the interval $[0, t]$.

    (c) The inventory at time t is $I(t) =$ starting inventory $+$ total production up to time $t -$ total demand up to time t. Since now $P(t) = 900 - Rt$,

    $$\begin{aligned} I(t) &= 1680 + \int_0^t (900 - Rs)\,ds - \int_0^t (800 - 10s)\,ds \\ &= (5 - \dfrac{R}{2})t^2 + 100t + 1680. \end{aligned}$$

    (d) When $t = 12$ the inventory is $3600 - 72R$; to make this zero, $R = 50$ must be true.

SECTION 5.3 THE FUNDAMENTAL THEOREM OF CALCULUS

16. (a) Let $F(x) = \int_1^x \frac{dt}{t}$. Then, using the chain rule, $\frac{d}{dx}F(ax) = aF'(ax) = a \cdot \frac{1}{ax} = \frac{1}{x}$.

   (b) Since $\ln(ax)$ has the same derivative as $\ln x$, it differs from that function only by a constant. Thus, $\ln(ax) = \ln x + C$.

   (c) Let $x = 1$. Then $\ln(ax) = \ln x + C \implies \ln(a) = C$ (since $\ln 1 = 0$).

5.4 Approximating Sums: The Integral as a Limit

1. Using 5 equal subintervals, the left sum approximation to $\int_{-5}^{5} g(x)\,dx$ is

$$2\Big(f(-5) + f(-3) + f(-1) + f(1) + f(3)\Big) = 4;$$

the right sum approximation is $2\Big(f(-3) + f(-1) + f(1) + f(3) + f(5)\Big) = 8;$

and, the midpoint sum approximation is $2\Big(f(-4) + f(-2) + f(0) + f(2) + f(4)\Big) = 7.$

Diagrams illustrating each of these sums appear below:

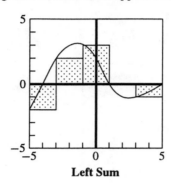
Left Sum

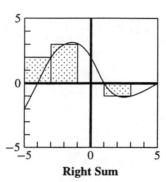
Right Sum

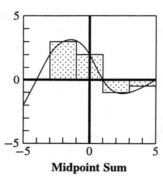
Midpoint Sum

2. (a) By counting squares, $\int_{0}^{8} f(x)\,dx < 6 + 5 + 3 + 6 + 7 + 6 + 5 + 3 = 41, 45.$

   (b) Yes. By counting squares one sees that the region below f and above the x-axis between $x = 1$ and $x = 7$ has area greater than 20.

   (c) $\int_{1}^{7} f(x)\,dx \approx f(3) \cdot 2 + f(5) \cdot 2 + f(7) \cdot 2 = 24$

   (d) $\int_{1}^{7} f(x)\,dx \approx f(1) \cdot 2 + f(3) \cdot 2 + f(5) \cdot 2 = 28$

   (e) $\int_{0}^{8} f(x)\,dx \approx 2\big(f(2) + f(4) + f(6) + f(8)\big) = 28$

   (f) $\int_{0}^{8} f(x)\,dx \approx 2\big(f(0) + f(2) + f(4) + f(6)\big) = 40$

3. $R_{10} = 2 \sum_{k=1}^{10} (2k)^2$

SECTION 5.4 APPROXIMATING SUMS: THE INTEGRAL AS A LIMIT

4. (a) $\dfrac{24}{N} \sum_{j=0}^{N-1} c(t_j) \cdot E(t_j)$, where $t_j = 24j/N$.

   (b) $\displaystyle\int_0^{24} c(t) E(t)\, dt$

5. left: $\displaystyle\int_0^5 \sqrt[3]{2x}\, dx \approx \dfrac{5}{10} \sum_{j=0}^{9} \sqrt[3]{2 \cdot j \cdot \dfrac{5}{10}} = \dfrac{1}{2} \sum_{j=0}^{9} \sqrt[3]{j}$

   right: $\displaystyle\int_0^5 \sqrt[3]{2x}\, dx \approx \dfrac{5}{10} \sum_{j=1}^{10} \sqrt[3]{2 \cdot j \cdot \dfrac{5}{10}} = \dfrac{1}{2} \sum_{j=1}^{10} \sqrt[3]{j}$

   midpoint: $\displaystyle\int_0^5 \sqrt[3]{2x}\, dx \approx \dfrac{5}{10} \sum_{j=0}^{9} \sqrt[3]{2 \cdot (j+0.5) \cdot \dfrac{5}{10}} = \dfrac{1}{2} \sum_{j=0}^{9} \sqrt[3]{j+0.5}$

6. left: $\displaystyle\int_0^5 \sqrt{3x}\, dx \approx \dfrac{5}{N} \sum_{k=1}^{N} \sqrt{3 \cdot (k-1) \cdot \dfrac{5}{N}} = \dfrac{5}{N} \sum_{j=0}^{N-1} \sqrt{3 \cdot j \cdot \dfrac{5}{N}}$

   right: $\displaystyle\int_0^5 \sqrt{3x}\, dx \approx \dfrac{5}{N} \sum_{k=1}^{N} \sqrt{3 \cdot k \cdot \dfrac{5}{N}} = \dfrac{5}{N} \sum_{j=0}^{N-1} \sqrt{3 \cdot (j+1) \cdot \dfrac{5}{N}}$

   midpoint: $\displaystyle\int_0^5 \sqrt{3x}\, dx \approx \dfrac{5}{N} \sum_{k=1}^{N} \sqrt{3 \cdot (k-0.5) \cdot \dfrac{5}{N}} = \dfrac{5}{N} \sum_{j=0}^{N-1} \sqrt{3 \cdot (j+0.5) \cdot \dfrac{5}{N}}$

7. $\dfrac{2}{100} \sum_{k=1}^{100} \sin\left(\dfrac{2k}{100}\right) \approx \displaystyle\int_0^2 \sin x\, dx = -\cos 2 + \cos 0 \approx 1.41615$. (This is a right sum approximation to the integral.)

8. Let $S = \dfrac{2}{10} \sum_{k=1}^{40} \cos\left(\dfrac{2k-1}{10}\right) = \dfrac{2}{10}\left(\cos(1/10) + \cos(3/10) + \cos(5/10) + \cdots + \cos(79/10)\right)$. Staring reveals that each of the subintervals has length $2/10 = 1/5$; their endpoints are $0, 2/10, 4/10, 6/10, \ldots, 78/10, 8$. Writing the sum S in the form

$$S = \dfrac{2}{10}\left(\cos(1/10) + \cos(3/10) + \cos(5/10) + \cdots + \cos(79/10)\right)$$

shows that S is the midpoint approximation M_{40} for the integral $\int_0^8 \cos(x)\, dx$. This integral is easy to evaluate: $\int_0^8 \cos(x)\, dx = \sin(x)\Big]_0^8 = \sin 8 \approx 0.98936$.

9. (a) The left sum approximation to $\int_2^5 f(x)\, dx$ with $n = 3$ equal subintervals is $f(2) \cdot (3-2) + f(3) \cdot (4-3) + f(4) \cdot (5-4) = 0.21 + 0.28 + 0.36 = 0.85$.

   (b) The trapezoid sum approximation is

   $$\dfrac{1}{2}(f(2) + f(3)) + \dfrac{1}{2}(f(3) + f(4)) + \dfrac{1}{2}(f(4) + f(5)) = 0.965$$

   (c)

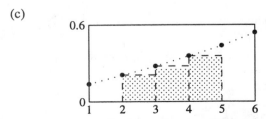
Left sum approximation to $\int_2^5 f(x)\, dx$

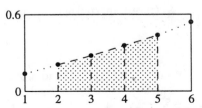
Trapezoid sum approximation to $\int_2^5 f(x)\, dx$

(d) A Riemann sum approximation of an integral $\int_a^b f(x)\,dx$ has the form $\sum_{j=1}^n f(c_j)\,\Delta x_j$ where $a = x_0 < x_1 < \cdots < x_n = b$ is a partition of the interval $[a, b]$ into n subintervals, $\Delta x_j = (x_j - x_{j-1})$ is the width of the jth subinterval, and c_j is a point in the jth subinterval. Thus, $f(2) \cdot 2 + f(4) \cdot 3$ is the sum obtained when the interval $[1, 6]$ is partitioned into the two unequal subintervals $[1, 3]$ and $[3, 6]$ and the integrand is evaluated at the points $c_1 = 2$ and $c_2 = 4$. See the picture below.

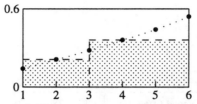

A Riemann sum approximation

10. (a) $\int_0^5 r(t)\,dt$

    (b) $\int_0^5 r(t)\,dt \approx 32e^{0.05} + 32e^{0.1} + 32e^{0.15} + 32e^{0.2} + 32e^{0.25} \approx 186.36$

    (c) Each term corresponds approximates the amount of oil consumed in one year.

    (d) $\int_0^5 32e^{0.05t}\,dt = \dfrac{32}{0.05} e^{0.05t}\Big]_0^5 = 640\left(e^{0.25} - 1\right) \approx 181.78$

11. $\displaystyle\lim_{n\to\infty} \frac{1}{n}\sum_{j=1}^n \left(\frac{j}{n}\right)^3 = \int_0^1 x^3\,dx = \frac{1}{4}$

12. $\displaystyle\lim_{n\to\infty} \frac{2}{n}\sum_{j=1}^n \left(\frac{2j}{n}\right)^3 = \int_0^2 x^3\,dx = 4$

13. (a) $\displaystyle\sum_{k=1}^4 \frac{5}{k(k+1)}(2.3 + (k-1)\cdot 0.5)$

    (b) No. The endpoints of the subintervals are $x_0 = 0$, $x_1 = 5/2$, $x_2 = 10/3$, $x_3 = 15/4$, and $x_4 = 4$. The sampling points are $c_1 = 2.3$, $c_2 = 2.8$, $c_3 = 3.3$, and $c_4 = 3.8$. However, since c_2 and c_3 lie in the same subinterval, the sum is not a Riemann sum.

14. (a) $L_4 = 6$, $R_4 = 6$, $M_4 = 5$, $T_4 = 6$; exact answer is $16/3$.

    (b) $L_4 = -8$, $R_4 = 8$, $M_4 = 0$, $T_4 = 0$; exact answer is 0.

    (c) $L_4 = 2$, $R_4 = 6$, $M_4 = 4$, $T_4 = 4$; exact answer is 4.

    (d) $L_4 \approx -0.785$, $R_4 = 0.785$, $M_4 = 0$, $T_4 = 0$; exact answer is 0

15. (a) One way is to connect the dots.

    (b) To estimate the total distance traveled:

    (i) A trapezoid approximating sum, 6 subdivisions:

    $$T_6 = \frac{1}{6}(42/2 + 38 + 36 + 57 + 0 + 55 + 51/2) = 38.75$$

    (ii) A left approximating sum, 6 subdivisions:

    $$L_6 = \frac{1}{6}(42 + 38 + 36 + 57 + 0 + 55) = 38$$

SECTION 5.4 APPROXIMATING SUMS: THE INTEGRAL AS A LIMIT

(iii) A midpoint approximating sum, 3 subdivisions

$$M_6 = \frac{1}{3}(38 + 57 + 55) = 50$$

The trapezoid rule answer might be best.

(c) To plot a plausible distance graph, one might estimate the distances covered over each 10-minute period (using trapezoids) and add them up:

Distance estimates over one hour							
time (min)	0	10	20	30	40	50	60
speed (mph)	42	38	36	57	0	55	51
total distance (miles)	0	6.66	12.83	20.58	25.33	29.92	38.75

Here's a graph of the resulting distance graph:

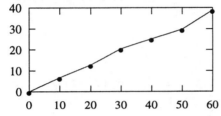

A possible distance function (min. vs. miles)

16. (a) $\displaystyle\int_5^{20} f(x)\,dx$

 (b) $\displaystyle\int_0^{15} f(x)\,dx$

 (c) $\displaystyle\int_{2.5}^{17.5} f(x)\,dx$

17. (a) $\displaystyle\int_2^{10} f(x)\,dx$

 (b) $\displaystyle\int_0^8 f(x)\,dx$

 (c) $\displaystyle\int_1^9 f(x)\,dx$

5.5 Approximating Sums: Interpretations and Applications

1. (b) The area is approximately $\displaystyle\frac{1}{5}\sum_{i=0}^{4}\left(\frac{i}{5}-\frac{i^2}{25}\right)=\frac{5}{32}=0.15625$.

   (c) Here's the picture:

   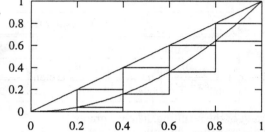

   (d) The area is $\displaystyle\int_0^1 (x-x^2)\,dx = \frac{1}{2}x^2 - \frac{1}{3}x^3\Big]_0^1 = \frac{1}{6}$.

2. Area $= \displaystyle\int_1^3 (\sin x - \cos x)\,dx = (-\cos x - \sin x)\Big|_1^3 = (\cos 1 + \sin 1) - (\cos 3 + \sin 3) \approx 2.2306$

3. (a) $\displaystyle\int_0^1 \left(1-\sqrt{1-x}\right)dx + \int_1^2 dx + \int_2^3 \left(1-(x-2)\right) = 1/3 + 1 + 1/2 = 11/6$

   (b) $\displaystyle\int_0^1 \left(1-\sqrt{1-x}\right)dx + \int_1^2 (2-x)\,dx = 1/3 + 1/2 = 5/6$

   (c) $\displaystyle\int_1^3 \left(1-(x-2)\right)dx = \int_1^3 (3-x)\,dx = 3x - \frac{1}{2}x^2\Big]_1^3 = 2$

   (d) (a) + (b) = 8/3

4. (a) The area is $\displaystyle\int_0^1 2\sqrt{1-x}\,dx = \int_{-1}^1 (1-y^2)\,dy = \frac{4}{3}$.

   (b) The area of the triangular region is $\displaystyle\int_1^3 (2-x)+1\,dx = \int_{-1}^1 (3-(y+2))\,dy = 2$.

   (c) The region is a square with area 6.

   (d) The area of Region 2 is (c) − (b) − (a) = 6 − 2 − 4/3 = 8/3.

5. Area $= \displaystyle\int_{-1}^1 (1-x^4)\,dx = \frac{8}{5}$.

6. Area $= \displaystyle\int_{-1}^0 (x^3-x)\,dx + \int_0^1 (x-x^3)\,dx = 2\int_0^1 (x-x^3)\,dx = \frac{1}{2}$

7. Area $= \displaystyle\int_0^1 (x^2-x^3)\,dx = \frac{1}{12}$

8. Area $= \displaystyle\int_{-1}^2 [(x+1)-(x^2-1)]\,dx = \frac{9}{2}$

SECTION 5.5 APPROXIMATING SUMS: INTERPRETATIONS AND APPLICATIONS

9. Area $= \int_{-3}^{2} [(2-y) - (y^2 - 4)]\, dy = \int_{-4}^{0} 2\sqrt{x+4}\, dx + \int_{0}^{5} \left[(2-x) + \sqrt{x+4} \right] dx = \dfrac{125}{6}$

10. Area $= \int_{0}^{4} \sqrt{x}\, dx = \dfrac{16}{3}$

11. Area $= \int_{0}^{1} (\sqrt{x} - x^2)\, dx = \dfrac{1}{3}$

12. Area $= \int_{-\frac{1}{4}}^{1} \left[(2-x) - \dfrac{9}{4x+5} \right] dx = \dfrac{65}{32} - \dfrac{9}{4} \ln(9) + \dfrac{9}{4} \ln(4) \approx 0.20666$

13. Area $= \int_{-1}^{1} \left[(2 - x^2) - \dfrac{9}{4x^2 + 5} \right] dx = \dfrac{10}{3} - \dfrac{9}{\sqrt{5}} \arctan\left(\dfrac{2}{\sqrt{5}} \right) \approx 0.39624$

14. Area $= \int_{-\frac{\pi}{4}}^{\frac{\pi}{6}} [(2 + \cos x) - \sec^2 x]\, dx \approx 2.2478$

15. Area $= \int_{0}^{1} e^x\, dx = e - 1$

16. Area $= \int_{-1}^{0} (2^x - 5^x)\, dx + \int_{0}^{1} (5^x - 2^x)\, dx = \dfrac{16}{5 \ln 5} - \dfrac{1}{2 \ln 2} \approx 1.2669$

6.1 Antiderivatives: The Idea

1. $\int \left(3x^5 + 4x^{-2}\right) dx = \frac{1}{2}x^6 - 4/x + C$

2. $\int \frac{dx}{3x} = \frac{1}{3}\ln|x| + C$

3. $\int \frac{dx}{4\sqrt{1-x^2}} = \frac{1}{4}\arcsin x + C$

4. $\int \frac{3}{x^2+1} dx = 3\arctan x + C$

5. $\int 3e^{4x} dx = \frac{3}{4}e^{4x} + C$

6. $\int (2\sin(3x) - 4\cos(5x)) dx = -\frac{2}{3}\cos(3x) - \frac{4}{5}\sin(5x)$

7. $\int 4\sec^2(3x) dx = \frac{4}{3}\tan(3x) + C$

8. $\int \left(1+\sqrt{x}\right)^2 dx = \int (1 + 2\sqrt{x} + x) dx = x + \frac{4}{3}x^{3/2} + \frac{1}{2}x^2 + C$

9. $\int (x+1)^2 \sqrt[3]{x}\, dx = \int \left(x^{7/3} + 2x^{4/3} + x^{1/3}\right) dx = \frac{3}{10}x^{10/3} + \frac{6}{7}x^{7/3} + \frac{3}{4}x^{4/3} + C$

10. $\int \frac{(3-x)^2}{x} = \int \frac{9 - 6x + x^2}{x} dx = 9\ln|x| - 6x + \frac{1}{2}x^2 + C$

11. $\int \frac{(1-x)^3}{\sqrt{x}} = \int \left(x^{-1/2} - 3x^{1/2} - x^{5/2}\right) dx = 2x^{1/2} - 2x^{3/2} + \frac{6}{5}x^{5/2} - \frac{2}{7}x^{7/2} + C$

12. $\int e^x(1-e^x) dx = \int \left(e^x - e^{2x}\right) = e^x - \frac{1}{2}e^{2x} + C$

13. **False.** $\left(e^{-\cos x} + C\right)' = e^{-\cos x} \sin x \neq e^{\sin x}$

14. **False.** $(\arctan(2x) + C)' = \frac{2}{1 + 4x^2}$

15. **True.** $(\tan x + C)' = \sec^2 x = 1 + \tan^2 x$

16. $(x\ln x - x + C)' = \ln x + \frac{x}{x} - 1 = \ln x$

17. $\left(x\arctan x - \frac{1}{2}\ln\left(1+x^2\right) + C\right)' = \arctan x + \frac{x}{1+x^2} - \frac{1}{2} \cdot \frac{2x}{1+x^2} = \arctan x$

18. $\left(\frac{1}{a}\arctan\left(\frac{x}{a}\right) + C\right)' = \frac{1}{a} \cdot \frac{1/a}{1 + (x/a)^2} = \frac{1}{a^2 + x^2}$

19. $(\ln|\sec x| + C)' = \frac{\sec x \tan x}{\sec x} = \tan x$

20. $(\ln|\sec x + \tan x| + C)' = \frac{\sec x \tan x + \sec^2 x}{\sec x + \tan x} = \frac{\sec x (\tan x + \sec^2 x)}{\sec x + \tan x} = \sec x$

Section 6.1 Antiderivatives: The Idea

21. $\left(\dfrac{1}{2}\ln\left|\dfrac{1+x}{1-x}\right|+C\right)' = \dfrac{1}{2}\cdot\dfrac{1-x}{1+x}\cdot\dfrac{2}{(1-x)^2} = \dfrac{1}{1-x^2}$

22. $\left(x\arcsin x + \sqrt{1-x^2}+C\right)' = \arcsin x + \dfrac{x}{\sqrt{1-x^2}} + \dfrac{-2x}{2\sqrt{1-x^2}} = \arcsin x$

23. (a) $\displaystyle\int \sin^2 x\, dx = \int \dfrac{1}{2}(1-\cos(2x))\,dx$
    $= \dfrac{1}{2}\left(\int 1\,dx - \int \cos(2x)\,dx\right)$
    $= \dfrac{1}{2}\left(x - \dfrac{1}{2}\sin(2x)\right) + C$
    $= \dfrac{x}{2} - \dfrac{1}{4}\sin(2x) + C$

    (b) $\sqrt{1-\cos(2x)} \neq \sqrt{2}\sin x$ when $\pi < x < 2\pi$.

24. $\sin^2 x = 1 - \cos^2 x \implies \sin^2 x + C_1 = (1-\cos^2 x) + C_1 = -\cos^2 x + (1+C_1) = -\cos^2 x + C_2$

25. $\dfrac{1}{2}\sin(2x) = \dfrac{1}{2}\cdot 2\cos x \sin x = \cos x \sin x$

26. $\dfrac{1}{2}\cos x \sin x = \dfrac{2\cos x \sin x}{4} = \dfrac{1}{4}\sin(2x)$

27. $\displaystyle\int \dfrac{x^2}{1+x^2}\,dx = x - \arctan x + C$

28. $\displaystyle\int \dfrac{4x^2+3x+2}{x+1}\,dx = 2x^2 - x + 3\ln|x+1| + C$

29. $\displaystyle\int \dfrac{x-1}{x+1}\,dx = x - 2\ln|x+1| + C$

30. $\displaystyle\int \dfrac{dx}{\sqrt{x-1}+\sqrt{x+1}} = -\tfrac{1}{3}(x-1)^{3/2} + \tfrac{1}{3}(x+1)^{3/2} + C$

31. $\displaystyle\int \tan^2 x\,dx = \tan x - x + C$

32. $\displaystyle\int \dfrac{dx}{1+\sin x}\,dx = \int \dfrac{1-\sin x}{\cos^2 x}\,dx = \int \left(\sec^2 x - \sec x \tan x\right) dx = \tan x - \sec x + C$

33. $\displaystyle\int \dfrac{dx}{1+9x^2} = \tfrac{1}{3}\arctan(3x) + C$

34. $\displaystyle\int \dfrac{6}{9+x^2}\,dx = 2\arctan(x/3) + C$

35. $\displaystyle\int \dfrac{2}{\sqrt{1-9x^2}}\,dx = \tfrac{2}{3}\arcsin(3x) + C$

36. $\displaystyle\int \dfrac{8}{\sqrt{4-x^2}}\,dx = 4\arcsin(x/2) + C$

6.2 Antidifferentiation by Substitution

1. $\int (4x+3)^{-3} dx = -\dfrac{1}{8(4x+3)^2} + C$

2. $\int x\sqrt{1+x^2}\, dx = \dfrac{1}{3}\left(1+x^2\right)^{3/2} + C$

3. $\int e^{\sin x} \cos x\, dx = e^{\sin x} + C$

4. $\int \dfrac{(\ln x)^3}{x} dx = \dfrac{(\ln x)^4}{4} + C$

5. $\int \dfrac{\arctan x}{1+x^2} dx = \dfrac{1}{2}(\arctan x)^2$

6. $\int \dfrac{\sin(\sqrt{x})}{\sqrt{x}} dx = -2\cos(\sqrt{x}) + C$

7. $\int \dfrac{e^{1/x}}{x^2} dx = \int \dfrac{e^{x^{-1}}}{x^2} dx = -e^{x^{-1}} + C = -e^{1/x} + C$

8. $\int x^3 \sqrt{9-x^2}\, dx = \sqrt{9-x^2}\left(\tfrac{1}{5}x^4 - \tfrac{3}{5}x^2 - \tfrac{54}{5}\right) + C$

9. $\int \dfrac{e^x}{1+e^{2x}} dx = \arctan(e^x) + C$

10. $\int \dfrac{x}{1+x} dx = \int \dfrac{(1+x)-1}{1+x} dx = x - \ln(1+x) + C$

11. $u = x^2$, $a = 4$, $b = 1$; $\quad \displaystyle\int_{-2}^{1} \dfrac{x}{1+x^4} dx = \dfrac{1}{2}\int_{4}^{1} \dfrac{1}{1+u^2} du = (\pi/4 - \arctan 4)/2 \approx -0.27021$

12. $u = 3x^2$, $a = 3\pi/2$, $b = 3\pi$; $\quad \displaystyle\int_{-\sqrt{\pi/2}}^{\sqrt{\pi}} x\cos(3x^2)\, dx = \dfrac{1}{6}\int_{3\pi/2}^{3\pi} \cos u\, du = \dfrac{1}{6}$

13. $u = 2x^2 + 1$, $a = 1$, $b = 19$; $\quad \displaystyle\int_{0}^{3} \dfrac{x}{(2x^2+1)^3} dx = \dfrac{1}{4}\int_{1}^{19} u^{-3} du = \dfrac{45}{361}$

14. $u = x^3/4$, $a = 1/4$, $b = 2$; $\quad \displaystyle\int_{1}^{2} x^2 e^{x^3/4} dx = \dfrac{4}{3}\int_{1/4}^{2} e^u du = \dfrac{4}{3}\left(e^2 - e^{1/4}\right)$

15. $\left(\dfrac{1}{2}\arctan(x^2) + C\right)' = \dfrac{1}{2}\dfrac{2x}{1+(ix^2)^2} = \dfrac{x}{1+x^4}$

16. $\left(\dfrac{(2x+3)^{3/2}}{6} - \dfrac{3\sqrt{2x+3}}{2} + C\right)' = \dfrac{\sqrt{2x+3}}{2} - \dfrac{3}{2\sqrt{2x+3}} = \dfrac{x}{\sqrt{2x+3}}$.

17. (a) If $u = 1 + \sqrt{x}$, then $du = \tfrac{1}{2}x^{-1/2} dx$ and $2(u-1)\, du = dx$. Thus,
$$\int \dfrac{dx}{1+\sqrt{x}} \to \int \dfrac{2(u-1)\, du}{u} = 2u - 2\ln|u| + C \to 2(1+\sqrt{x}) - 2\ln(1+\sqrt{x}) + C$$
(b)

SECTION 6.2 ANTIDIFFERENTIATION BY SUBSTITUTION

18. Let $u = x/a$. Then, $du = \frac{1}{a} dx$ so $\displaystyle\int \frac{dx}{\sqrt{a^2 - x^2}} = \frac{1}{a}\int \frac{dx}{\sqrt{1-(x/a)^2}} \to \int \frac{du}{\sqrt{1-u^2}} =$
    $\arcsin u + C \to \arcsin(x/a) + C$

19. If $u = a + b/x$, then $du = -\frac{b}{x^2} dx$. Therefore,
    $\displaystyle\int \frac{dx}{ax^2 + bx} = \int \frac{dx}{x^2(a+b/x)} \to -\frac{1}{b}\int \frac{du}{u} = -\frac{1}{b}\ln|u| + C \to -\frac{1}{b}\ln|a+b/x| + C =$
    $-\frac{1}{b}\ln|ax+b| + \frac{1}{b}\ln|x| + C.$

20. If $u = e^x$, then $\frac{1}{u} du = dx$. Therefore,
    $\displaystyle\int \frac{dx}{1+e^x} = \int \frac{du}{u(1+u)} = \int \left(\frac{1}{u} - \frac{1}{1+u}\right) du = \ln u - \ln|1+u| + C = \ln(e^x) - \ln|1+e^x| + C =$
    $x - \ln(1+e^x) + C.$

21. If $\int_0^{12} g(x)\,dx = \pi$, then $\int_0^4 g(3x)\,dx = \frac{\pi}{3}$. **Reason:** Make the substitutions $u = 3x$ and $du = 3\,dx$ in the second integral: $\displaystyle\int_0^4 g(3x)\,dx = \frac{1}{3}\int_0^4 3 \cdot g(3x)\,dx = \frac{1}{3}\int_0^{12} g(u)\,du = \frac{1}{3}\pi$

22. (a) When $u = \sec x$, $du = \sec x \tan x\, dx$ so
    $I = \displaystyle\int \sec^2 x \tan x\, dx = \int \sec x \cdot \sec x \tan x\, dx \to \int u\, du = \frac{1}{2}u^2 + C \to \frac{1}{2}\sec^2 x + C.$

    (b) When $u = \tan x$, $du = \sec^2 x\, dx$ so
    $I = \displaystyle\int \sec^2 x \tan x\, dx = \int \tan x \cdot \sec^2 x\, dx \to \int u\, du = \frac{1}{2}u^2 + C \to \frac{1}{2}\tan^2 x + C.$

    (c) Since $\sec^2 x = 1 + \tan^2 x$, the expressions in parts (a) and (b) differ from each other by a constant (1/2). Because an antiderivative is unique only up to an additive constant (i.e., two antiderivatives of the same function may differ by a constant), there is no paradox.

23. Let $u = 2x + 3$. Then, $\displaystyle\int \cos(2x+3)\,dx = \frac{1}{2}\int \cos u\, du = \frac{1}{2}\sin u + C = \frac{1}{2}\sin(2x+3) + C.$

24. Let $u = 2 - 3x$. Then $du = -3\,dx$. Therefore, $\displaystyle\int \sin(2-3x)\,dx \to -\frac{1}{3}\int \sin u\, du = \frac{1}{3}\cos u + C \to$
    $\frac{1}{3}\cos(2-3x) + C$

25. Let $u = 3x - 2$. Then $du = 3\,dx$ and $\frac{1}{3}du = dx$. Therefore, $\displaystyle\int (3x-2)^4\, dx \to \frac{1}{3}\int u^4\, du = \frac{1}{15}u^5 + C \to$
    $\frac{1}{15}(3x-2)^5 + C.$

26. Let $u = 1 - x^2$. Then, $\displaystyle\int x\cos(1-x^2)\,dx = -\frac{1}{2}\sin(1-x^2) + C.$

27. Let $u = 1 + x^4$. Then, $\displaystyle\int \frac{2x^3}{1+x^4}\,dx = \frac{\ln(1+x^4)}{2} + C.$

28. Let $u = 3x + 2$. Then, $\displaystyle\int x(3x+2)^4\, dx = \frac{1}{9}(u-2)u^4\, du = \frac{1}{54}u^6 - \frac{2}{45}u^5 + C =$
    $\frac{1}{54}(3x+2)^6 - \frac{2}{45}(3x+2)^5 + C.$

29. Let $u = 1 - 2x$. Then, $du = -2\,dx$ and $-\frac{1}{2}du = dx$. Therefore, $\displaystyle\int \frac{dx}{1-2x} \to -\frac{1}{2}\int \frac{du}{u} = -\frac{1}{2}\ln|u| + C \to$
    $-\frac{1}{2}\ln|1-2x| + C.$

30. Let $u = 3x - 2$. Then, $du = 3\,dx$ and $\frac{1}{3}\,du = dx$. Therefore, $\int \sqrt{3x-2}\,dx \to \frac{1}{3}\int \sqrt{u}\,du = \frac{2}{9}u^{3/2} + C \to \frac{2}{9}(3x-2)^{3/2} + C$.

31. Let $u = 3 - 2x$. Then $x = (3-u)/2$, $du = -2\,dx$, and $-\frac{1}{2}\,du = dx$. Therefore,

$$\int x\sqrt{3-2x}\,dx \to -\frac{1}{4}\int (3-u)u^{1/2}\,du = -\frac{1}{4}\int \left(3u^{1/2} - u^{3/2}\right)du$$
$$= \frac{1}{10}u^{5/2} - \frac{1}{2}u^{3/2} + C$$
$$\to \frac{1}{10}(3-2x)^{5/2} - \frac{1}{2}(3-2x)^{3/2} + C.$$

32. Let $u = \ln x$. Then, $\int \frac{\ln x}{x}\,dx = \frac{1}{2}(\ln x)^2 + C$.

33. Let $u = 1 + x^{-1}$. Then, $\int \frac{\sqrt{1+x^{-1}}}{x^2}\,dx = -\frac{2}{3}\left(1+x^{-1}\right)^{3/2} + C$.

34. Let $u = x^2$. Then, $\int xe^{x^2}\,dx = \frac{1}{2}e^{x^2} + C$.

35. Let $u = x^4 - 1$. Then, $\int x^3\left(x^4 - 1\right)^2 dx = \frac{1}{12}\left(x^4 - 1\right)^3 + C$.

36. Let $u = 1 + x^2$. Then $x^2 = u - 1$ and $du = 2x\,dx$. Therefore, $\int \frac{x^3}{1+x^2}\,dx = \int \frac{x^2 \cdot x}{1+x^2}\,dx \to \frac{1}{2}\int \frac{u-1}{u}\,du = \frac{1}{2}u - \frac{1}{2}\ln|u| + C \to \frac{1}{2}(1+x^2) - \frac{1}{2}\ln(1+x^2) + C$.

37. Let $u = x^3$. Then, $\int \frac{x^2}{1+x^6}\,dx = \frac{1}{3}\arctan\left(x^3\right) + C$.

38. Let $u = x^2$. Then, $\int \frac{x}{\sqrt{1-x^4}}\,dx = \frac{1}{2}\arcsin(x^2) + C$.

39. Let $u = 4x^3 + 5$. Then, $\int x^2\sqrt{4x^3+5}\,dx = \frac{1}{18}\left(4x^3+5\right)^{3/2} + C$.

40. Let $u = 1 + x^2$. Then, $\int \frac{x}{\sqrt{1+x^2}}\,dx = \sqrt{1+x^2} + C$.

41. $\int \frac{x+4}{x^2+1}\,dx = \int \frac{x}{x^2+1}\,dx + 4\int \frac{dx}{x^2+1} = \int \frac{x}{x^2+1}\,dx + 4\arctan x = \frac{1}{2}\ln\left(1+x^2\right) + 4\arctan x + C$.
(The substitution $u = 1 + x^2$ is used in the last step.)

42. Let $u = 1 - x^2$. Then, $\int x\left(1-x^2\right)^{15} dx = -\frac{1}{32}\left(1-x^2\right)^{16} + C$.

43. Let $u = x^2 + 3x + 5$. Then, $\int \frac{2x+3}{(x^2+3x+5)^4}\,dx = -\frac{1}{3}\left(x^2+3x+5\right)^{3} + C$.

44. Let $u = x^2 + 4x + 5$. Then, $\int (x+2)\left(x^2+4x+5\right)^6 dx = \frac{1}{14}\left(x^2+4x+5\right)^7 + C$.

45. Let $u = 3x^2 + 6x + 5$. Then, $\int \frac{x+1}{\sqrt[3]{3x^2+6x+5}}\,dx = \frac{1}{4}\left(3x^2+6x+5\right)^{2/3} + C$.

Section 6.2 Antidifferentiation by Substitution

46. Let $u = 1 + e^{2x}$. Then, $\int \dfrac{e^{2x}}{(1+e^{2x})^3}\,dx = -\dfrac{1}{4}\left(1+e^{2x}\right)^{-2} + C$.

47. Let $u = 2e^x + 3$. Then, $\int \dfrac{e^x}{(2e^x+3)^2}\,dx = -\dfrac{1}{2}(2e^x+3)^{-1} + C$.

48. Let $u = \sqrt{x}$. Then, $\int \dfrac{e^{\sqrt{x}}}{\sqrt{x}}\,dx = 2e^{\sqrt{x}} + C$.

49. Let $u = x + 1$. Then, $\int \dfrac{2x+3}{(x+1)^2}\,dx = \int \dfrac{2(u-1)+3}{u^2}\,du = 2\ln|x+1| - (x+1)^{-1} + C$.

50. Let $u = x - 3$. Then, $\int \dfrac{x^2}{x-3}\,dx = \int \dfrac{(u+3)^2}{u}\,du = \int \left(u + 6 + \dfrac{9}{u}\right)\,du$
$= \dfrac{1}{2}(x-3)^2 + 6(x-3) + 9\ln(x-3) + C$.

51. Let $u = \cos x$. Then, $\int \tan x\,dx = -\ln|\cos x| + C$.

52. Let $u = \sec x$. Then, $\int \sec x \tan x\,dx = \int du = \sec x + C$.

53. Let $u = \arcsin x$. Then, $\int \dfrac{\arcsin x}{\sqrt{1-x^2}}\,dx = \dfrac{1}{2}(\arcsin x)^2 + C$.

54. Let $u = 1 + \sec x$. Then, $\int \sec x \tan x \sqrt{1+\sec x}\,dx = \dfrac{2}{3}(1+\sec x)^{3/2} + C$.

55. Let $u = 3x^2 + 4$. Then, $\int \dfrac{5x}{3x^2+4}\,dx = \dfrac{5}{6}\ln\left(3x^2+4\right) + C$.

56. Let $u = \sin x$. Then, $\int \dfrac{\cos x}{1+\sin^2 x}\,dx = \arctan(\sin x) + C$.

57. Let $u = \tan x$. Then, $\int \dfrac{\sec^2 x}{\sqrt{1-\tan^2 x}}\,dx = \arcsin(\tan x) + C$.

58. Let $u = \cos x$. Then, $\int \tan^2 x \csc x\,dx = \int \dfrac{\sin^2 x}{\cos^2 x} \cdot \dfrac{1}{\sin x}\,dx = \int \dfrac{\sin x}{\cos^2 x}\,dx \to -\int \dfrac{du}{u^2} = \dfrac{1}{u} + C \to \sec x + C$.

59. Let $u = x^2$. Then, $\int x \tan\left(x^2\right)\,dx = \dfrac{1}{2}\int \tan u\,du = \dfrac{1}{2}\int \dfrac{\sin u}{\cos u}\,du = -\dfrac{1}{2}\ln|\cos u| + C = -\dfrac{1}{2}\ln\left|\cos\left(x^2\right)\right| + C$.

60. Let $u = x^3$. Then, $\int x^2 \sec^2\left(x^3\right)\,dx = \dfrac{1}{3}\tan\left(x^3\right) + C$.

61. Let $u = \sin x$. Then, $\int \dfrac{\cos x}{\sin^4 x}\,dx = -\dfrac{1}{3\sin^3 x} + C = -\dfrac{1}{3}\csc^3 x + C$.

62. Let $u = x^5 + 6$. Then, $\int x^4 \sqrt[3]{x^5+6}\,dx = \dfrac{3\left(x^5+6\right)^{4/3}}{20} + C$.

63. Let $u = \ln(\cos x)$. Then $du = -\frac{\sin x}{\cos x} dx = -\tan x\, dx$. Therefore, $\int \ln(\cos x) \tan x\, dx \to -\int u\, du$
$= -\frac{1}{2}u^2 + C \to -\frac{1}{2}\left(\ln(\cos x)\right)^2 + C$.

64. Let $u = 1 + \sqrt{x}$. Then $du = \frac{1}{2}x^{-1/2} dx$ and $dx = 2(u-1)\, du$. Therefore, $\int \frac{(1+\sqrt{x})^3}{\sqrt{x}} dx \to$
$\int \frac{u^3}{u-1} \cdot 2(u-1)\, du = 2\int u^3\, du = \frac{1}{2}u^4 + C \to \frac{1}{2}(1+\sqrt{x})^4 + C$.

65. Let $u = sqrtx + 2$. Then $du = \frac{1}{2}x^{-1/2} dx$ and $dx = 2(u-2)du$. Therefore, $\int \frac{dx}{\sqrt{x}\left(\sqrt{x}+2\right)^3} \to$
$2\int \frac{du}{u^3} = -\frac{1}{u^2} + C \to -\left(\sqrt{x}+2\right)^{-2} + C$.

66. Let $u = x^2 + 2$. Then $du = 2x\, du$. Therefore, $\int x^3\sqrt{x^2+2}\, dx = \int x^2\sqrt{x^2+2} \cdot x\, dx \to$
$\frac{1}{2}\int (u-2)\sqrt{u}\, du = \frac{1}{2}\int \left(u^{3/2} - 2u^{1/2}\right) du = \frac{1}{5}u^{5/2} - \frac{2}{3}u^{3/2} + C \to \frac{2}{5}(x^2+2)^{5/2} - \frac{2}{3}(x^2+2)^{3/2} + C$

67. Let $u = 1 + \sqrt{x}$. Then $du = \frac{1}{2}x^{-1/2} dx$ and $2(u-1)\, du = dx$. Therefore, $\int \sqrt{1+\sqrt{x}}\, dx \to$
$\int u^{1/2} \cdot 2(u-1)\, du = \frac{4}{5}u^{5/2} - \frac{2}{3}u^{3/2} + C \to \frac{4}{5}(1+\sqrt{x})^{5/2} - \frac{2}{3}(1+\sqrt{x})^{3/2} + C$.

68. Let $u = e^x + 1$. Then $du = e^x dx$. Therefore, $\int \frac{e^x}{e^{2x}+2e^x+1} dx = \int \frac{e^x}{(e^x+1)^2} dx \to \int \frac{du}{u^2} =$
$-\frac{1}{u} + C \to -\frac{1}{e^x+1} + C$.

69. First, note that $\int \frac{e^{\tan x}}{1-\sin^2 x} dx = \int \frac{e^{\tan x}}{\cos^2 x} dx = \int \sec^2 x\, e^{\tan x} dx$. Now, let $u = \tan x$. Then,
$\int \frac{e^{\tan x}}{1-\sin^2 x} dx = e^{\tan x} + C$.

70. Let $u = 1 + x^2$. Then, $\int_0^2 \frac{x}{(1+x^2)^3} dx = -\frac{1}{4}\left(1+x^2\right)^{-2}\bigg]_0^2 = \frac{6}{25}$

71. Let $u = 8 - x$. Then, $\int_{-19}^8 \sqrt[3]{8-x}\, dx = -\frac{3}{4}(8-x)^{4/3}\bigg]_{-19}^8 = \frac{243}{4}$

72. Let $u = \ln x$. Then, $\int_1^e \frac{\sin(\ln x)}{x} dx = -\cos(\ln x)\bigg]_1^e = 1 - \cos 1$

73. Let $u = \ln x$. Then, $\int_e^{4e} \frac{dx}{x\sqrt{\ln x}} = 2\sqrt{\ln x}\bigg]_e^{4e} = 2\sqrt{1+2\ln 2} - 2$

74. Let $u = \sin x$. Then, $\int_0^\pi \sin^3 x \cos x\, dx = \frac{1}{4}\sin^4 x\bigg]_0^\pi = 0$

75. Let $u = \cos x$. Then, $\int_{-\pi/2}^\pi e^{\cos x} \sin x\, dx = -e^{\cos x}\bigg]_{-\pi/2}^\pi = 1 - e^{-1}$

SECTION 6.2 ANTIDIFFERENTIATION BY SUBSTITUTION

76. Let $u = x^2$. Then $du = 2x\, dx$. Therefore, $\int_0^1 x\sqrt{1-x^4}\, dx = \dfrac{1}{2}\int_0^1 \sqrt{1-u^2}\, du = \dfrac{1}{2}\cdot\dfrac{\pi}{4} = \dfrac{\pi}{8}$.

[NOTE: $\int_0^1 \sqrt{1-u^2}\, du$ is the area of a quarter-circle with radius 1/2.]

77. Let $u = \pi - x$. Then $x = \pi - u$ and $du = -dx$. Therefore,

$$\int_0^\pi xf(\sin x)\, dx = -\int_\pi^0 (\pi - u)f(\sin(\pi - u))\, du$$

$$= \int_0^\pi (\pi - u)f(\sin(\pi - u))\, du$$

$$= \int_0^\pi (\pi - u)f(\sin u)\, du$$

$$= \pi\int_0^\pi f(\sin u)\, du - \int_0^\pi uf(\sin u)\, du$$

This implies that $2\int_0^\pi xf(\sin x)\, dx = \pi\int_0^\pi f(\sin x)\, dx$ or, equivalently, that

$\int_0^\pi xf(\sin x)\, dx = \dfrac{\pi}{2}\int_0^\pi f(\sin x)\, dx$.

78. Let $u = a + b - x$. Then $du = -dx$. Therefore, $\int_a^b f(a+b-x)\, dx = -\int_b^a f(u)\, du = \int_a^b f(u)\, du$.

79. Let $u = 1 - x$. Then, $\int_0^1 x^n(1-x)^m\, dx = \int_1^0 (1-u)^n u^m\, (-du) = \int_0^1 u^m(1-u)^n\, du$.

80. Let $u^2 = 2x + 1$. Then, $dx = u\, du$ and

$$\int x\sqrt{2x+1}\, dx = \dfrac{1}{2}\int (u^2-1)u^2\, du = \dfrac{1}{2}\left(\dfrac{u^5}{5} - \dfrac{u^3}{3}\right) + C = \dfrac{(2x+1)^{5/2}}{10} - \dfrac{(2x+1)^{3/2}}{6} + C.$$

81. First, note that $\int \dfrac{dx}{\sqrt{x} + \sqrt[3]{x}} = \int \dfrac{dx}{\sqrt[3]{x}(\sqrt[6]{x} + 1)}$. Now, let $u^6 = x$ and $w = u + 1$ to obtain

$$\int \dfrac{dx}{\sqrt{x} + \sqrt[3]{x}} = 6\int \dfrac{u^5}{u^2(u+1)} = 6\int \dfrac{u^3}{u+1}$$

$$= 6\int \dfrac{(w-1)^3}{w}\, dw = 6\int \left(w^2 - 3w + 3 - w^{-1}\right) dw$$

$$= 2w^3 - 9w^2 + 18w - 6\ln w + C$$

$$= 2x^{1/2} - 3x^{1/3} + 6x^{1/6} - 6\ln\left(x^{1/6} + 1\right) + C.$$

82. The "proof" is flawed because $u = x^{-2}$ is not defined when $x = 0$.

83. $\sqrt{1 - \sin^2 x} = -\cos x$ if $\pi/2 \le x \le \pi$. The "proof" incorrectly replaces $\sqrt{1 - \sin^2 x}$ by $\cos x$ for all x in the interval $[0, \pi]$.

6.3 Integral Aids: Tables and Computers

1. $\int \dfrac{dx}{3+2e^{5x}} = \dfrac{1}{3}x - \dfrac{1}{15}\ln\left(3+2e^{5x}\right).$ [Use formula #58.]

2. $\int \dfrac{dx}{x(2x+3)} = \dfrac{1}{3}\ln\left|\dfrac{x}{2x+3}\right|.$ [Use formula #23.]

3. $\int \dfrac{dx}{x^2(3-x)} = -\dfrac{1}{3x} - \dfrac{1}{9}\ln\left|\dfrac{3-x}{x}\right| = \dfrac{1}{9}\ln\left|\dfrac{x}{3-x}\right| - \dfrac{1}{3x}.$ [Use formula #24.]

4. $\displaystyle\int_{e}^{4e} x^2 \ln x\, dx = \dfrac{64}{3}\ln(4e)e^3 - \dfrac{22}{3}e^3.$ [Use formula #56.]

5. $\int \tan^3(5x)\, dx = \dfrac{1}{10}\tan^2(5x) - \dfrac{1}{5}\ln|\cos(5x)|.$ [Use formulas #50 and #7.]

6. $\int \dfrac{dx}{x^2\sqrt{2x+1}} = -\dfrac{\sqrt{2x+1}}{x} - \int \dfrac{dx}{x\sqrt{2x+1}} = -\dfrac{\sqrt{2x+1}}{x} - \ln\left|\dfrac{\sqrt{2x+1}-1}{\sqrt{2x+1}+1}\right|.$ [Use the substitution $u = \sqrt{2x+1}$.]

7. $\int x \sin(2x)\, dx = \dfrac{1}{4}\sin(2x) - \dfrac{x}{2}\cos(2x).$ [Use formula #46.]

8. $\int x^2 e^{3x}\, dx = \dfrac{e^{3x}}{3}\left(x^2 - \dfrac{2}{9}x + \dfrac{2}{27}\right).$ [Use formula #53 (twice).]

9. $\int e^{2x}\cos(3x)\, dx = \dfrac{e^{2x}}{13}(2\cos(3x) + 3\sin(3x)).$ [Use formula #55.]

10. $\int \dfrac{2x+3}{4x+5}\, dx = 2\int \dfrac{x}{4x+5}\, dx + 3\int \dfrac{dx}{4x+5} = 2\left(\dfrac{x}{4} - \dfrac{5}{16}\ln|4x+5|\right) + \dfrac{3}{4}\ln|4x+5| = \dfrac{x}{2} + \dfrac{1}{8}\ln|4x+5|.$ [Use formulas #21 and #20.]

11. $\int \dfrac{dx}{4-x^2}\, dx = \dfrac{1}{4}\ln\left|\dfrac{2+x}{2-x}\right|.$ [Use formula #30.]

12. $\int x^2\sqrt{1-3x}\, dx = -\dfrac{2}{21}(1-3x)^{3/2}\left(x^2 + \dfrac{36}{135}x + \dfrac{8}{135}\right).$ [Use the substitution $u = 1 - 3x$ or $u = \sqrt{1-3x}$.]

13. $\int \dfrac{4x+5}{(2x+3)^2}\, dx = 4\int \dfrac{x}{(2x+3)^2}\, dx + 5\int \dfrac{dx}{(2x+3)^2} = \dfrac{1}{2(2x+3)} + \ln|2x+3|.$ [Use formulas #22 and #19.]

14. $\int \dfrac{dx}{x\sqrt{3x-2}} = \dfrac{2}{\sqrt{2}}\arctan\left(\dfrac{3x-2}{2}\right).$ [Use formula #29.]

15. $\int \dfrac{dx}{4x^2-1} = \dfrac{1}{4}\ln\left|\dfrac{2x-1}{2x+1}\right|.$ [Use formula #32.]

16. $\int \dfrac{dx}{(4x^2-9)^2} = -\dfrac{x}{18(4x^2-9)} - \dfrac{1}{216}\ln\left|\dfrac{2x-3}{2x+3}\right|.$ [Use formulas #33 and #32.]

17. $\int \dfrac{x+2}{2+x^2}\, dx = \int \dfrac{x}{2+x^2}\, dx + 2\int \dfrac{dx}{2+x^2} = \dfrac{1}{2}\ln\left(x^2+2\right) + \sqrt{2}\arctan(x\sqrt{2}/2).$ [Use formulas #35 and #31.]

18. $\int \dfrac{3}{\sqrt{6x+x^2}}\,dx = 3\int \dfrac{dx}{\sqrt{(x+3)^2-9}} = 3\ln\left|x+3+\sqrt{6x+x^2}\right|.$
[Complete the square, then make the substitution $u = x+3$ in formula #38.]

19. $\int \dfrac{5}{4x^2+20x+16}\,dx = \dfrac{5}{4}\int \dfrac{dx}{x^2+5x+4}\,dx = \dfrac{5}{4}\int \dfrac{dx}{(x+5/2)^2-9/4}\,dx = \dfrac{5}{12}\ln\left|\dfrac{x+1}{x+4}\right|.$ [Use formula #32.]

20. $\int \dfrac{dx}{\sqrt{x^2+2x+26}} = \int \dfrac{dx}{\sqrt{(x+1)^2+25}} = \ln\left|x+1+\sqrt{x^2+2x+26}\right|.$ [Make the substitution $u = x+1$ in formula #38.]

21. Making the substitution $u = x^2$ together with formula #47 (or #49),
$$\int x^3 \cos(x^2)\,dx \to \dfrac{1}{2}\int u\cos u\,du = \dfrac{1}{2}(u\sin u + \cos u) \to \dfrac{1}{2}\left(x^2\sin\left(x^2\right)+\cos\left(x^2\right)\right)$$

22. $\int \dfrac{x}{\sqrt{x^2+4x+3}}\,dx = \int \dfrac{x}{\sqrt{(x+2)^2-1}}\,dx$
$\to \int \dfrac{u-2}{\sqrt{u^2-1}}\,du = \int \dfrac{u}{\sqrt{u^2-1}}\,dx - 2\int \dfrac{du}{\sqrt{u^2-1}}$
$= \sqrt{u^2-1} - 2\ln\left|u+\sqrt{u^2-1}\right|$
$\to \sqrt{x^2+4x+3} - 2\ln\left|x+2+\sqrt{x^2+4x+3}\right|$

23. $\int \dfrac{dx}{(x^2+3x+2)^2} = \int \dfrac{dx}{((x+3/2)^2-1/4)^2} = -\dfrac{2x+3}{x^2+3x+2} - 2\ln\left|\dfrac{x+1}{x+2}\right|.$ [Use formula #33.]

24. $\int x^2(\cos x + 3\sin x)\,dx = \int x^2\cos x\,dx + 3\int x^2\sin x\,dx =$
$x^2\sin x - 2\sin x + 2x\cos x - 3x^2\cos x + 6\cos x + 6x\sin x.$
[Use formulas #48 and #49.]

25. $\int \dfrac{e^x}{e^{2x}-2e^x+5}\,dx \to \int \dfrac{du}{u^2-2u+5} = \int \dfrac{du}{(u-1)^2+4} \to \int \dfrac{dw}{w^2+4} = \dfrac{1}{2}\arctan(w/2) \to$
$\dfrac{1}{2}\arctan\left(\dfrac{e^x-1}{2}\right) + C.$
[Use formulas #13 or #31.]

26. $\int \cos x \sin x \sin^2\left(2\cos^2 x + 1\right)\,dx \to -\dfrac{1}{4}\int \sin^2 u\,du = \dfrac{1}{16}\sin(2u) - \dfrac{1}{8}u \to$
$\dfrac{1}{16}\sin\left(4\cos^2 x + 2\right) - \dfrac{1}{4}\cos^2 x - \dfrac{1}{8}$
[Use formula #39 or #40.]

27. $\int \sqrt{x^2+4x+1}\,dx = \int \sqrt{(x+2)^2-3}\,dx = \dfrac{1}{2}\left((x+2)\sqrt{x^2+4x+1} - 3\ln\left|x+2+\sqrt{x^2+4x+1}\right|\right).$
[Complete the square, then use formula #36.]

28. $\int \dfrac{\cos x}{3\sin^2 x - 11\sin x - 4}\,dx \to \dfrac{1}{3}\int \dfrac{du}{u^2 - 11u/3 - 4/3} = \dfrac{1}{3}\int \dfrac{du}{(u-11/6)^2 - 169/36} =$
$\dfrac{1}{13}\ln\left|\dfrac{u-4}{u+1/3}\right| \to \dfrac{1}{13}\ln\left|\dfrac{\sin x - 4}{\sin x + 1/3}\right|.$
[Use formula #32.]

29. $\displaystyle\int \frac{\cos x \sin x}{(\cos x - 4)(3\cos x + 1)} dx = \int \frac{\cos x \sin x}{3\cos^2 x - 11\cos x - 4} dx$

$\displaystyle\rightarrow -\int \frac{u}{3u^2 - 11u - 4} du$

$\displaystyle = -\frac{1}{3}\int \frac{u}{u^2 - 11u/3 - 4/3} du = -\frac{1}{3}\int \frac{u}{(u - 11/6)^2 - 169/36} du$

$\displaystyle\rightarrow -\frac{1}{3}\int \frac{w + 11/6}{w^2 - (13/6)^2} dw$

$\displaystyle = -\frac{4}{13}\ln|w - 13/6| - \frac{1}{39}\ln|w + 13/6|$

$\displaystyle\rightarrow -\frac{4}{13}\ln|u - 4| - \frac{1}{39}\ln|u + 1/3|$

$\displaystyle\rightarrow -\frac{4}{13}\ln|\cos x - 4| - \frac{1}{39}\ln|\cos x + 1/3|$

[Use formula #32.]

30. $\displaystyle\int \frac{e^{2x}}{\sqrt{e^{2x} - e^x + 1}} dx \rightarrow \int \frac{u}{\sqrt{u^2 - u + 1}} du = \int \frac{u}{\sqrt{(u - 1/2)^2 + 3/4}} du$

$\displaystyle\rightarrow \int \frac{w + 1/2}{\sqrt{w^2 + 3/4}} dw = \sqrt{w^2 + 3/4} + \frac{1}{2}\ln\left|w + \sqrt{w^2 + 3/4}\right|$

$\displaystyle\rightarrow \sqrt{u^2 - u + 1} + \frac{1}{2}\ln\left|u - 1/2 + \sqrt{u^2 - u + 1}\right|$

$\displaystyle\rightarrow \sqrt{e^{2x} - e^x + 1} + \frac{1}{2}\ln\left|e^x - 1/2 + \sqrt{e^{2x} - e^x + 1}\right|$

[Use formula #38.]

31. $\displaystyle\int x \sin(3x + 4)\, dx \rightarrow \frac{1}{9}\int (u - 4)\sin u\, du$

$\displaystyle = \frac{1}{9}(\sin u - u\cos u + 4\cos u)$

$\displaystyle\rightarrow \frac{1}{9}(\sin(3x + 4) - (3x + 4)\cos(3x + 4) + 4\cos(3x + 4))$

[Use formula #46 or #48.]

Section 7.1 The Idea of Approximation

7.1 The Idea of Approximation

1. (a) $I = \int_1^4 \dfrac{dx}{\sqrt{x}} = 2\sqrt{x}\Big]_1^4 = 4 - 2 = 2$

    (b) $L_3 = \dfrac{4-1}{3}(f(1) + f(2) + f(3)) = \dfrac{1}{\sqrt{1}} + \dfrac{1}{\sqrt{2}} + \dfrac{1}{\sqrt{3}} \approx 2.28446;$
    $R_3 = \dfrac{4-1}{3}(f(2) + f(3) + f(4)) = \dfrac{1}{\sqrt{2}} + \dfrac{1}{\sqrt{3}} + \dfrac{1}{\sqrt{4}} \approx 1.78446$

    (c) $|I - L_3| \approx 0.28446;\ |I - R_3| \approx 0.21554$

    (d) Yes — the actual approximation errors are less than $|f(4) - f(1)|\dfrac{4-1}{3} = \left|\dfrac{1}{\sqrt{4}} - 1\right| = \dfrac{1}{2}$ (the bound given in Theorem 1).

    (e) The approximation error made by T_3 is less than that allowed by Theorem 1: $T_3 = (L_3 + R_3)/2 \approx 2.03446$ so $|I - T_3| \approx 0.03446 \le 0.25$ (the error bound for the trapezoid rule estimate).

    (f) The value of n must be chosen so that $|f(4) - f(1)|\dfrac{4-1}{n} = \dfrac{3}{2n} \le 0.005$. This will be true if $n \ge 300$.

2. (a) $I = \dfrac{x^4}{4}\Big]_0^1 = \dfrac{1}{4}.$

    (b) $L_4 = (f(0) + f(1/4) + f(1/2) + f(3/4)) \cdot \dfrac{1}{4} = \left(0^3 + (1/4)^3 + (1/2)^3 + (3/4)^3\right) \cdot \dfrac{1}{4} = \dfrac{9}{64}.$
    Similarly, $R_4 = (f(1/4) + f(1/2) + f(3/4) + f(1)) \cdot \dfrac{1}{4} = \left((1/4)^3 + (1/2)^3 + (3/4)^3 + 1^3\right) \cdot \dfrac{1}{4} = \dfrac{25}{64}.$

    (c) $|I - L_4| = \left|\dfrac{1}{4} - \dfrac{9}{64}\right| = \dfrac{7}{64}.\ |I - R_4| = \left|\dfrac{1}{4} - \dfrac{25}{64}\right| = \dfrac{9}{64}.$

    (d) The error bounds are the same for both L_4 and R_4:
    $$|I - L_4| \le \dfrac{|f(1) - f(0)| \cdot (1 - 0)}{4} = \dfrac{1}{4}.$$
    The actual errors are less (as they should be).

    (e) $T_4 = \dfrac{L_4 + R_4}{2} = \dfrac{17}{64}$; the actual error is only 1/64—much less than what the theorem predicts.

    (f) For L_n and R_n, the theorem says we're OK if $\dfrac{|f(1) - f(0)| \cdot (1 - 0)}{n} = \dfrac{1}{n} \le 0.005$, i.e., if $n \ge 200$.

3. (a) A reasonable guess is around 1.

    (b) $L_{20} \approx 0.961197;\ R_{20} \approx 0.811215$. They overestimate and underestimate, respectively, since the function is decreasing on the interval of integration.

    (c) For L_n and R_n, the theorem says we're OK if $\dfrac{|f(3) - f(0)| \cdot (3 - 0)}{n} \approx \dfrac{3}{n} \le 0.005$, i.e., if $n \ge 600$.

4. For $I = \int_1^3 x\,dx$, $f(x) = x$, $a = 1$, $b = 3$, so
$$|f(b) - f(a)|\dfrac{(b-a)}{n} = 2 \cdot \dfrac{2}{n} = \dfrac{4}{n} \le 0.000005 \iff n \ge 800{,}000.$$

5. For $\int_1^2 x^2\,dx$ we need $n \ge 600{,}000$.

6. For $\int_1^4 \sqrt{x}\,dx$ we need $n \ge 600{,}000$.

7. For $\int_1^2 x^{-1}\,dx$ we need $n \geq 100{,}000$.

8. For $\int_2^3 \sin x\,dx$ we need $n \geq 153{,}636$.

9. $n \geq 800{,}000$

10. $n \geq 600{,}000$

11. $n \geq 600{,}000$

12. $n \geq 100{,}000$

13. $n \geq 153{,}636$

14. $n \geq 400{,}000$

15. $n \geq 300{,}000$

16. $n \geq 300{,}000$

17. $n \geq 50{,}000$

18. $n \geq 76{,}818$

19. (a) No. Since f is decreasing over the interval of integration, L_n overestimates I for any n.

   (b) By the theorem, $|I - L_{16}| \leq \dfrac{|f(1) - f(0)| \cdot (1 - 0)}{16} = \dfrac{3}{16}$.

   (c) The idea is that since f is monotone, we can find the *difference* between L_{16} and R_{16}. That is,
   $$|L_{16} - R_{16}| = \dfrac{|f(1) - f(0)| \cdot (1 - 0)}{16} = \dfrac{3}{16}.$$
   Moreover, since f is decreasing, we know that $L_n > R_n$, so $R_{16} = L_{16} + \dfrac{3}{16} = 5.5047$.

   (d) By definition, $T_{16} = R_{16} + L_{16})/2 = (5.3172 + 5.5047)/2 = 5.41095$. By the theorem, $|I - T_{16}| \leq \dfrac{|f(1) - f(0)| \cdot (1 - 0)}{2 \cdot 16} = \dfrac{3}{32}$.

20. (a) No. For any n, R_n underestimates I because the integrand — $1/(1+f(x))$ — is decreasing on the interval of integration.

   (b) Applying Theorem 1,
   $$|I - R_{10}| \leq \left|\dfrac{1}{10} - \dfrac{1}{3}\right| \dfrac{(6-1)}{10} = \dfrac{1}{3}.$$

   (c) $L_{10} = R_{10} - \dfrac{1}{2}\left(\dfrac{1}{10} - \dfrac{1}{3}\right) = 20.08536 + \dfrac{1}{3} \approx 20.41869$

21. A picture similar to that in the proof of Theorem 1 makes this clear.

22. For a *decreasing* function: $R_n \leq M_n \leq L_n$. (A picture should make this clear.)

23. The average value of f over the interval $[1, 5]$ is $\dfrac{1}{4}\int_1^5 f(x)\,dx$. Therefore, $T_n/4$ will estimate the average value of f over the interval $[1, 5]$ within 0.01 if n is chosen such that $\left|\int_1^5 f(x)\,dx - T_n\right| \leq 0.04$. Theorem 1 guarantees that this inequality holds for all $n \geq 22$. Thus, $T_{22}/4 \approx 0.9028$ approximates the average value of f over the interval $[1, 5]$ within 0.01.

SECTION 7.1 THE IDEA OF APPROXIMATION

24. (a) $I = \int_0^\pi \sin x \, dx = -\cos x \Big]_0^1 = 2.$

    (b) $|I - L_4| = |2 - \pi(\sqrt{2} + 1)| \approx 1.8961.$

    (c) Theorem 1 doesn't apply here, because the integrand is *not* monotone.

25. Notice that the integrand is *not* monotone over the interval $[0, 2]$. Thus before using Theorem 1 we should break the interval of integration two pieces, on each of which the integrand *is* monotone. A look at the graph shows that the function is *increasing* on $[0, \sqrt{\pi/2}]$ and *decreasing* on $[\sqrt{\pi/2}, 2]$. Thus we can approximate answers on each of these intervals, say with error less than 0.005, and add the results. The respective answers are about

    $$\int_0^{\sqrt{\pi/2}} \sin(x^2) \, dx \approx 0.549276; \qquad \int_{\sqrt{\pi/2}}^2 \sin(x^2) \, dx \approx 0.255652;$$

    a good estimate to I, therefore, is around 0.80 or 0.81. (More advanced methods can be used to show that the "true" answer is about 0.80485.)

26. The integrand is decreasing on the interval $(2, \pi)$ and increasing on the interval $(\pi, 5)$. NOTE: Do with a single sum? Idea is to use minimum and maximum values of integrand to computer bound on the approximation error.

27. The statement **must** be true. The condition $f'(x) > 0$ for all x in $[3, 8]$ means that f is increasing, so *all* left sums underestimate I.

28. The statement **cannot** be true. From the information given, we may infer that f is decreasing on the interval $[3, 8]$. Therefore, *all* left sums overestimate I.

29. The statement **may** be true. $|I - L_{1000}| \leq |-4 - 2| \dfrac{5}{1000} = 0.03.$

30. The statement **may** be true.

31. The statement **must** be true. Because f is monotone, Theorem 1 applies. It says that if f is monotone on $[3, 8]$, $f(3) = 5$, and $f(8) = 1$, then $|I - T_{1000}| \leq |f(8) - f(3)| \dfrac{5}{2000} = \dfrac{20}{2000} = 0.01 < 0.05.$

32. The statement **may** be true. However, for some monotone functions the L_n error may be less than what the bound guarantees (i.e., $|I - L_n| < 0.1$ may be true even though $n < 200$).

33. The statement **may** be true. If the function f is not monotone, it may happen that the actual approximation error made by R_{20} is greater than that of R_{10}.

34. The key idea is that for equally-spaced points $a = x_0 < x_1 < x_2 < \cdots < x_n = b$,

    $$L_n = f(x_0)\Delta x + f(x_1)\Delta x + f(x_2)\Delta x + \cdots + f(x_{n-1})\Delta x,$$

    while

    $$R_n = f(x_1)\Delta x + f(x_1)\Delta x + f(x_2)\Delta x + \cdots + f(x_n)\Delta x,$$

    where $\Delta x = (b-a)/n$. Thus, by algebra,

    $$R_n - L_n = f(x_n)\Delta x - f(x_0)\Delta x = (f(b) - f(a))\Delta x = (f(b) - f(a))\dfrac{b-a}{n}$$

    or, equivalently, $R_n = L_n + [f(b) - f(a)] \cdot \dfrac{(b-a)}{n}.$

35. Recall that $T_n = (L_n + R_n)/2$. Into this, substitute the expression for R_n derived in the previous exercise.

36. If $f(a) = f(b)$, then $f(b) - f(a) = 0$. In this case, $R_n = L_n$ (see Exercise 34), so $T_n = L_n$ (see Exercise 35).

37. (a) $T_{10} = \frac{1}{2}(L_{10} + R_{10}) = 9.495$. Using equation 6.1.1, $|I - T_{10}| \leq \frac{1}{2}|R_{10} - L_{10}| = 0.0818$.

   (b) Since f is an increasing function on the interval of integration, $L_{10} \leq I \leq R_{50}$ is true. Thus, the midpoint of the interval from L_{10} to R_{50}, $\frac{1}{2}(L_{10} + R_{50})$, cannot be further from I than one-half the length of the interval, $R_{50} - L_{10}$.

38. A picture shows that for any linear function f, T_n commits zero error in approximating $\int_a^b f(x)\,dx$.

39. As the picture in the proof of Theorem 1 shows, the "exact" integral I must lie somewhere *between* L_n and R_n, i.e., in an interval of length

$$|R_n - L_n| = |f(b) - f(a)|\,\Delta x = |f(b) - f(a)|\frac{(b-a)}{n}.$$

Since T_n is the *midpoint* of this same interval, I must lie within a distance of *half* the interval's width from T_n. (Draw L_n, R_n, and T_n on a number line to understand all this.)

40. The idea is that for any monotone function, M_n lies somewhere *between* L_n and R_n. Since both L_n and R_n satisfy an inequality of the desired type, so must M_n. (Think about it—draw a picture on the number line.)

41. (a) By the Theorem's formula, $|I - L_4| \leq \dfrac{9 \cdot 11}{n}$; thus we want $\dfrac{9 \cdot 11}{n} \leq 0.005$, or $n \geq 19{,}800$. A lot!

   (b) By the Theorem's formula, $|I - L_4| \leq \dfrac{8 \cdot 1}{n}$; thus we want $\dfrac{8}{n} \leq 0.004$, or $n \geq 2{,}000$.

   (c) By the Theorem's formula, $|I - L_4| \leq \dfrac{1 \cdot 10}{n}$; thus we want $\dfrac{10}{n} \leq 0.001$, or $n \geq 10{,}000$.

   (d) Adding the estimates produced in the previous two parts does the trick.

   (e) The estimate in part (d) requires significantly less computational effort.

42. (a) $n = 500$

   (b) $n = 219$

   (c) $n = 55$

   (d) The sum of the estimates from parts (b) and (c) approximates I within $0.009 + 0.001 = 0.01$.

   (e) The estimate computed in part (d) requires only about half as much computation effort as the estimate in part (a) — 274 versus 500 function evaluations.

43. The point of (iv), in each case, is to see that the ACTUAL errors committed are no more than the theoretical bounds guaranteed by Theorem 1. Each exercise has many parts; here they are, tabulated.

   (a) $I = \int_1^2 x^2\,dx = 7/3$.

n	4	8	16	32	64	128
L_n	1.9688	2.1485	2.2403	2.2867	2.3103	2.3220
$I - L_n$	0.3645	0.1848	0.0930	0.0466	0.0230	0.0113
Thm 1 bound	0.7500	0.3750	0.1875	0.0937	0.0468	0.0234

   (b) $I = \int_1^4 \sqrt{x}\,dx = 14/3$.

n	4	8	16	32	64	128
L_n	4.2802	4.4764	4.5724	4.6199	4.6435	4.6560
$I - L_n$	0.3865	0.1903	0.0943	0.0468	0.0232	0.0107
Thm 1 bound	0.75000	0.37500	0.18750	0.0937	0.0468	0.0234

(c) $I = \int_1^2 x^{-1}\,dx = \ln 2 \approx 0.6932$.

n	4	8	16	32	64	128
L_n	0.75953	0.72538	0.70900	0.70103	0.69709	0.69514
$I - L_n$	-0.0664	-0.0322	-0.01585	-0.0079	-0.0039	-0.0020
Thm 1 bound	-0.1250	-0.0625	-0.0312	-0.0156	-0.0078	-0.0039

(d) $I = \int_2^3 \sin x\,dx = -\cos 3 + \cos 2 \approx 0.57384$.

n	4	8	16	32	64	128
L_n	0.6669	0.6211	0.5976	0.5858	0.5798	0.5768
$I - L_n$	-0.0930	-0.0472	-0.0238	-0.0120	-0.0600	-0.0301
Thm 1 bound	-0.19205	-0.09602	-0.04801	-0.02400	-0.01200	-0.00600

45. The answers can be read from the tables found in Exercise 43. The point is to see that for the Left rule, *doubling* n roughly *halves* the error committed. Thus the error committed with $n = 256$ should be roughly $1/2$ that committed with $n = 128$.

7.2 More on Error: Left and Right Sums and the First Derivative

1. $I = 1$; $L_{10} = 1.000$; $R_{10} = 1.000$; $|I - L_{10}| = 0$; $|I - R_{10}| = 0$; $K_1 = 0$; $\dfrac{K_1(b-a)^2}{2n} = 0$

2. $I = 4.$; $L_{10} = 3.800$; $R_{10} = 4.200$; $|I - L_{10}| = 0.200$; $|I - R_{10}| = 0.200$; $K_1 = 1$; $\dfrac{K_1(b-a)^2}{2n} = 0.2000$

3. $I = 2.333$; $L_{10} = 2.185$; $R_{10} = 2.485$; $|I - L_{10}| = 0.148$; $|I - R_{10}| = 0.152$; $K_1 = 4$; $\dfrac{K_1(b-a)^2}{2n} = 0.2000$

4. $I = 4.667$; $L_{10} = 4.515$; $R_{10} = 4.815$; $|I - L_{10}| = 0.152$; $|I - R_{10}| = 0.148$; $K_1 = 0.5$; $\dfrac{K_1(b-a)^2}{2n} = 0.2250$

5. $I = 0.6931$; $L_{10} = 0.7187$; $R_{10} = 0.6687$; $|I - L_{10}| = 0.0256$; $|I - R_{10}| = 0.0244$;
   $K_1 = 1$; $\dfrac{K_1(b-a)^2}{2n} = 0.0500$

6. $I = 0.5738$; $L_{10} = 0.6118$; $R_{10} = 0.5350$; $|I - L_{10}| = 0.0380$; $|I - R_{10}| = 0.0388$;
   $K_1 = 1$; $\dfrac{K_1(b-a)^2}{2n} = 0.0500$

7. The condition $-4 \le f'(x) \le 3$ if $1 \le x \le 2$ implies that K_1 must be a number greater than $|-4| = 4$. Thus, Theorem 2 guarantees that $|I - L_n| \le 4/(2n)$.

8. For $\int_0^3 e^{-x^2}\,dx$ we consider $f'(x) = -\exp(-x^2)2x$. On $[0, 3]$, $|f'(x)| \le 0.86$, so $K_1 = 0.86$ works. Hence we need $n \ge 0.86 \cdot 3^2 \cdot 100 = 774$.

9. For $\int_0^2 \sin(x^2)\,dx$, $K_1 = 4$ works, so any $n \ge 4 \cdot 2^2 \cdot 100 = 1600$ will do.

10. For $\int_0^1 (1+x^2)^{-1}\,dx$, $K_1 = 0.7$ works so any $n \ge 0.7 \cdot 1^2 \cdot 100 = 70$ will do.

11. For $\int_1^{10} \sin(1/x)\,dx$, $K_1 = \cos 1$ works so any $n \ge \cos 1 \cdot 9^2 \cdot 100 \approx 4376.45$ will do.

12. For $I = \int_1^4 e^x/x\,dx$, we can use $K_1 = 11$, so Theorem 2 says that $|I - L_n| \le \dfrac{11 \cdot 3^2}{2n} = \dfrac{99}{2n} < \dfrac{50}{n}$. Theorem 1 also applies, since the integrand is monotone. It says that
    $$|I - L_n| \le \frac{|f(4) - f(1)| \cdot (4-1)}{n} \approx \frac{33}{n}.$$
    Here, in other words, Theorem 1 does better than Theorem 2.

13. (a) $I = \pi$ because the integral gives the area of the northeast quadrant of a circle of radius 2.

    (b) $L_{10} \approx 3.3045$; $|I - L_{10}| \approx 0.1629$.

    (c) Theorem 2 doesn't give a good bound here because we can't compute K_1—$f'(x)$ is unbounded on the interval $(0, 2)$.

    (d) Theorem 1 says that
    $$|L_{10} - I| \le \frac{|f(2) - f(0)| \cdot 2}{10} = 0.4.$$
    This number is larger, as it should be, than the actual error committed.

14. Since f' is negative on $[a, b]$, f is decreasing. Therefore $R_n \le I \le L_n$.

SECTION 7.2 MORE ON ERROR: LEFT AND RIGHT SUMS AND THE FIRST DERIVATIVE   39

15. The information given implies that f is increasing and concave down on the interval $[a, b]$. A sketch shows that the approximation error made by L_n includes all of the area corresponding to the approximation error made by T_n and more.

    An algebraic proof of this result is also possible. Since f is (strictly) increasing and (strictly) concave down on the interval of integration, $L_n < I < R_n$ and $I - T_n < 0$. Thus,

    $$(I - L_n) - (I - T_n) = T_n - L_n = \tfrac{1}{2}(R_n - L_n) > 0$$

    which implies that $|I - T_n| < |I - L_n|$.

17. (a) From the graph it is apparent that $|f'(x)| \leq 9$ if $0 \leq x \leq 4$. Thus, the error bound inequality

    $$|I - L_n| \leq \frac{9 \cdot 4^2}{2n} \leq 0.0001$$

    implies that any value of $n \geq 720{,}000$ will do the trick.

    (b) **No.** Since the value of $f'(x)$ is negative for all x in the interval of integration, f is decreasing over this interval. It follows that R_n underestimates I (i.e., $R_n < I$).

    (c) Since $F' = f$, we need a bound on $|f(x)|$ if $2 \leq x \leq 4$. From the graph, $|f'(x)| \leq 4$ on $2 \leq x \leq 4$ so $f(2) = 0 \implies |f(x)| \leq 4(x - 2)$ if $2 \leq x \leq 4$. Therefore, $|f(x)| \leq 8$ over the interval $[2, 4]$ and so $|L_n - \int_2^4 F(x)\,dx| \leq \frac{8 \cdot 2^2}{2n} \leq 0.01$ if $n \geq 1600$.

18. Any linear function (i.e., of the form $f(x) = ax + b$) will do.

19. Use a picture similar to that in Example 3.

20. We'll tabulate a lot of information first:

Exercise	I	T_{10}	M_{10}	$I - T_{10}$	$I - M_{10}$
1. $\int_2^3 1\,dx$	1.	1.0000	1.0000	0	0
2. $\int_1^3 x\,dx$	4.	4.0000	4.0000	0	0
3. $\int_1^2 x^2\,dx$	2.3333	2.3350	2.3327	-0.0017	0.0006
4. $\int_1^4 \sqrt{x}\,dx$	4.6667	4.6650	4.6677	0.0017	-0.0010
5. $\int_1^2 x^{-1}\,dx$	0.69315	0.69378	0.69284	-0.00063	0.00031
6. $\int_2^3 \sin x\,dx$	0.57384	0.57339	0.57410	0.00045	-0.00026

    (a) The approximation errors made by T_{10} and M_{10} are generally considerably less than those made by L_{10} and R_{10}.

    (b) Approximation errors made by T_{10} are generally about twice the size of, and opposite in sign to, those made by M_{10}.

    (c) T_{10} makes no approximation error on the first two—in both cases, the first derivative is constant.

    (d) T_{10} underestimates if the integrand is concave down, i.e., for integrands with *negative* second derivative.

    (e) T_{10} overestimates if the integrand is concave up, i.e., for integrands with *positive* second derivative.

    (f) M_{10} makes no approximation error on the first two—in both cases, the first derivative is constant.

    (g) M_{10} underestimates if the integrand is concave up, i.e., for integrands with *positive* second derivative.

    (h) M_{10} overestimates if the integrand is concave down, i.e., for integrands with *negative* second derivative.

21. (a) Since f is increasing, the approximation error made on each subinterval is nonnegative. Furthermore, this approximation error is less than or equal to $K_1 (\Delta x)^2/2$ (the area of a triangle of width Δx and height $K_1 \Delta x$). Since there are n subintervals,

$$0 \le I - L_n \le \sum_{i=1}^{n} \frac{K_1(\Delta x)^2}{2} = \sum_{i=1}^{n} \frac{K_1(b-a)^2}{2n^2} = \frac{K_1(b-a)^2}{2n}.$$

(b) Since f is increasing, the approximation error made on each subinterval is nonpositive. Now, an argument similar to that in part (a) leads to the desired result.

(c) When $f(x) = x$,

$$L_n = \left(\frac{b-a}{n}\right) \sum_{i=0}^{n-1} (a+i(b-a)/n) = \left(\frac{b-a}{n}\right) (na + (n-1)(b-a)) = a(b-a) + \frac{(b-a)^2}{2} - \frac{(b-a)^2}{2n}.$$

Therefore, $I - L_n = \left(\frac{b^2}{2} - \frac{a^2}{2}\right) - a(b-a) - \frac{(b-a)^2}{2} + \frac{(b-a)^2}{2n} = \frac{(b-a)^2}{2n}.$

(d) Taking the *average* of L_n and R_n will produce a better estimate: $(L_n + R_n)/2$. This is the Trapezoid rule.

23. $|I - T_n| = |(I - L_n)/2 + (I - R_n)/2| \le \frac{1}{2}|I - L_n| + \frac{1}{2}|I - L_n| \le \frac{K_1(b-a)^2}{2n}.$

24. The statement is true—the larger the derivative is on an interval, the worse L_n behaves there.

7.3 Trapezoid Sums, Midpoint Sums, and the Second Derivative

1. (a) $M_2 = \frac{1}{2}(f(1/4) + f(3/4)) = \frac{e^{1/16} + e^{9/16}}{2} \approx 1.409$.

   $T_2 = \frac{1}{4}(f(0) + 2f(1/2) + f(1)) = \frac{1 + 2e^{1/4} + e^1}{4} \approx 1.571$.

   (b) In all parts below, we'll use $K_1 = 6$, $K_2 = 16.31$. Then

   $L_{10} \approx 1.381$. Error bound: $|L_{10} - I| \leq \frac{K_1 \cdot 1}{20} = \frac{6}{20} = 0.3$.

   $R_{10} \approx 1.553$. Error bound: same as L_{10}.

   $M_{10} \approx 1.460$. Error bound: $|M_{10} - I| \leq \frac{K_2 \cdot 1^3}{24 \cdot 100} = \frac{16.31}{2400} \approx 0.007$.

   $T_{10} \approx 1.467$. Error bound: $|T_{10} - I| \leq \frac{K_2 \cdot 1^3}{12 \cdot 100} = \frac{16.31}{1200} \approx 0.0136$.

   L_{10} and M_{10} *under*estimate the exact value; the others underestimate.

   (c) We want $|M_n - I| \leq \frac{K_2(b-a)^3}{24n^2} = \frac{16.31(1)^3}{24n^2} \leq 0.0005$. The last inequality holds if $n^2 \geq \frac{16.31}{24(0.0005)} \approx 1359.17$, i.e., if $n \geq \sqrt{\frac{4000}{3}} \approx 36.87$. In particular, $n = 37$ is OK so the estimate $M_{37} \approx 1.46248$ is good to 3 decimal places.

2. For $I = \int_0^1 \sin x \, dx$, $K_2 = 1$ will do. (We could also use $K_2 = \sin 1 \approx 0.84$, but $K_1 = 1$ makes things simpler.) Thus

   $$|T_n - I| \leq \frac{1 \cdot 1^3}{12n^2} \leq 10^{-10} \iff n^2 \geq \frac{10^{10}}{12} \iff n \geq \frac{10^5}{\sqrt{12}} \approx 28{,}868.$$

   T_n computed with this (gigantic) value of n slightly *overestimates* the integral, because the integrand is concave down on $[0, 1]$.

3. Since $|f''(x)| \leq 2.5$ if $0 \leq x \leq 5$, $\left|T_n - \int_a^b f(x)\,dx\right| \leq 2.5 \cdot 5^3/12n^2$. The expression on the right is less than 0.001 if $n \geq 162$.

4. $K_2 = 0$; $I = 1$; $T_{10} = 1.0000$; $|I - T_{10}| = 0$; $\frac{K_2(b-a)^3}{12n^2} = 0$; $M_{10} = 1.0000$; $|I - M_{10}| = 0$; $\frac{K_2(b-a)^3}{24n^2} = 0$

5. $K_2 = 0$; $I = 4$; $T_{10} = 4.0000$; $|I - T_{10}| = 0$; $\frac{K_2(b-a)^3}{12n^2} = 0$; $M_{10} = 4.0000$; $|I - M_{10}| = 0$; $\frac{K_2(b-a)^3}{24n^2} = 0$

6. $K_2 = 2$; $I = 2.3333$; $T_{10} = 2.3350$; $|I - T_{10}| = 0.0017$; $\frac{K_2(b-a)^3}{12n^2} = 0.0017$; $M_{10} = 2.3325$; $|I - M_{10}| = 0.000\overline{83}$; $\frac{K_2(b-a)^3}{24n^2} = 0.000\overline{83}$

7. $K_2 = 0.25$; $I = 4.6667$; $T_{10} = 4.6648$; $|I - T_{10}| = 0.00187$; $\frac{K_2(b-a)^3}{12n^2} = 0.0056$; $M_{10} = 4.6676$; $|I - M_{10}| = 0.0009$; $\frac{K_2(b-a)^3}{24n^2} = 0.0028$

8. $K_2 = 2$; $I = 0.6932$; $T_{10} = 0.6938$; $|I - T_{10}| = 0.0006$; $\frac{K_2(b-a)^3}{12n^2} = 0.0017$; $M_{10} = 0.6928$; $|I - M_{10}| = 0.00031$; $\frac{K_2(b-a)^3}{24n^2} = 0.0008$

9. $K_2 = 1$; $I = 0.57384$; $T_{10} = 0.57337$; $|I - T_{10}| = 0.00047$; $\dfrac{K_2(b-a)^3}{12n^2} = 0.00083$; $M_{10} = 0.57408$;
$|I - M_{10}| = 0.00024$; $\dfrac{K_2(b-a)^3}{24n^2} = 0.00042$

10. For $\int_0^3 e^{-x^2}\,dx$, $K_2 = 2$ works. Therefore, $n^2 \geq \dfrac{2 \cdot (3-0)^3}{24 \cdot 0.005} = 450 \implies n \geq 22$ will do.

11. For $\int_0^2 \sin(x^2)\,dx$, $K_2 = 10.80156$ works. Therefore, $n^2 \geq \dfrac{10.80156 \cdot 2^3}{24 \cdot 0.005} = 720.104 \implies n \geq 27$ will do.

12. For $\int_0^1 (1+x^2)^{-1}\,dx$, $K_2 = 2$. Therefore, $n \geq \dfrac{2 \cdot 1^3}{24 \cdot 0.005} \approx 16.66667 \implies n \geq 5$ will do.

13. For $\int_1^{10} \sin(1/x)\,dx$, $K_2 = 0.425$ works. Therefore, $n^2 \geq \dfrac{0.425 \cdot 9^3}{24 \cdot 0.005} = 2581.875 \implies n \geq 51$ will do.

14. Let $I = \int_a^b f(x)\,dx$. Since f is increasing on $[a, b]$, the left rule *underestimates* I, the right rule *overestimates*, and the other two lie in between. Since f is concave up on the interval $[a, b]$, $M_n \leq T_n$ must be true. Thus, $L_n \leq M_n \leq T_n \leq R_n$ so $L_n = 8.52974$, $M_n = 9.71090$, $T_n = 9.74890$, and $R_n = 11.04407$.

15. Because the function is monotone increasing and concave down, L_{30} and T_{30} underestimate I; M_{30} and R_{30} overestimate I. Thus we have $L_{30} < T_{30} < I < M_{30} < R_{30}$.

16. Since $F''(x) = f'(x) \geq 0$ on $[a, b]$, T_{100} overestimates the value of I.

17. The graph shows that f'' is positive—hence f is *concave up*—over the interval of integration.

    (a) $L_{100} \leq R_{100}$ could be true or false, depending on whether f is increasing or decreasing. The graph of f'' doesn't tell which is the case.

    (b) $T_{200} \leq M_{200}$ is false: If the graph of f is concave up, T_n must *overestimate* I and M_n must *underestimate*. Thus $T_n > M_n$ for any n.

    (c) $M_{50} \leq L_{50}$ could be true or false, depending on whether f is increasing or decreasing. The graph of f'' doesn't tell which is the case.

18. (a) **No.** Since f' is an increasing function on the interval of integration, f is concave upwards on this interval. It follows that M_n *under*estimates I (i.e., $M_n < I$).

    (b) Since $F'' = f'$, the graph shows that we can take $K_2 = 5.5$. Thus, any $n \geq 5$ will work.

19. (a) **Must** be true. Since f is concave upwards on the interval of integration, any trapezoidal rule estimate overestimates I (i.e., $I - T_n < 0$).

    (b) **Cannot** be true. Since f is decreasing on the interval of integration $M_n > R_n$ for every n. [If f is decreasing on an interval $[a, b]$, then $m = (a+b)/2$ is the midpoint of the interval and $f(m) > f(b)$.]

    (c) **Cannot** be true. Because f is decreasing and concave upwards on the interval of integration, both L_n and T_n overestimate I for any n. Also, $T_n < L_n$ since $R_n < L_n$ (f is decreasing) and $T_n = (L_n + R_n)/2$. Together, these results imply that $L_n - I > T_n - I > 0 \iff |I - L_n| > |I - T_n|$.

20. (a) $R_4 = 0.75 \cdot (26.522 + 48.755 + 68.328 + 86.790) = 172.79625$.
    $|I - R_4| \leq 0.75 \cdot |86.790 - 2.0000| = 63.5925$.

    (b) $T_4 = \frac{1}{2}(L_4 + R_4) = \frac{1}{2} \cdot 0.75 \cdot (2 - 86.790) + R_4 = 141.00$.
    From the graph, it is apparent that $|f''(x)| < 6$ if $-1 \leq x \leq 2$ so, taking $K_2 = 6$, we have
    $$|I - T_4| \leq \dfrac{6 \cdot 3^3}{12 \cdot 4^2} = 0.84375.$$

SECTION 7.3 TRAPEZOID SUMS, MIDPOINT SUMS, AND THE SECOND DERIVATIVE

21. If $f(x) = x^2$, then f is concave up, so M_n underestimates I and the approximation error is as bad as Theorem 3 allows.

22. If $f(x) = -x^2$, then f is concave down, so M_n overestimates I and the approximation error is as bad as Theorem 3 allows.

23. If $f(x) = -x^2$, then f is concave down, so T_n underestimates I and the approximation error is as bad as Theorem 3 allows.

24. If $f(x) = x^2$, then f is concave up, so T_n overestimates I and the approximation error i as bad as Theorem 3 allows.

25. (a) The error bounds for M_n and M_{10n} are

$$|I - M_n| \leq \frac{K_2(b-a)^3}{24n^2}; \qquad |I - M_{10n}| \leq \frac{K_2(b-a)^3}{24 \cdot (10n)^2} = \frac{K_2(b-a)^3}{24 \cdot 100n^2}.$$

The second bound has an extra factor of 100 in the denominator. This means that using ten times as many subintervals in M_n gives about *two* extra decimal places of accuracy.

(b) The error bounds for L_n and L_{10n} are

$$|I - L_n| \leq \frac{K_1(b-a)^2}{2n}; \qquad |I - L_{10n}| \leq \frac{K_1(b-a)^2}{2 \cdot 10n} = \frac{K_1(b-a)^2}{20n}.$$

Here the second bound has an extra factor of 10 in the denominator, using ten times as many subintervals in L_n gives only *one* extra decimal place of accuracy.

26. (a) Since $f'' \geq 0$, the integrand is concave up, so M_n underestimates. In other words, $0 \leq I - M_n$. The second inequality comes from the error bound formula.

(b) Since $f'' \geq 0$, the integrand is concave up, so T_n underestimates. In other words, $I - T_n \leq 0$. The first inequality comes from the error bound formula.

(c) The first two parts show that M_n and T_n often commit errors of opposite sign: one overestimates, the other underestimates. Thus I typically lies *between* T_n and M_n.

The error bound formulas also show (see the denominators) that the worst-case M_n error is *half* the worst-case T_n error. Thus it's reasonable to use, as a better estimate of I, the number 2/3 of the way from T_n to M_n, i.e., $\frac{2}{3}M_n + \frac{1}{3}T_n$.

This estimate *is* used in practice—it's called **Simpson's rule**.

27. (a) Here's a table of values for the various integrals. (The final column is sensitive to roundoff errors—so results may vary.)

Exercise	I	T_{10}	M_{10}	$I - T_{10}$	$I - M_{10}$	$\dfrac{I - M_{10}}{I - T_{10}}$
5. $\int_1^2 x^2\, dx$	2.3333	2.3350	2.3325	−0.00167	0.00083	−0.5000
6. $\int_1^4 \sqrt{x}\, dx$	4.6667	4.6648	4.6676	0.001871	−0.000934	−0.4992
7. $\int_1^2 x^{-1}\, dx$	0.6931	0.6938	0.6928	−0.000624	0.000312	−0.4995
8. $\int_2^3 \sin x\, dx$	0.5738	0.5734	0.5741	0.000478	−0.000239	−0.5001

The results show the general pattern: the T_{10} error is about *twice* the M_{10} error, and the errors are *opposite in sign*.

(b) $I = \int_0^1 \sqrt{x}\, dx = 2/3$; $T_{10} \approx 0.6605$; $M_{10} \approx 0.6684$; $I - T_{10} \approx 0.006157$, $I - M_{10} \approx -0.001717$, $(I - M_{10})/(I - T_{10}) \approx -0.27889$.

The surprise here is that this time the last quantity is not near -0.5 as in each case in the previous part. The difference is that the graph of this integrand is vertical at the left endpoint of integration. This fouls up the usual error estimates.

7.4 Simpson's Rule

1. For $I = \int_a^b dx = x\big]_a^b = b - a$. We need to show that S_2 has the *same* value as I. Here goes:

$$S_2 = \frac{b-a}{6}(f(a) + 4f((a+b)/2) + f(b)) = \frac{b-a}{6}(1 + 4 \cdot 1 + 1) = b - a.$$

2. For $I = \int_a^b x\, dx = \frac{x^2}{2}\big]_a^b = \frac{b^2-a^2}{2}$. We need to show that S_2 has the *same* value as I. Here goes:

$$\begin{aligned} S_2 &= \frac{b-a}{6}(f(a) + 4f((a+b)/2) + f(b)) \\ &= \frac{b-a}{6}(a + 4((a+b)/2) + b) \\ &= \frac{b-a}{6}(3a + 3b) \\ &= \frac{b^2 - a^2}{2}. \end{aligned}$$

3. For $I = \int_a^b x^2\, dx = \frac{x^3}{3}\big]_a^b = \frac{b^3 - a^3}{3}$. We need to show that S_2 has the *same* value as I. Here goes:

$$S_2 = \frac{b-a}{6}(f(a) + 4f((a+b)/2) + f(b)) = \frac{b-a}{6}\left(a^2 + 4\frac{(a+b)^2}{4} + b^2\right)$$

$$= \frac{b-a}{6}\left(2a^2 + 2ab + 2b^2\right) = \frac{b-a}{3}\left(a^2 + ab + b^2\right) = \frac{b^3 - a^3}{3}.$$

4. $I = \int_a^b x^3\, dx = \frac{x^4}{4}\big]_a^b = \frac{b^4 - a^4}{4}$. We need to show that S_2 has the same value. Considerable algebra is involved. Here goes:

$$\begin{aligned} S_2 &= \frac{b-a}{6}(f(a) + 4f((a+b)/2) + f(b)) \\ &= \frac{b-a}{6}\left(a^3 + 4(a+b)^3/8 + b^3\right) \\ &= \frac{b-a}{6}\left(a^3 + (a^3 + 3ab^2 + 3a^2b + b^3)/2 + b^3\right) \\ &= \frac{b-a}{12}\left(3a^3 + 3ab^2 + 3a^2b + 3b^3\right) \\ &= \frac{b-a}{4}\left(a^3 + ab^2 + a^2b + b^3\right) = \frac{b^4 - a^4}{4}. \end{aligned}$$

5. Let $I = \int_a^b x^4\, dx = \frac{b^5}{5} - \frac{a^5}{5}$. By comparison, $S_2 = \frac{b-a}{6}\left(a^4 + 4((b+a)/2)^4 + b^4\right)$.

Some careful algebra now shows: $|I - S_2| = \left|\frac{b^5 - a^5}{5} - \frac{(b-a)}{6}\left(a^4 + \frac{(b+a)^4}{4} + b^4\right)\right| = \frac{(b-a)^5}{120}$.

The error bound is: $|S_2 - I| \leq \frac{K_4(b-a)^5}{180n^4} = \frac{24(b-a)^5}{180 \cdot 2^4} = \frac{(b-a)^5}{120}$. Thus S_2 does commit the maximum possible error.

6. (a) Subdividing [0, 2] into 4 pieces creates partition points at 0, 1/2, 1, 3/2, and 2. Therefore:

$$S_4 = \frac{2-0}{3 \cdot 4}\left(f(0)+4f(1/2)+2f(1)+4f(3/2)+f(2)\right) = \frac{1}{6}\left(e^0 + 4e^{1/4} + 2e^1 + 4e^{9/4} + e^4\right) \approx 17.353.$$

(b) Calculating $f^{(4)}$ and plotting it over [0, 2] suggests we use something like $K_4 = 25{,}200$. Therefore

$$\text{error} \leq \frac{25{,}200 \cdot 2^5}{180 \cdot 4^4} = \frac{35}{2} = 17.5.$$

(That's a lot of error!)

(c) To assure error no more than 10^{-5} we'd need

$$\frac{25{,}200 \cdot 2^5}{180 n^4} < 10^{-5}.$$

Solving this for n gives $n \geq 145.5$, so $n = 146$ works. ($S_{146} \approx 16.45263$, by the way.)

7. (a) $I = \int_0^1 \cos(100x)\,dx = \left.\dfrac{\sin(100x)}{100}\right]_0^1 = \dfrac{\sin 100}{100} \approx -0.0050637.$

(b) $S_{10} \approx 0.036019$; thus the actual approximation error is $|I - S_{10}| \approx 0.041083 < 0.05.$

(c) From the error bound formula, with $K_4 = 100^4 = 10^8$ and $n = 10$, we get

$$|\text{error}| \leq \frac{K_4 \cdot 1^5}{180 \cdot 10^8} = \frac{10^8}{180 \cdot 10^4} \approx 55.556.$$

This bound is large because K_4 is so large.

8. (a) $S_4 = \frac{1}{6}(\cos 5 + 4\cos 5.5 + 2\cos 6 + 4\cos 6.5 + \cos 7) \approx 1.6165.$

(b) Since the fourth derivative of $\cos x$ is $\cos x$, we may take $K_4 = 1$. Then,

$$|I - S_4| \leq \frac{1 \cdot 2^5}{180 \cdot 4^4} = \frac{1}{1440} \approx 0.0006944.$$

(c) $I = \int_5^7 \cos x\,dx = \sin 7 - \sin 5 \approx 1.6159.$ Thus, S_4 *over*estimates I.

9. Let $I = \int_1^7 f(x)\,dx$. Since f is positive on the interval of integration, all five estimates are positive numbers. Since f is increasing on the interval of integration, $L_{100} \leq I \leq R_{100}$. Since f is concave down on the interval of integration, $T_{100} \leq I \leq M_{100}$. Furthermore, since T_{100} is the average of L_{100} and R_{100}, $L_{100} \leq T_{100} \leq R_{100}$. Also, S_{100} is a weighted average of T_{100} and M_{100}, so $T_{100} \leq S_{100} \leq M_{100}$. Therefore,

$$L_{100} \leq T_{100} \leq S_{100} \leq M_{100} \leq R_{100}.$$

10. The key idea is that the error committed by S_n is inversely proportional to the *fourth* power of n. Thus, e.g., multiplying n by 10 diminishes the error by a factor of 10^4.

    (a) If S_{10} commits error ≤ 0.01, then S_{100} makes error $\leq 0.01/10^4 = 10^{-6}$.

    (b) If S_{10} commits error ≤ 0.01, then S_{1000} makes error $\leq 0.01/100^4 = 10^{-10}$.

11. $S_4 = \frac{1}{3}(2M_2 + T_2) = 141.425.$

    When $-1 \leq x \leq 2$, $\left|f^{(4)}(x)\right| \leq 8 = K_4$, so $|I - S_4| \leq \dfrac{8 \cdot 2^5}{180 \cdot 4^4} = 0.0421875.$

SECTION 7.4   SIMPSON'S RULE   47

12. By examining graphs of the first, second, and fourth derivatives of the integrand, we determine that $K_1 \geq 1$, $K_2 \geq 3.25$, and $K_4 \geq 47.5$. Thus, to achieve the desired accuracy, L_n or R_n requires $n \geq 2000$; T_n requires $n \geq 45$; M_n requires $n \geq 33$; and, S_n requires $n \geq 10$. [NOTES: The integrand is not monotonic over the interval of integration, so Theorem 1 cannot be used to produce an error bound. Also, this integral cannot be evaluated using the Fundamental Theorem of Calculus because no elementary antiderivative exists.]

    Sample answers: $M_{33} = 0.45148$, $S_{10} = 0.45131$.

13. For $I = \int_0^1 \sin(\sin(x))\,dx$ we can use $K_2 = 1$ and $K_4 = 3.8$.

    (a) For the midpoint rule we have M_n error $\leq \dfrac{1 \cdot 1}{24n^2} = \dfrac{1}{24n^2}$. Setting $n = 4$, $n = 8$, and $n = 16$ gives, respectively: M_4 error $\leq \dfrac{1}{384} \approx 0.0026$; M_8 error $\leq \dfrac{1}{4 \cdot 384} \approx 0.00065$; M_{16} error $\leq \dfrac{1}{16 \cdot 384} \approx 0.000163$.

    (b) For Simpson's rule we have S_n error $\leq \dfrac{1 \cdot 3.8}{180 n^4} = \dfrac{3.8}{180 n^4}$. Setting $n = 4$, $n = 8$, and $n = 16$ gives, respectively: S_4 error $\leq \dfrac{3.8}{180 \cdot 4^4} \approx 8.2465 \times 10^{-5}$; S_8 error $\leq \dfrac{3.8}{180 \cdot 8^4} \approx 5.1541 \times 10^{-6}$; S_{16} error $\leq \dfrac{3.8}{180 \cdot 16^4} \approx 3.2213 \times 10^{-7}$.

14. (a) We'll do L_4 explicitly; the others are similar.

    $$L_4 = \frac{1}{4}\bigl(f(0) + f(0.25) + f(0.5) + f(0.75)\bigr) = 1.1485.$$

    Similar calculations show:

    $$R_4 = 1.1805; \quad M_2 = 1.1345; \quad T_4 = 1.1645; \quad S_4 = 1.1545.$$

    (b) Upward concavity means that T_4 must *overestimate* and M_2 must *underestimate*. Therefore I has to lie in the interval $[1.1345, 1.1645]$. (The S_4 estimate is consistent with this.)

15. (a) We're given that $f^{(4)}(x) = \bigl(\cos^4 x - 6\cos^2 x \sin x - 4\cos^2 x + 3\sin^2 x + \sin x\bigr) \cdot e^{\sin x}$. Recall that (1) sines and cosines always lie in $[-1, 1]$; and (2) the absolute value of a sum is no more than the sum of the absolute values. It follows that for x in $[0, 1]$,

    $$|f^{(4)}(x)| \leq (1 + 6 + 4 + 3 + 1)e^1 = 15e \approx 40.8 < 41.$$

    (b) The graph of $f^{(4)}$ on $[0, 1]$ gives a better bound—it shows that $|f^{(4)}| < 11$.

    (c) Using $K_4 = 11$, we see that get $|S_n - I| \leq 0.001$ if $\dfrac{11 \cdot (200\pi)^5}{180 n^4} \leq 0.001$. Solving this inequality for n gives $n \geq 8795$; thus $n = 8796$ is OK. (This is a huge number—somewhat impractical for real calculations.)

    (d) $I = 100 \int_0^{2\pi} f(x)\,dx$ because f is *periodic*; it *repeats* itself every 2π. On $[-50\pi, 150\pi]$, it repeats itself 100 times.

    (e) For $\int_0^{2\pi} f(x)\,dx$, the Simpson error formula says error $\leq \dfrac{K_4 \cdot (2\pi)^5}{180 n^4} = \dfrac{11 \cdot (2\pi)^5}{180 n^4} \leq 0.00001$; solving for n gives $n \geq 87.9$. Thus $n = 88$ will do. (In fact, $S_{88} \approx 7.9549265$; this is guaranteed correct within 0.00001.)

    (f) To find the original integral, we can multiply our estimate for $\int_0^{2\pi} f(x)\,dx$ (found with S_{88}) by 100; the error won't be more than $100 \cdot 0.00001 = 0.001$. (The result: $\int_{-50\pi}^{150\pi} f(x)\,dx \approx 795.493$ is correct to within 0.001.)

8.1 Introduction

1. We know that $r_B(t) = A + B(t-35) + C(t-35)^2$ for some constants A, B, and C. We'll show that the only possible values for A, B, and C are $A = 13$, $B = 0$, and $C = -1/100$.

   Note first that
   $$r_B(t) = A + B(t-35) + C(t-35)^2 \implies r'_B(t) = B + 2C(t-35).$$
   From the problem we know that $r_B(35) = 13$, $r_B(60) = 27/4$, and $r'_B(35) = 0$. This leads to three simple conditions on A, B, and C:
   $$r_B(35) = A = 13; \quad r'_B(35) = B = 0; \quad r_B(60) = A + B \cdot 25 + C \cdot 25^2 = 27/4.$$
   Thus $A = 13$, $B = 0$, and (from the last equation) $C = -1/100$.

2. We want to find a quadratic function $r_J(t)$ such that
   $$r_J(5) = 4; \quad r_J(15) = 9; \quad r_J(25) = 12; \quad r_J(35) = 13; \quad r_J(45) = 12; \quad r_J(55) = 9.$$
   Since we're looking for a *quadratic* function, any three of these data are enough—a quadratic function is determined by any three points on its graph.

   There are several ways to proceed. One way is to write $r_J(t) = A + Bt + Ct^2$ and use any three equations above to solve for A, B, and C. The result is
   $$r_J(t) = \frac{3}{4} + \frac{7}{10}t - \frac{1}{100}t^2.$$

   Another approach (which gives the same result!) is to notice the "symmetry" of the values around $t = 35$. If we then write $r_J(t) = A + B(t-35) + C(t-35)^2$, then
   $$r_J(35) = 13 \implies A = 13;$$
   $$r_J(25) = 12 \implies 12 = 13 - 10B + 100C;$$
   $$r_J(45) = 12 \implies 12 = 13 + 10B + 100C.$$
   Solving the last two equations gives $B = 0$ and $C = -1/100$. Thus $r_J(t) = 13 - (t-35)^2/100$. (Expanding the formula in powers of t gives $r_J(t) = 3/4 + 7t/10 - t^2/100$, as above.) In fact, r_J and r_B have the *same* formula, so their integrals from $t = 0$ to $t = 60$ must be equal. The calculation is easy: $\int_0^{60} r_J(t)\,dt = 585$. The bottom line is that if we estimate the harvest total using quadratic functions, our estimates for Brown and Jones are the *same*—585 bushels over the hour.

3. (a) The appropriate integral is $\int_5^{55} r_B(t)\,dt = \int_5^{55}\left(13 - \frac{(t-35)^2}{100}\right)dt = \frac{1600}{3}$. Thus Brown harvests about 533.33 bushels from $t = 5$ to $t = 55$.

   (b) We want to estimate the integral $\int_5^{55} r_J(t)\,dt$ using T_5, the trapezoid rule with 5 subdivisions, each of length 10. Here's the result:
   $$\left(\frac{r_J(5) + r_J(15)}{2} + \frac{r_J(15) + r_J(25)}{2} + \cdots + \frac{r_J(35) + r_J(45)}{2}\right) \cdot 10 = 525.$$
   By this estimate, therefore, Jones harvests 525 bushels over the period.

4. (a) The curve appears to have length around 1.5. (Note that a diagonal line from $(0,0)$ to $(1,1)$ has length $\sqrt{2}$.)

   (b) The shaded region appears to have area around 1.5, since it resembles a trapezoid with base 1 and altitudes 1 and 2.

SECTION 8.1 INTRODUCTION 49

(c) The midpoint rule error bound formula says that $|M_n - I| \le \dfrac{K_2(b-a)^3}{24n^2}$. Here we can use $n = 20$, $a = 0$, $b = 1$, and $K_2 = 4$. (To find K_2, plot the second derivative of $y = \sqrt{1+4x^2}$ over the interval $[0, 1]$.) This gives $|M_{20} - I| \le \dfrac{4}{24 \cdot 20^2} = \dfrac{1}{2400} \approx 0.0004$.

(d) We want to find $I = \int_0^1 \sqrt{1+4x^2}\, dx$ symbolically. One way is to write $\sqrt{1+4x^2} = 2\sqrt{1/4 + x^2}$; now we can use Formula #38 in the table of integrals, with $p = 1/2$. The result:

$$\int \sqrt{1+4x^2}\, dx = 2 \int \sqrt{1/4 + x^2}\, dx = x\sqrt{x^2 + 1/4} + \frac{\ln(x + \sqrt{x^2 + 1/4})}{4}.$$

Therefore,

$$\int_0^1 \sqrt{1+4x^2}\, dx = \left. x\sqrt{x^2+1/4} + \frac{\ln(x+\sqrt{x^2+1/4})}{4} \right|_0^1 = \frac{\sqrt{5}}{2} + \frac{\ln(2+\sqrt{5})}{4} \approx 1.4789.$$

5. (a) The linear function $f(x) = bx/a$ does the job. The length of its graph from $x = 0$ to $x = a$ is given by the integral

$$\int_0^a \sqrt{1 + f'(x)^2}\, dx = \int_0^a \sqrt{1 + b^2/a^2}\, dx = a\sqrt{1 + b^2/a^2} = \sqrt{a^2 + b^2}.$$

(An easier way to find the answer is to use the distance formula in the plane.)

(b) If $f(x) = x^2 + 1$, then $f'(x) = 2x$, so we want the integral $I = \int_0^1 \sqrt{1 + f'(x)^2}\, dx = \int_0^1 \sqrt{1+4x^2}\, dx$. We estimated this same integral in Example 2, where we got $M_{20} \approx 1.479$.

(c) For each of the two curves in question we have $dy/dx = \cos x$, so the length integral is the same in each case: $I = \int_0^\pi \sqrt{1 + f'(x)^2}\, dx = \int_0^\pi \sqrt{1 + \cos^2 x}\, dx$. For this integral, $M_{20} \approx 3.820$—that's a good estimate for the length of both curves.

(d) If $g(x) = f(x) + C$, for any constant C, then $g'(x) = f'(x)$, so the length of the g-graph from $x = a$ to $x = b$ is $\int_a^b \sqrt{1 + g'(x)^2}\, dx = \int_a^b \sqrt{1 + f'(x)^2}\, dx$. The last quantity is independent of C. Geometrically, the idea is that adding a constant C to f raises or lowers the graph of f, but *doesn't change its length*.

6. The integral has the same *numerical* value in each part: $\int_0^3 f(t)\, dt = \int_0^3 \left(5t^2 - 20t + 50\right) dt = 105$. Differences among parts have to do only with interpretation.

(a) Meaning: the car traveled 105 miles during the three hours from midnight to 3 am.

(b) Meaning: the car traveled 105 feet in the first 3 seconds after midnight.

(c) Meaning: the car's velocity increased by 105 feet per minute over the first 3 minutes after midnight.

7. (a) $I = \int_0^1 \sqrt{1+x}\, dx = \left. \dfrac{2}{3}(1+x)^{3/2} \right|_0^1 = \dfrac{4\sqrt{2} - 2}{3} \approx 1.219.$

(b) I is the area under the curve $y = \sqrt{1+x}$ from $x = 0$ to $x = 1$. (The region in question is more or less trapezoidal, with base 1 and altitudes 1 and $\sqrt{2}$.)

(c) We want a function f for which
$$I = \int_0^1 \sqrt{1+x}\, dx = \int_0^1 \sqrt{1+f'(x)^2}\, dx.$$
Let's look, therefore, for an f for which $f'(x)^2 = x$, or $f'(x) = \sqrt{x}$. *Any* antiderivative of $\sqrt{x}$ will do; let's use $f(x) = 2x^{3/2}/3$. Plotting this f over $[0, 1]$ gives a graph whose length appears to be around 1.2, as the previous part suggests.

(d) Any antiderivative of $\sqrt{x}$, i.e., any function of the form $f(x) = \dfrac{2}{3}x^{3/2} + C$, where C is a constant. (There are other possibilities, too. For instance, we could use $f(x) = -2x^{3/2}/3$, the *opposite* of the function in the previous part.

8.2 Finding Volumes by Integration

1. $V = \int_0^8 \pi \left(x^3\right)^2 dx = \dfrac{8^7 \pi}{7}$

2. $V = \int_0^1 \pi \left(1^2 - \left(x^4\right)^2\right) dx = \dfrac{8\pi}{9}$

3. $V = \int_0^2 \pi \left((x+6)^2 - \left(x^3\right)^2\right) dx = \dfrac{1688\pi}{21}$

4. $V = \int_0^1 \pi \left(\left(x^2\right)^2 - \left(x^3\right)^2\right) dx = \dfrac{2\pi}{35}$

5. $V = \int_0^2 \pi \left(4^2 - \left(y^2\right)^2\right) dy = \dfrac{128\pi}{5}$

6. $V = \int_0^1 \pi \left(\left(\sqrt{y}\right)^2 - \left(y^2\right)^2\right) dy = \dfrac{3\pi}{10}$

7. $V = \int_0^4 \pi \left(\sqrt{y}\right)^2 dy - \int_1^4 \pi \left(\log_2 y\right)^2 dy = \pi \left(\dfrac{16}{\ln 2} - 8 - \dfrac{6}{(\ln 2)^2}\right) \approx 8.15214$

8. $V = \int_0^e \pi 1^2 dy - \int_1^e \pi (\ln y)^2 dy = 2\pi$

9. At height y above the base, a cross section parallel to the base of the cone is a circle of radius $\dfrac{r}{h}(h - y)$. Thus, the volume of the cone is

$$V = \int_0^h \pi \left(\dfrac{r}{h}(h-y)\right)^2 dy = \dfrac{\pi r^2}{h^2} \int_{-h}^0 u^2 \, du = \dfrac{\pi r^2 h}{3}.$$

[The substitution $u = h - y$ was used to evaluate the integral.]

Alternatively, if the vertex of the cone is placed at the origin and the center of the base is placed on the positive x-axis, the area of a cross-section parallel to the base is $A(x) = \pi(rx/h)^2$. Thus, the volume of the cone is

$$V = \dfrac{\pi r^2}{h^2} \int_0^h x^2 \, dx = \dfrac{\pi r^2 h}{3}.$$

10. At height y above the base, a cross section parallel to the base of the pyramid is a square with edge length $\ell - \dfrac{\ell}{h}y$. Thus, the volume of the pyramid is

$$V = \int_0^h \ell^2 \left(1 - \dfrac{y}{h}\right)^2 dy = h\ell^2 \int_0^1 u^2 \, du = \tfrac{1}{3} h\ell^2.$$

11. (a) $\pi \int_0^4 ((6-y) - (-2))^2 \, dy - \pi \int_0^4 (\sqrt{y} - (-2))^2 \, dy$

    (b) $\pi \int_0^2 (x^2 - (-1))^2 \, dx + \pi \int_2^6 ((6-x) - (-1))^2 \, dx$

12. $V = \int_0^1 \pi \left(1^2 - \left(1 - \sqrt{x}\right)^2\right) dx = \dfrac{5\pi}{6}$

13. $V = \int_{-1}^2 \pi \left(((x+1) + 1)^2 - \left((x^2 - 1) + 1\right)^2\right) dx = \dfrac{72\pi}{5}$

14. $V = \int_{-4}^{0} \pi \left(\left(2+\sqrt{x+4}\right)^2 - \left(2-\sqrt{x+4}\right)^2 \right) dx + \int_{0}^{5} \pi \left(\left(2+\sqrt{x+4}\right)^2 - (2-(2-x))^2 \right) dx = \frac{128\pi}{3} + \frac{123\pi}{2} = \frac{625\pi}{6}$

15. $V = \int_{0}^{1} \pi \left(\left(\sqrt{x}+2\right)^2 - \left(x^2+2\right)^2 \right) dx = \frac{49\pi}{30}$

16. The area of an isosceles right triangle with hypotenuse h is $h^2/4$. Thus, the volume of the solid is $\int_{1}^{4} \frac{dx}{4x^2} = \frac{3}{16}$.

17. A circle of circumference c has area $\frac{c^2}{4\pi}$. Thus, since 1 foot = 12 inches, the volume of the pole is approximately
$$S_6 = \frac{60 \cdot 12}{3 \cdot 6 \cdot 4\pi} \left(1 \cdot 16^2 + 4 \cdot 14^2 + 2 \cdot 10^2 + 4 \cdot 5^2 + 2 \cdot 3^2 + 4 \cdot 2^2 + 1 \cdot 1^2\right) = \frac{13750}{\pi} \approx 4376.8 \text{ in}^3.$$

18. $V = \pi \int_{0}^{\pi/4} \left(1 - \tan^2 y\right) dy = \pi \left(y - (\tan y - y)\right)\Big]_{0}^{\pi/4} = \pi \left(\frac{\pi}{2} - 1\right) \approx 1.7932$.

19. The radius of the glass (in inches) at height h (in inches) is $r = 1 + 0.1h$ when $0 \le h \le 5$. Thus, the volume of the glass is $V = \pi \int_{0}^{5} (1 + 0.1h)^2 \, dh = \frac{95\pi}{12} \approx 24.87 \text{ in}^3$.

20. From the information given, the radius of the Earth is $r = C/2\pi \approx 3{,}963$ miles.

    (a) $V = \pi \int_{\sqrt{2}r/2}^{r} (r^2 - y^2) \, dy = \left(\frac{2}{3} - \frac{5\sqrt{2}}{12}\right) \pi r^3 \approx 1.1514 \times 10^{10} \text{ miles}^3$.

    (b) $V = \pi \int_{0}^{\sqrt{2}r/2} (r^2 - y^2) \, dy = \frac{5\sqrt{2}\pi r^3}{12} \approx 1.1522 \times 10^{11} \text{ miles}^3$.

21. The volume of water in the balloon is $V = \pi \int_{-3}^{1} (9 - y^2) \, dy = 80\pi/3$ cubic inches.

22. **Method 1:** Cross-sections parallel to the ends of the tank have constant area:
$$A(x) = 2 \int_{-4}^{2} \sqrt{4^2 - y^2} \, dy = 2 \left(\frac{1}{2} y \sqrt{4^2 - y^2} + 8 \arcsin\left(\frac{y}{4}\right)\right)\Big]_{-4}^{2} = 4\sqrt{3} + \frac{32\pi}{3}.$$

    Thus, the volume of gasoline in the tank is $V = \int_{0}^{25} A(x) \, dx = 100\sqrt{3} + \frac{800\pi}{3} \approx 1011.0$ cubic feet.

    Method 2: The cross-section parallel to the surface at height y has area $A(y) = 25 \cdot 2\sqrt{4^2 - y^2}$. Thus, the volume of gasoline in the tank is $V = \int_{-4}^{2} A(y) \, dy = 50 \int_{-4}^{2} \sqrt{4^2 - y^2} \, dy$.

    Method 3: The volume of the tank is 400π cubic feet. The volume of the "empty" portion at the top is $25 \int_{-\sqrt{12}}^{\sqrt{12}} \left(\sqrt{4^2 - x^2} - 2\right) dx$ cubic feet. Now the volume of the gasoline in the tank can be found by subtraction.

23. (a) $V = \pi \int_{-3}^{-1} \arctan^2 x \, dx$

    (b) For any n, L_n overestimates V since $\left(\arctan^2 x\right)' = \frac{2 \arctan x}{1 + x^2} < 0$ over the interval of integration (i.e., the integrand is a decreasing function).

SECTION 8.2 FINDING VOLUMES BY INTEGRATION

24. The cross-sectional area at height h is $V'(h)$. Thus, the cross-sectional area of the object 1 inch above its base is $1.5 + \cos 1$ square inches.

25. Cross-sections parallel to the ends of the tank have area:

$$A(x) = 2\int_{-6}^{3} \sqrt{9 - \tfrac{y^2}{4}}\, dy = 4\int_{-3}^{3/2} \sqrt{9 - u^2}\, du = 2\left(u\sqrt{9 - u^2} + 9\arcsin\left(\tfrac{u}{3}\right)\right)\Big|_{-3}^{3/2} = \frac{9\sqrt{3}}{2} + 12\pi.$$

Thus, the volume of fuel oil in the tank is $V = \int_0^{10} A(x)\, dx = 45\sqrt{3} + 120\pi \approx 454.93$ cubic feet.

27. (a) The area of the annulus is $\pi(r + \Delta r)^2 - \pi r^2 = \pi(2r + \Delta r)\Delta r$.

(b) The area of a circle of radius R is is the area of a circle of radius r and $n - 1$ concentric annuli with thickness Δr. The area of the circle of radius r is $\pi(\Delta r)^2$ and the area of the kth annulus is $\pi(2r_k + \Delta r)\Delta r$. Therefore, since $r_0 = 0$, the area of the circle of radius R is $\displaystyle\sum_{k=0}^{n-1} \pi(2r_k + \Delta r)\Delta r = \pi R^2$.

(c) Observe that $\displaystyle\lim_{n\to\infty} \sum_{k=0}^{n-1} 2\pi r_k \Delta r = \int_0^R 2\pi r\, dr$ — the sum on the left is a left sum approximation to the integral on the right. Therefore, since $\displaystyle\lim_{n\to\infty} \sum_{k=0}^{n-1} (\Delta r)^2 = \lim_{n\to\infty} \sum_{k=0}^{n-1} \left(\frac{R}{n}\right)^2 = \lim_{n\to\infty} \frac{R^2}{n} = 0$,

$$\lim_{n\to\infty} \sum_{k=0}^{n-1} \pi(2r_k + \Delta r)\Delta r = \int_0^R 2\pi r\, dr.$$

(d) This is an alternate algebraic representation of the sum in part (b). The area of a circle of radius R is written as the sum of the area of a circle and the areas of $n - 1$ annuli, each with thickness Δr.

29. (a) Since f is a decreasing function, each left cylindrical shell overestimates the volume it encloses. Similarly, each right shell underestimates the volume it encloses. Therefore, the left and right sums overestimate and underestimate, respectively, the volume V.

(b) $\displaystyle\lim_{n\to\infty} \sum_{k=1}^{n} \pi(2x_k - \Delta x)f(x_k)\Delta x = \lim_{n\to\infty} \sum_{k=0}^{n-1} \pi(2x_k + \Delta x)f(x_k)\Delta x = \int_a^b 2\pi x f(x)\, dx = V.$

(c) The roles of the left and right sums in parts (a) and (b) are interchanged when f is an increasing function. Otherwise, the ideas are the same.

31. (a) The racetrack principle implies that $f(z) \geq f(c) - K(z - c)$. Since $d \geq z$, it follows that $f(z) \geq f(c) - K(z - d)$. The result that $f(z) \leq f(c) + K(d - c)$ can be derived using a similar argument.

(b) Using part (a), we find that $f(x_k) - K\Delta x \leq f(z) \leq f(x_k) + K\Delta x$ when $x_k \leq z \leq x_{k+1}$. Thus, $\pi(2x_k - \Delta x)(f(x_k) - K\Delta x)\Delta x$ underestimates the volume of the "shell" obtained by rotating the area under the graph of f between x_k and x_{k+1}. Similarly, $\pi(2x_k + \Delta x)(f(x_k) + K\Delta x)\Delta x$ overestimates the volume of the "shell" obtained by rotating the area under the graph of f between x_k and x_{k+1}.

(c) Since $\displaystyle\lim_{n\to\infty} (\Delta x)^2 = 0$, $\displaystyle\lim_{n\to\infty} (\Delta x)^3 = 0$, and f is bounded on the interval $[a, b]$,

$$\lim_{n\to\infty} \sum_{k=0}^{n-1} \pi(2x_k - \Delta x)(f(x_k) - K\Delta x)\Delta x = \lim_{n\to\infty} \sum_{k=0}^{n-1} \pi(2x_k + \Delta x)(f(x_k) + K\Delta x)\Delta x = \int_a^b 2\pi x f(x)\, dx.$$

32. (a) $V = \pi \int_0^2 \left(8^2 - y^6\right) dy = \dfrac{768\pi}{7}$

(b) $V = 2\pi \int_0^8 x^{4/3}\, dx = \dfrac{768\pi}{7}$

33. (a) Solving the equation $y = x\sqrt{1-x^2}$ for x^2 yields $x^2 = \left(1 \pm \sqrt{1+4y^2}\right)/2$. Thus, the outer boundary of the region is the curve $x = \left(1 + \sqrt{1+4y^2}\right)^{1/2}$ and the inner boundary of the region is the curve $x = \left(1 - \sqrt{1+4y^2}\right)^{1/2}$ when $0 \le y \le 1/2$. Therefore, the volume of the solid of revolution is

$$V = \int_0^{1/2} \frac{\pi}{2}\left(1 + \sqrt{1-4y^2}\right) dy - \int_0^{1/2} \frac{\pi}{2}\left(1 - \sqrt{1-4y^2}\right) dy.$$

(b) Using the method of cylindrical shells, $V = \int_0^1 \pi x f(x)\, dx = \int_0^1 2\pi x^2 \sqrt{1-x^2}\, dx$.

(c) The integral in part (b) can be evaluated using the substitutions $u = 1 - x^2$, $w = u - 1/2$ and the table of integrals in the back of the textbook:

$$\int_0^1 2\pi x^2 \sqrt{1-x^2}\, dx = \pi \int_0^1 \sqrt{(1/2)^2 - (u - 1/2)^2}\, du = \pi \int_{-1/2}^{1/2} \sqrt{(1/2)^2 - w^2}\, dw = \dfrac{\pi^2}{8}.$$

35. (a) Let $p(r)$ be the population density at a distance r from the center of the city. Since p is a linear function it can be written in the form $p(r) = ar + b$. Now, $p(0) = K$ and $p(R) = 0$, so $b = K$ and $a = -K/R$. Thus, $p(r) = -Kr/R + K = K(1 - r/R)$.

(b) The population of the city is $\int_0^R p(r)\, dr = \dfrac{KR}{2}$.

8.3 Arclength

1. The integral is $\int_0^1 \sqrt{1+1}\,dx = \sqrt{2}$. This result is (of course) the same as that given by the usual distance formula.

2. The length of the line segment is $\int_0^1 \sqrt{1+m^2}\,dx = \sqrt{1+m^2}$. We'd get the same answer using the distance formula, since the line in question joins the points $(0, b)$ and $(1, m+b)$.

3. (a) Here's a picture. Since the line connecting the endpoints of the curve C has length
$\sqrt{(3-1)^2 + (109/12 - 7/12)^2} = \sqrt{305/4} \approx 8.7321$, the length of the curve is slightly more than this.

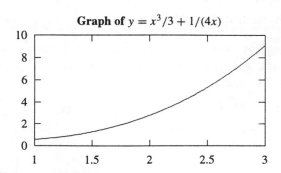

Graph of $y = x^3/3 + 1/(4x)$

(b) By the arclength formula, the length is the integral:

$$\int_1^3 \sqrt{1+(f')^2} = \int_1^3 \sqrt{1 + \left(x^2 - \frac{1}{4x^2}\right)^2}\,dx = \int_1^3 \sqrt{1 + x^4 - \frac{1}{2} + \frac{1}{16x^4}}\,dx$$
$$= \int_1^3 \sqrt{x^4 + \frac{1}{2} + \frac{1}{16x^4}}\,dx = \int_1^3 \sqrt{\left(x^2 + \frac{1}{4x^2}\right)^2}\,dx$$
$$= \int_1^3 \left(x^2 + \frac{1}{4x^2}\right)\,dx = \left.\frac{x^3}{3} - \frac{1}{4x}\right]_1^3$$
$$= \frac{53}{6} \approx 8.833333333.$$

4. $\text{length} = \int_1^4 \sqrt{1 + \left(\tfrac{1}{2}(e^x + e^{-x})'\right)^2}\,dx$
$= \int_1^4 \sqrt{1 + \left(\tfrac{1}{2}(e^x - e^{-x})\right)^2}\,dx$
$= \int_1^4 \tfrac{1}{2}(e^x + e^{-x})\,dx = \left.\tfrac{1}{2}(e^x - e^{-x})\right|_1^4$
$= \left(e^4 - e^{-4} - e^1 + e^{-1}\right)/2 \approx 26.11472.$

5. The shortest distance between two points A and B is a line. If the curve passing through the points A and B is not a line, then any polygonal path from A to B through intermediate points on the curve will be longer than the line segment connecting the two points.

6. $\text{length} = \int_0^1 \sqrt{1 + (2x)^2}\,dx = \frac{1}{4}\left(4\sqrt{17} + \ln\left(4 + \sqrt{17}\right) - 2\sqrt{5} - \ln\left(2 + \sqrt{5}\right)\right) \approx 3.1678$

7. length $= \int_0^1 \sqrt{1+e^{2x}}\,dx = \frac{1}{2}\int_1^{e^2} \frac{\sqrt{1+u}}{u} = \frac{1}{2}\int_1^{e^2} \frac{1+u}{u\sqrt{1+u}} = \left(\sqrt{1+u} + \frac{1}{2}\ln\left|\frac{\sqrt{1+u}-1}{\sqrt{1+u}-1}\right|\right)\Big|_1^{e^2}$
≈ 2.003497110.

8. The length of the curve $y = x^2$ between $x = 1$ and $x = \alpha$ is $\int_1^\alpha \sqrt{1+4x^2}\,dx = \alpha\sqrt{1+4\alpha^2}/2 + \frac{1}{4}\ln\left(2\alpha + \sqrt{1+4\alpha^2}\right) - \sqrt{5}/2 - \frac{1}{4}\ln\left(2+\sqrt{5}\right)$. Thus, the length of the curve is greater than 8 and less than 9 when $\alpha = 3$.

   Alternative Solution: Since $2x \le \sqrt{1+4x^2} \le \sqrt{5}x$ for all $x \ge 1$, the length of the curve $y = x^2$ between $x = 1$ and $x = \alpha$ is greater than $\alpha^2 - 1$ and less than $\sqrt{5}(\alpha^2-1)/2$. Therefore, the length of the curve is greater than 8 and less than 9 when $\alpha = 3$.

9. (a) The length of the curve $y = f(x)$ from $x = a$ to $x = b$ is $\int_a^b \ell(x)\,dx$, where $\ell(x) = \sqrt{1+(f'(x))^2}$. Now, $\ell'(x) = f'(x)f''(x)/\sqrt{1+(f'(x))^2}$ so $\ell'(x) < 0$ on interval $[a,b]$ because $f'(x) > 0$ and $f''(x) < 0$ when $a \le x \le b$. Thus, $\ell(x)$ is a decreasing function on the interval $[a,b]$ and, therefore, L_n *over*estimates the value of the arclength integral.

   (b) Let $\ell(x) = \sqrt{1+(g'(x))^2}$. Then, $\left|L_{10} - \int_0^1 \ell(x)\,dx\right| \le \frac{K_1}{20}$, where K_1 is a number such that $|\ell'(x)| \le K_1$ when $0 \le x \le 1$. Since $4 \le g'(x) \le 7$ and $-3 \le g''(x) \le -2$ when $0 \le x \le 1$,

   $$|\ell'(x)| = \left|\frac{g'(x)g''(x)}{\sqrt{1+(g'(x))^2}}\right| \le \left|\frac{7\cdot -3}{\sqrt{1+(4)^2}}\right| = \frac{21}{\sqrt{17}}.$$

   Therefore, the left sum estimate L_{10} approximates the arclength integral within $21/20\sqrt{17} \approx 0.25466$.

10. (a) $J = \int_0^4 \sqrt{1+(f'(x))^2}\,dx$, so

    $$\begin{aligned} L_4 &= \sqrt{1+(f'(0))^2} + \sqrt{1+(f'(1))^2} + \sqrt{1+(f'(2))^2} + \sqrt{1+(f'(3))^2} \\ &= \sqrt{1+(-9)^2} + \sqrt{1+(-11/2)^2} + \sqrt{1+(-3)^2} + \sqrt{1+(-3/2)^2} \\ &\approx 19.611. \end{aligned}$$

    (b) Let $\ell(x) = \sqrt{1+(f'(x))^2}$. Then $J = \int_0^4 \ell(x)\,dx$. Now, when $0 \le x \le 4$,

    $$\ell'(x) = \frac{f'(x)f''(x)}{\sqrt{1+(f'(x))^2}} < 0$$

    because $f'(x) < 0$ and $f''(x) > 0$. Thus, $\ell(x)$ is a decreasing function on the interval of integration and, therefore, L_4 *over*estimates J (i.e., $J < L_4$).

11. After 5 minutes, the bug has crawled 30 feet. Since the length of the curve $y = \frac{1}{3}(x^2+2)^{3/2}$ between $x = 1$ and $x = s$ is $s^3/3 + s - 4/3$, $x \approx 4.3271$ is the x-coordinate of the bug's position after 5 minutes. Thus, the bug is (approximately) at the point $(4.3271, 31.4470)$.

8.4 Work

1. (a) work $= 115 \times 0.06$ inch-pounds $= 6.9$ inch-pounds $= 0.575$ foot-pounds.

   (b) According to Hooke's Law, the force required to compress a spring x units is proportional to x. Since a force of 115 pounds compresses the spring by 0.06 inches, the constant of proportionality (the *spring constant* is $k = 115/0.06$. Therefore, a force of 175 pounds will compress the spring approximately 0.0913 inches and work $= 175 \times 0.0913$ inch-pounds $= 15.978$ inch-pounds $= 1.3315$ foot-pounds.

2. work $= \displaystyle\int_0^{200} 5 \cdot (200 - x)\, dx = 100{,}000$ foot-pounds.

3. Each parallel slice is a circular cylinder with cross-sectional area 25π and thickness Δx. Therefore,
$$\text{work} = \int_0^5 62.4 \cdot 25\pi \cdot (10 - x)\, dx = 58{,}500\pi \text{ foot-pounds} \approx 183{,}783 \text{ foot-pounds}.$$

4. For reasons as in Example 5, the necessary forces F_1 and F_2 for the two buckets are given, respectively, by
$$F_1(x) = 60 + 0.25 \cdot (60 - x) = 75 - \frac{x}{4};$$
$$F_2(x) = 50 + 0.25 \cdot (70 - x) = 67.5 - \frac{x}{4}.$$

   (The unit of force is pounds; x measures distance in feet from the bottom of the well.) To find the work in each case, we integrate:
$$W_1 = \int_0^{60} F_1(x)\, dx = \int_0^{60} \left(75 - \frac{x}{4}\right) dx = 4050 \text{ foot-pounds};$$
$$W_2 = \int_0^{70} F_2(x)\, dx = \int_0^{70} \left(67.5 - \frac{x}{4}\right) dx = 4112.5 \text{ foot-pounds}.$$

   Raising the *second* bucket takes a little more work.

5. Let x denote the number of inches of compression.

   (a) To find k, use the equation (implicit in the problem statement) $F(2) = 2k = 10$. It follows that $k = 5$ (and hence that $F(x) = 5x$.)

   (b) Compressing from 16 inches to 12 inches means compressing from $x = 2$ to $x = 6$. Thus the work done is
$$W = \int_2^6 F(x)\, dx = \int_2^6 5x\, dx = 80 \text{ inch-pounds}.$$

6. In from Example 2 the orbit had "altitude" 15,000 miles; that corresponds to a distance of 19,000 miles from earth's center. The current situation is almost the same—the only difference is that the orbit is now 20,000 miles from earth's center (or, equivalently, at "altitude" 16,000 miles). The reasoning in Example 2 shows that the work done (in mile-pounds) in moving from $x = 4000$ to $x = 20000$ is given by the integral:
$$\int_{4000}^{20000} \frac{1.6 \times 10^{10}}{x^2} = -\frac{1.6 \times 10^{10}}{x} \bigg]_{4000}^{20000} = 3{,}200{,}000 \text{ mile-pounds}.$$

   (Compare this to the result of Example 2, which showed that raising the satellite to 19,000 miles took 3,158,000 mile-pounds of work. One point of this problem is that the next 1000 miles of altitude comes cheaply—it "costs" only 42,000 mile-pounds of work.)

7. $\text{work} = \int_{-4}^{4} 42 \cdot 2\sqrt{16-y^2} \cdot 15 \cdot (y+17)\, dy$

$= -420\left(16-y^2\right)^{3/2} + 10{,}710y\sqrt{16-y^2} + 171{,}360\arcsin(y/4)\Big|_{-4}^{4}$

$= 171{,}360\pi \text{ foot-pounds} \approx 538{,}343 \text{ foot-pounds}$

8. Let's find an expression for $F(x)$, the force required to lift the bucket, water, and rope when the bucket is x feet from the bottom.

    Since the bucket loses 40 pounds of water in 75 feet of vertical travel, it loses $40/75 = 8/15$ pounds per foot. Thus, at height x, the bucket and water weigh $80 - 8x/15$ pounds. When the height is x, $75 - x$ feet of rope must be lifted; they weigh $(75-x) \cdot 0.65$ pounds. Putting it all together:

    $$F(x) = 80 - \frac{8x}{15} + (75-x) \cdot 0.65 \text{ pounds.}$$

    Thus the necessary work W is given by

    $$W = \int_0^{75} F(x)\, dx = \int_0^{75}\left(80 - \frac{8x}{15} + (75-x) \cdot 0.65\right) dx = 6328.125 \text{ foot-pounds.}$$

9. (a) $W = \int_0^2 F(x)\, dx = \int_0^2 40x\, dx = 80$ foot-pounds.

    (b) $W = \int_0^s F(x)\, dx = \int_0^s 40x\, dx = 20s^2$ foot-pounds.

    (c) By the problem and the previous part, $W = 20s^2 = 10000$. Solving this for s gives $s = 10\sqrt{5} \approx 22.3361$ feet.

10. (a) $\text{work} = \int_0^{10} kx\, dx = 50k \text{ foot} - \text{pounds}$

    (b) $\text{work} = \int_a^{a+10} kx\, dx = k(50 + 10a) \text{ foot} - \text{pounds}$

11. Let x denote the distance (in feet) that the spring is extended. Notice that since the chain weighs 20 pounds, we start with $x = 5$. (Draw a picture! Note, too, that other x-scales are possible.)

    The problem is to find $\int_5^7 F(x)\, dx$, where $F(x)$ is the net downward force necessary at a given x.

    Let's find a formula for $F(x)$; there are two main ingredients: (i) For any value of x, the spring exerts an *upward* force of $4x$ pounds. (ii) For a given value of x, the length of chain remaining above the floor is $10 - (x-5) = 15 - x$ feet. (A diagram should make this convincing.) Since the chain weighs 2 pounds per foot, this length of chain exerts a *downward* force of $2(15-x) = 30 - 2x$ pounds.

    Putting (i) and (ii) together means that the *net* downward force required for given x is $F(x) = 4x - (30-2x) = 6x - 30$ pounds. Thus the desired work is

    $$W = \int_5^7 F(x)\, dx = \int_5^7 (6x - 30)\, dx = 12 \text{ foot-pounds.}$$

12. In each case, the work done is given by the integral $W = \int_a^b F(x)\, dx = \int_a^b kf'(x)\, dx$.

    (a) If $k = 10$, $a = 0$, $b = 1$, $f(x) = x$, then $W = \int_0^1 10 \cdot dx = 10 \cdot x\Big]_0^1 = 10.$

(b) If $k = 10$, $a = 0$, $b = 1$, $f(x) = x^3$, then $W = \int_0^1 10 \cdot 3x^2 \, dx = 10 \cdot x^3 \Big]_0^1 = 10$.

(c) If $k = 10$, $a = 0$, $b = 1$, $f(x) = x^n$, and n is *any* positive integer, then $W = \int_0^1 10 \cdot nx^{n-1} \, dx = 10 \cdot x^n \Big]_0^1 = 10$.

(d) We got the same answer each time. Mathematically, the point is that

$$W = \int_a^b F(x) \, dx = \int_a^b kf'(x) \, dx = k \ f(x)]_a^b = k(f(b) - f(a)).$$

Physically, the point is that in every case, the work is always the product of the object's weight and the *vertical* distance through which it travels. In a sense, then, the particular curve along which it travels makes no difference.

8.5 Present Value

1. For any interest rate r, the present value of one $1 million payment, 23 years ahead, is $PV = 1,000,000 \cdot e^{-r \cdot 23}$.

    (a) If $r = 0.06$, then $PV = \$1,000,000 e^{-0.06 \cdot 23} \approx \$251,579$.

    (b) If $r = 0.08$, then $PV = \$1,000,000 e^{-0.08 \cdot 23} \approx \$155,817$.

    (c) To find the desired r we solve the equation $PV = 100,000 = 1,000,000 e^{-r \cdot 23}$ for r. The result: $r = (\ln 10)/23 \approx 0.10011$—just a bit above 10%.

2. For any interest rate r, the present value of one $1 million payment, 45 years ahead, is $PV = 1,000,000 \cdot e^{-r \cdot 45}$. If $r = 0.06$, the formula gives $PV \approx 67,206$; if $r = 0.08$, $PV \approx 27,324$.

    If $r = 0.2$, then the formula gives $PV \approx 123.41$. (This preposterous result shows why 45-year bonds paying 20% are not (legally) available.)

3. For any interest rate r (real or nominal) the present value of one $1 million payment, 23 years ahead, is $PV = 1,000,000 \cdot e^{-r \cdot 23}$.

    (a) If $r = 0.02$, then $PV = \$1,000,000 e^{-0.02 \cdot 23} \approx \$631,284$.

    (b) If $r = 0.04$, then $PV = \$1,000,000 e^{-0.04 \cdot 23} \approx \$398,519$.

    (c) To find Betty's r we solve the equation $PV = 200,000 = 1,000,000 e^{-r \cdot 23}$ for r. The result: $r = (\ln 5)/23 \approx 0.07$. Thus Betty needs to find a *real* interest rate—after inflation—of 7%.

4. For any interest rate r, the present value of one $1 million payment, 45 years ahead, is $PV = 1,000,000 \cdot e^{-r \cdot 45}$. If $r = 0.02$, the formula gives $PV \approx 406,570$; if $r = 0.04$, $PV \approx 165,299$.

    If $r = 0.15$, then the formula gives $PV \approx 1,170.88$. (This preposterous result shows why 45-year investments paying 15% real interest are either extremely risky or illegal.)

5. At any interest rate r, the present value formula for several future payments says, in this situation, that

$$PV = 40,000 e^{-18r} + 42,000 e^{-19r} + 44,000 e^{-20r} + 46,000 e^{-21r}.$$

    If $r = 0.06$, the formula (and some electronic help) give $PV \approx \$53,316.85$. If $r = 0.08$, $PV \approx \$36,119.66$.

6. Let $t = 0$ denote 1993. Then the payment stream function $p(t) = 43,000$ applies for $18 \leq t \leq 22$ (i.e., from 2011 to 2015). For a given r the present value is the integral

$$PV = \int_{18}^{22} 43,000 e^{-rt} \, dt = -43,000 \frac{e^{-rt}}{r} \bigg]_{18}^{22} = 43,000 \frac{e^{-18r} - e^{-22r}}{r}.$$

    Substituting $r = 0.06$ gives $PV \approx \$51,929.83$. If $r = 0.08$, then $PV \approx \$34,874.56$. (For comparison: the four-lump-payment "discrete" model gave $PV \approx \$51,614.92$ if $r = 0.06$ and $PV \approx \$35,001.42$ if $r = 0.08$.)

7. (a) p_I is the graph which peaks at $t = 180$. When $t = 180$, $p_I = 110$.

    (b) The function $\cos t$ has period 2π. The functions p_T and p_I both have period 360.

    (c) The constant 50 affects the amplitude of the graphs. The constant 60 shifts the graphs upward. The constants 180, 105, and π affect where each graph has its local maxima and minima.

8. (a) $\int_{10}^{20} 12,000 e^{-0.06t} \, dt \approx \$49,523$

    (b) $\int_{10}^{20} 12,000 e^{-0.08t} \, dt \approx \$37,115$

    (c) $\int_{10}^{20} 12,000 \, dt = \$120,000$

SECTION 8.5 PRESENT VALUE

(d) $\int_{10}^{20} 12{,}000 e^{-0.04t}\, dt \approx \$66{,}297$

(e) $\int_{10}^{20} 12{,}000 e^{-0.06t}\, dt \approx \$49{,}523$

9. The total return from the investment is \$40,000 paid continuously over an 8-year time interval. Since the present value of the return from the investment is $PV = \$5{,}000 \int_0^8 e^{-0.06t}\, dt \approx \$31{,}768$, this is a worthwhile investment.

10. Using the midpoint rule with 50 subdivisions (and Maple): $PV_F \approx 21068.89$, $PV_T \approx 21203.64$, and $PV_I \approx 21067.78$.

11. (a) $\left(\dfrac{e^{ax}}{a^2+b^2}(a\cos(bx)+b\sin(bx))\right)' = \dfrac{ae^{ax}}{a^2+b^2}(a\cos(bx)+b\sin(bx)) + \dfrac{e^{ax}}{a^2+b^2}(-ab\sin(bx)+b^2\cos(bx))$
$= e^{ax}\cos(bx)$

(b) $\int_0^{360}\left(50\cos\left(\pi\cdot\dfrac{t-180}{180}\right)+60\right)e^{-0.1t/360}\, dt = \left(216{,}000 - \dfrac{180{,}000}{1+400\pi^2}\right)\left(1 - e^{-1/10}\right) \approx 20{,}550.78$

NOTE: $\cos\left(\pi\cdot\dfrac{t-180}{180}\right) = \cos\left(\dfrac{\pi t}{180}-\pi\right) = -\cos\left(\dfrac{\pi t}{180}\right)$

13. The interest earned from the income received at time t is $p(t)e^{r(T-t)}$. Thus, the total income accrued between $t=0$ and $t=T$ is $\int_0^T p(t)e^{r(T-t)}\, dt = e^{rT}\int_0^T p(t)e^{-rt}\, dt = e^{rT}\, PV$.

8.6 Fourier Polynomials

1. (a) $f(x) = f'(x) = f''(x) = f'''(x) = e^x$ so $f(0) = f'(0) = f''(0) = f'''(0) = e^0 = 1$.
   Now, $p(0) = 1$; $p'(x) = 1 + x + x^2/2$ so $p'(0) = 1$; $p''(x) = 1 + x$ so $p''(0) = 1$; $p'''(x) = 1$ so $p'''(0) = 1$.

2. (a) Both $\sin(x)$ and $\cos(x)$ are 2π-periodic. Thus $f(x + 2\pi) = \cos(k(x + 2\pi)) = \cos(kx + 2k\pi) = \cos(kx) = f(x)$. The function g behaves the same way.

   (b) Each summand of a trigonometric polynomial is 2π-periodic. Therefore, so is the polynomial itself.

   (c) The smallest period of $p(x) = \cos(4x) + \sin(8x)$ is $\pi/2$, because $p(x + \pi/2) = \cos(4(x + \pi/2)) + \sin(8(x + \pi/2)) = \cos(4x + 2\pi) + \sin(8x + 4\pi) = \cos(4x) + \sin(8x) = p(x)$.

3. Graphs show that the function $\sin(kx)$ is odd, so the integral is zero. For $\cos(kx)$, the symmetry about $x = 0$ gives the same result. The other part is a straightforward symbolic calculation.

4. (a) $f(x) = \cos x$, $f'(x) = -\sin x$, $f''(x) = -\cos x$, $f'''(x) = \sin x$, and $f^{(4)}(x) = \cos x$. Thus, $f(0) = 1$, $f'(0) = 0$, $f''(0) = -1$, $f'''(0) = 0$, and $f^{(4)}(0) = 1$.
   $p(x) = 1 - x^2/2 + x^4/24$, $p'(x) = -x + x^3/6$, $p''(x) = -1 + x^2/2$, $p'''(x) = x$, and $p^{(4)}(x) = 1$. Thus, $p(0) = 1$, $p'(0) = 0$, $p''(0) = -1$, $p'''(0) = 0$, and $p^{(4)}(0) = 1$.

5. All summands give zero integrals except the constant term.

6. (a) Let $u = \sin(mx)$. Then, $\int_{-\pi}^{\pi} \cos(mx)\sin(mx)\,dx = \frac{1}{m}\int_0^0 u\,du = 0$.

   (b) $\int_{-\pi}^{\pi} \cos^2(mx)\,dx = \left(\frac{x}{2} + \frac{\sin(2mx)}{4m}\right)\Big]_{-\pi}^{\pi} = \pi$

   (c) $\int_{-\pi}^{\pi} \sin^2(mx)\,dx = \left(\frac{x}{2} - \frac{\sin(2mx)}{4m}\right)\Big]_{-\pi}^{\pi} = \pi$

7. The integrand is odd because it's the product of an odd function and an even function. Integrating any odd function over $[-\pi, \pi]$ gives zero.

8. $\int_{-\pi}^{\pi} x \sin(kx)\,dx = \left(\frac{1}{k^2}\sin(kx) - \frac{x}{k}\cos(kx)\right)\Big]_{-\pi}^{\pi} = -\frac{2\pi}{k}\cos(k\pi)$ so $\frac{1}{\pi}\int_{-\pi}^{\pi} x \sin(kx)\,dx = \frac{-2\cos(k\pi)}{k}$.

9. (b) $a_0 = \frac{1}{2\pi}\int_{-\pi}^{\pi} f(x)\,dx = \frac{1}{2\pi}\int_0^{\pi} dx = \frac{1}{2}$

   $a_k = \frac{1}{\pi}\int_{-\pi}^{\pi} f(x)\cos(kx)\,dx = \frac{1}{\pi}\int_0^{\pi} \cos(kx)\,dx = \frac{\sin(kx)}{k\pi}\Big]_0^{\pi} = 0$

   $b_k = \frac{1}{\pi}\int_{-\pi}^{\pi} f(x)\sin(kx)\,dx = \frac{1}{\pi}\int_0^{\pi} \sin(kx)\,dx = -\frac{\cos(kx)}{k\pi}\Big]_0^{\pi} = \frac{1 - \cos(k\pi)}{k\pi}$.

   Thus, $b_{2m} = 0$ and $b_{2m+1} = \frac{2}{(2m+1)\pi}$ for $m = 1, 2, 3, \ldots$.

   (c) $q_1(x) = \frac{1}{2} + \frac{2\sin x}{\pi}$;

   $q_3(x) = \frac{1}{2} + \frac{2\sin x}{\pi} + \frac{2\sin(3x)}{3\pi}$;

   $q_5(x) = \frac{1}{2} + \frac{2\sin x}{\pi} + \frac{2\sin(3x)}{3\pi} + \frac{2\sin(5x)}{5\pi}$;

   $q_7(x) = \frac{1}{2} + \frac{2\sin x}{\pi} + \frac{2\sin(3x)}{3\pi} + \frac{2\sin(5x)}{5\pi} + \frac{2\sin(7x)}{7\pi}$

10. (b) $a_0 = \dfrac{\pi}{2}$ and $a_k = \dfrac{2}{\pi k^2}\bigl(1 - \cos(k\pi)\bigr)$. Thus, $a_{2m} = 0$ and $a_{2m-1} = 4/\pi(2m-1)^2$ for $m = 1, 2, 3, \ldots$.
    $b_k = 0$

   (c) $q_1(x) = \dfrac{\pi}{2} + \dfrac{4\cos x}{\pi}$;

   $q_3(x) = \dfrac{\pi}{2} + \dfrac{4\cos x}{\pi} + \dfrac{4\cos(3x)}{9\pi}$;

   $q_5(x) = \dfrac{\pi}{2} + \dfrac{4\cos x}{\pi} + \dfrac{4\cos(3x)}{9\pi} + \dfrac{4\cos(5x)}{25\pi}$;

   $q_7(x) = \dfrac{\pi}{2} + \dfrac{4\cos x}{\pi} + \dfrac{4\cos(3x)}{9\pi} + \dfrac{4\cos(5x)}{25\pi} + \dfrac{4\cos(7x)}{49\pi}$

9.1 Integration by Parts

1. $\int xe^{2x}\,dx = \dfrac{xe^{2x}}{2} - \dfrac{e^{2x}}{4} + C \quad [du = dx,\ v = e^{2x}/2]$

2. $\int x\sin(3x)\,dx = \dfrac{\sin(3x)}{9} - \dfrac{x\cos(3x)}{3} + C \quad [du = dx,\ v = -\tfrac{1}{3}\cos(3x)]$

3. $\int x\sec^2 x\,dx = x\tan x + \ln|\cos x| + C \quad [du = dx,\ v = \tan x]$

4. $\int \sqrt{x}\ln x\,dx = \dfrac{2}{3}x^{3/2}\ln x - \dfrac{4}{9}x^{3/2} + C \quad [du = x^{-1}\,dx,\ v = \tfrac{2}{3}x^{3/2}]$

5. $\int x\sqrt{1+x}\,dx = \dfrac{2}{3}x(1+x)^{3/2} - \dfrac{4}{15}(1+x)^{5/2} + C \quad [du = dx,\ v = \tfrac{2}{3}(1+x)^{3/2}]$

6. $\int \arcsin x\,dx = x\arcsin(x) + \sqrt{1-x^2} + C \quad [du = (1-x^2)^{-1/2}\,dx,\ v = x]$

7. $\int_0^\pi x\cos(2x)\,dx = \dfrac{1}{2}x\sin(2x) + \dfrac{1}{4}\cos(2x)\Big]_0^\pi = 0. \quad [M_2 = 0]$

8. $\int_0^1 xe^{-x}\,dx = -(1+x)e^{-x}\Big]_0^1 = 1 - 2e^{-1} \approx 0.26424. \quad [M_2 \approx 0.27449]$

9. $\int_1^e x\ln x\,dx = \dfrac{1}{2}x^2\ln x - \dfrac{1}{4}x^2\Big]_1^e = \dfrac{1}{4}\left(e^2 + 1\right) \approx 2.09726. \quad [M_2 \approx 2.0670]$

10. $\int_{\pi/4}^{\pi/2} x\csc^2 x\,dx = -x\cot x + \ln(\sin x)\Big]_{\pi/4}^{\pi/2} = \pi/4 - \ln\left(\sqrt{2}/2\right) = \pi/4 + \ln\left(\sqrt{2}\right) \approx 1.13197.$
    $[M_2 \approx 1.1188]$

11. $\int_{-1}^{\sqrt{2}/2} x^2 \arctan x\,dx = \dfrac{1}{3}x^3\arctan x - \dfrac{1}{6}x^2 + \dfrac{1}{6}\ln\left(1+x^2\right)\Big]_{-1}^{\sqrt{2}/2}$
    $= \dfrac{\sqrt{2}}{12}\arctan(\sqrt{2}/2) + \dfrac{1}{12} + \dfrac{1}{6}\ln(3/4) - \dfrac{\pi}{12} \approx -0.15388. \quad [M_2 \approx -0.12765]$

12. $\int_1^4 e^{3x}\cos(2x)\,dx = \dfrac{3}{13}e^{3x}\cos(2x) + \dfrac{2}{13}e^{3x}\sin(2x)\Big]_1^4 \approx 19{,}307. \quad [M_2 \approx 24{,}861]$

13. Let $u = x$ and $dv = \cos^2 x\,dx = \tfrac{1}{2}(1 + \cos(2x))\,dx$. Then $du = dx$ and $v = \tfrac{1}{2}\left(x + \tfrac{1}{2}\sin(2x)\right)$. Therefore,
    $\int x\cos^2 x\,dx = \dfrac{1}{2}x^2 + \dfrac{1}{4}x\sin(2x) - \dfrac{1}{2}\int\left(x + \dfrac{\sin(2x)}{2}\right)dx = \dfrac{1}{4}x^2 + \dfrac{1}{4}x\sin(2x) + \dfrac{1}{8}\cos(2x) + C.$

14. $\int x\sin x\cos x\,dx = \dfrac{1}{4}\cos x\sin x - \dfrac{1}{2}x\cos^2 x + \dfrac{1}{4}x + C \quad [u = x,\ dv = \sin x\cos x\,dx,\ v = \tfrac{1}{2}\cos^2 x]$

15. (a) Let $u = \sin x$ and $dv = \sin x\,dx$. Then, $du = \cos x\,dx$ and $v = -\cos x$ so
    $\int \sin^2 x\,dx = -\sin x\cos x + \int \cos^2 x\,dx.$

Section 9.1 Integration by Parts

(b) $\int \sin^2 x \, dx = -\sin x \cos x + \int \cos^2 x \, dx = -\sin x \cos x + \int \left(1 - \sin^2 x\right) dx.$

Therefore, $2 \int \sin^2 x \, dx = x - \sin x \cos x$

16. $\int x^2 \cos x \, dx = x^2 \sin x - 2 \sin x + 2x \cos(x) + C$

    Let $u = x^2$, $dv = \cos x \, dx$. Then, $\int x^2 \cos x \, dx = x^2 \sin x - 2 \int x \sin x \, dx$. The remaining antiderivative can be found using integration by parts again.

17. $\int \arccos x \, dx = x \arccos x - \sqrt{1 - x^2} + C \quad [u = \arccos x, \, dv = dx]$

18. $\int \sqrt{x} \ln\left(\sqrt[3]{x}\right) dx = \frac{1}{3} \int \sqrt{x} \ln x \, dx = \frac{2}{9} x^{3/2} \ln x - \frac{4}{27} x^{3/2} + C. \quad [u = \ln x, \, dv = \sqrt{x} \, dx]$

19. $\int (\ln x)^2 \, dx = x (\ln x)^2 - 2x \ln x + 2x + C \quad [u = (\ln x)^2, \, dv = dx \text{ or } u = \ln x, \, dv = \ln x \, dx]$

20. $\int \frac{\ln x}{x^2} dx = -\frac{\ln x}{x} - \frac{1}{x} + C \quad [u = \ln x, \, dv = x^{-2} dx]$

21. $\int x e^x \sin x \, dx = \frac{1}{2}(1 - x)e^x \cos x + \frac{1}{2} x e^x \sin x + C. \quad [u = x, \, dv = e^x \sin x \, dx]$

22. $\int \arctan(1/x) \, dx = x \arctan(1/x) + \frac{1}{2} \ln\left(x^2 + 1\right) + C. \quad [u = \arctan(1/x), \, dv = dx]$

23. $\int x \arctan x \, dx = \frac{1}{2}\left(x^2 \arctan x - x + \arctan x\right) + C$

    Use integration by parts with $u = \arctan x$ and $dv = x \, dx$ to show that $\int x \arctan x \, dx = \frac{1}{2}\left(x^2 \arctan x - \int x^2/(1 + x^2) \, dx\right)$. The last antiderivative can be found using the algebraic identity

    $$\frac{x^2}{1 + x^2} = \frac{(1 + x^2) - 1}{1 + x^2} = 1 - \frac{1}{1 + x^2}$$

24. $\int e^{\sqrt{x}} \, dx = 2 e^{\sqrt{x}} \sqrt{x} - 2 e^{\sqrt{x}} + C$

    First substitute $w = \sqrt{x}$, $w^2 = x$, $2w \, dw = dx$. This produces the new integral $\int 2w e^w \, dw$. Now use parts, with $u = w$, $dv = e^w \, dw$. The answer follows directly.

25. $\int x^5 \sin\left(x^3\right) dx = \frac{\sin\left(x^3\right)}{3} - \frac{x^3 \cos\left(x^3\right)}{3} + C$

    First substitute $w = x^3$, $dw = 3x^2 \, dx$, to get the new integral $\frac{1}{3} \int w \sin w \, dw$. Now use parts, with $u = w$, $dv = \sin w \, dw$.

26. $\int \sin(\ln x) \, dx \frac{x}{2}(\sin(\ln x) - \cos(\ln x)) + C$

    First substitute $w = \ln x$, $x = e^w$, $dx = e^w \, dw$. This gives the new integral $\int \sin(\ln x) \, dx = \int e^w \sin w \, dw$. The new integral is done (with integration by parts, twice) in the book.

27. $\int \sqrt{x} e^{-\sqrt{x}} \, dx = -2 e^{-\sqrt{x}}\left(x + 2\sqrt{x} + 2\right) + C$

    First substitute $w = \sqrt{x}$, $w^2 = x$, $2w \, dw = dx$. This gives the new integral $2 \int w^2 e^{-w} \, dw$. Finding the latter integral requires using integrations by parts twice.

28. $\int \sin(\sqrt{x})\, dx = \sin(\sqrt{x}) - \sqrt{x}\cos(\sqrt{x}) + C$
First substitute $w = \sqrt{x}$, then use integration by parts with $u = w$ and $dv = \sin w\, dw$.

29. $\int \dfrac{\arctan(\sqrt{x})}{\sqrt{x}}\, dx = 2\sqrt{x}\arctan(\sqrt{x}) - \ln(1+x) + C$
Substitute $w = \sqrt{x}$, then use integration by parts with $u = \arctan w$ and $dv = dw$.

30. $\int \sqrt{x}\arctan(\sqrt{x})\, dx = \dfrac{2}{3}x^{3/2}\arctan(\sqrt{x}) - \dfrac{1}{3}(1+x) + \dfrac{1}{3}\ln(1+x) + C$
Substitute $w = \sqrt{x}$, then use integration by parts with $u = \arctan w$ and $dv = w^2\, dw$.

31. Using integration by parts with $u = \arcsin x$ and $dv = x^2\, dx$,

$$\int x^2 \arcsin x\, dx = \dfrac{1}{3}x^3 \arcsin x - \dfrac{1}{3}\int \dfrac{x^3}{\sqrt{1-x^2}}\, dx.$$

To find the remaining antiderivative, use the substituion $w = 1 - x^2$:

$$\int \dfrac{x^3}{\sqrt{1-x^2}}\, dx \to -\dfrac{1}{2}\int \dfrac{1-w}{\sqrt{w}}\, dw = -\dfrac{1}{2}\int \dfrac{dw}{\sqrt{w}} + \dfrac{1}{2}\int \sqrt{w}\, dw$$

$$= -\sqrt{w} + \dfrac{1}{3}w^{3/2} + C \to -\sqrt{1-x^2} + \dfrac{1}{3}(1+x^2)^{3/2} + C$$

Therefore, $\int x^2 \arcsin x\, dx = \dfrac{1}{3}x^3 \arcsin x + \dfrac{1}{3}\sqrt{1-x^2} - \dfrac{1}{9}(1+x^2)^{3/2} + C$.

32. $f(x) = x^4/4$. Integration by parts with $u = \cos x$ and $dv = x^3\, dx$ leads to $\int x^3 \cos x\, dx = \dfrac{1}{4}x^4 \cos x + \int \dfrac{1}{4}x^4 \sin x\, dx$.

33. (a) $I_1 = \int x(\ln x)^1\, dx$. Let $u = \ln x$, $dv = x\, dx$; $du = \dfrac{1}{x}\, dx$; $v = \dfrac{x^2}{2}$. Then

$$I_1 = \int x(\ln x)^1\, dx = \dfrac{x^2}{2}\ln x - \int \dfrac{x}{2}\, dx = \dfrac{x^2}{2}\ln x - \dfrac{x^2}{4} + C$$

(b) $I_n = \int x(\ln x)^n\, dx$. If we let $u = (\ln x)^n$; $dv = x\, dx$; $du = u(\ln x)^{n-1}\cdot \dfrac{1}{x}\, dx$; $v = \dfrac{x^2}{2}$. then we get

$$I_n = \dfrac{(\ln x)^n x^2}{2} - \dfrac{n}{2}\int x(\ln x)^{n-1}\, dx = \dfrac{x^2}{2}(\ln x)^n - \dfrac{n}{2}I_{n-1}$$

(c) By reduction formula plus the fact that $I_1 = \dfrac{x^2}{2}\ln x - \dfrac{x^2}{4} + C$, we have

$$I_2 = \dfrac{x^2}{2}(\ln x)^2 - I_1 = \dfrac{x^2}{2}(\ln x)^2 - \left(\dfrac{x^2}{2}\ln x - \dfrac{x^2}{4}\right) + C = \dfrac{x^2}{2}\left((\ln x)^2 - \ln x + \dfrac{1}{2}\right) + C.$$

Similarly, $I_3 = \dfrac{x^2}{2}(\ln x)^3 - \dfrac{3}{2}I_2 = \dfrac{x^2}{2}\left((\ln x)^3 - \dfrac{3}{2}(\ln x)^2 + \dfrac{3}{2}\ln x - \dfrac{3}{4}\right) + C$.

(d) This is an immediate consequence of the Fundamental Theorem of Calculus.

SECTION 9.1 INTEGRATION BY PARTS

(e) Carrying out the differentiation on the right side of the identity

$$(\text{Right Side})' = x \cdot (\ln x)^n + \frac{x^2}{2} \cdot n(\ln x)^{n-1} \cdot \frac{1}{x} - \frac{n}{2} x (\ln x)^{n-1}$$
$$= x \cdot (\ln x)^n \quad \text{(after a bit of algebra.)}$$

34. (a) $\left(\int x^n e^x \, dx \right)' = x^n e^x$, $(x^n e^x)' = \left(nx^{n-1} + x^n \right) e^x$, $\left(\int x^{n-1} e^x \, dx \right)' = x^{n-1} e^x$.

    (b) Let $u = x^n$ and $dv = e^x \, dx$. Then, $du = nx^{n-1} \, dx$ and $v = e^x$.

35. Let $u = (\ln x)^n$ and $dv = dx$.

36. An integration by parts with $u = (x^2 + a^2)^r$ and $dv = dx$ shows that

$$\int (x^2 + a^2)^r = x(x^2 + a^2)^r - 2r \int x^2 (x^2 + a^2)^{r-1} \, dx$$
$$= x(x^2 + a^2)^r - 2r \left(\int (x^2 + a^2)^r \, dx - a^2 \int (x^2 + a^2)^{r-1} \, dx \right).$$

A bit of algebra now produces the desired result.

37. (a) $-\frac{1}{2} \cos(x^2) + C$

    (b) $-\frac{1}{2} x^2 \cos(x^2) + \frac{1}{2} \sin(x^2) + C$

    (c) $\frac{1}{2} x^2 \sin(x^2) + \frac{1}{2} \cos(x^2) + C$

    (d) $\int x^2 \cos(x^2) \, dx = \frac{1}{2} x \sin(x^2) - \frac{1}{2} \int \cos(x^2)$. Since the expression on the left side of the equals sign is not elementary, neither is the expression on the right side.

38. Integration by parts with $u = x^{n-1}$ and $dv = x \sin(x^2) \, dx$, leads to the reduction formula

$$\int x^n \sin(x^2) \, dx = -x^{n-1} \frac{\cos(x^2)}{2} + \frac{n-1}{2} \int x^{n-2} \cos(x^2) \, dx.$$

Similarly, integration by parts with $u = x^{n-1}$ and $dv = x \cos(x^2) \, dx$, leads to the reduction formula

$$\int x^n \cos(x^2) \, dx = x^{n-1} \frac{\sin(x^2)}{2} - \frac{n-1}{2} \int x^{n-2} \sin(x^2) \, dx.$$

Thus, when n is an odd integer, the reduction formulas can be used to find an elementary antiderivative of $x^n \sin(x^2)$. However, when n is even, the reduction formulas lead to an expression involving $\int \cos(x^2) \, dx$ or $\int \sin(x^2) \, dx$. Since neither $\cos(x^2)$ nor $\sin(x^2)$ has an elementary antiderivative, neither does $x^n \sin(x^2)$.

39. An integration by parts with $u = f(x)$ and $dv = \sin x \, dx$ shows that

$$\int_0^\pi f(x) \sin x \, dx = -f(x) \cos x \Big]_0^\pi + \int_0^\pi f'(x) \cos x \, dx = f(\pi) + f(0) + \int_0^\pi f'(x) \cos x \, dx.$$

An integration by parts with $u = \sin x$ and $dv = f''(x) \, dx$ shows that

$$\int_0^\pi f''(x) \sin x \, dx = f'(x) \sin x \Big]_0^\pi - \int_0^\pi f'(x) \cos x \, dx = -\int_0^\pi f'(x) \cos x \, dx.$$

Combining these results, we have

$$6 = \int_0^\pi f(x)\sin x\,dx + \int_0^\pi f''(x)\sin x\,dx = f(\pi) + f(0) = f(\pi) + 2.$$

From this it follows that $f(\pi) = 4$.

40. Using integration by parts (twice), we can show that

$$\int f(x)\cos x\,dx = f(x)\sin x - \int f'(x)\sin x\,dx = f(x)\sin x + f'(x)\cos x - \int f''(x)\cos x\,dx.$$

Thus,
$$\int_{-\pi/2}^{3\pi/2} f(x)\cos x\,dx = \left(f(x)\sin x + f'(x)\cos x\right)\Big]_{-\pi/2}^{3\pi/2} - \int_{-\pi/2}^{3\pi/2} f''(x)\cos x\,dx$$
$$= -f(3\pi/2) + f(-\pi/2) - 4$$
$$= -\int_{-\pi/2}^{3\pi/2} f'(x)\,dx - 4$$
$$= -5.$$

9.2 Partial Fractions

1. (b) $\int \dfrac{5x+7}{(x+1)(x+2)}\,dx = 2\ln|x+1| + 3\ln|x+2| + C$

2. (b) $\int \dfrac{2}{x^2-1}\,dx = \int \left(\dfrac{1}{x-1} - \dfrac{1}{x+1}\right) dx = \ln|x-1| - \ln|x+1| + C$

3. (a) Using partial fractions, $\dfrac{x^2+3x-1}{x(x+1)(x-2)} = \dfrac{1}{2}\cdot\dfrac{1}{x} - \dfrac{1}{x+1} + \dfrac{3}{2}\cdot\dfrac{1}{x-2}$. Thus, $A = 1/2$, $B = 1$, and $C = 3/2$.

   (b) $\int \dfrac{x^2+3x-1}{x(x+1)(x-2)}\,dx = \dfrac{1}{2}\ln|x| - \ln|x+1| + \dfrac{3}{2}\ln|x-2| + C.$

4. (a) $A = -\tfrac{1}{4}$, $B = \tfrac{5}{4}$, $C = 0$.

   (b) $\int \dfrac{x^2-1}{x(x^2+4)}\,dx = -\dfrac{1}{4}\ln|x| + \dfrac{5}{8}\ln\left|x^2+4\right| + K$

5. Let $I = \int \dfrac{6}{(x-2)(x^2-1)}\,dx$. Partial fractions work, again:

   $$\dfrac{6}{(x-2)(x^2-1)} = \dfrac{6}{(x-2)(x-1)(x+1)} = \dfrac{A}{x-2} + \dfrac{B}{x-1} + \dfrac{C}{x+1}$$

   Solving gives $A = 2$, $B = -3$, and $C = 1$, so $I = 2\ln|x-2| - 3\ln|x-1| + \ln|x+1| + K$.

6. (a) $\dfrac{x^2}{(x+1)^3} = \dfrac{1}{x+1} - \dfrac{2}{(x+1)^2} + \dfrac{1}{(x+1)^3}$ (i.e., $A = 1$, $B = -2$, and $C = 1$).

   (b) $\int \dfrac{x^2}{(x+1)^3}\,dx = \ln|x+1| + \dfrac{2}{x+1} - \dfrac{1}{2(x+1)^2}$

7. (a) $\dfrac{x^2+x}{(x^2+1)^2} = \dfrac{1}{x^2+1} + \dfrac{x-1}{(x^2+1)^2}$ (i.e., $A = 0$, $B = 1$, $C = 1$, $D = -1$)

   (b) $\int \dfrac{x^2+x}{(x^2+1)^2}\,dx = \dfrac{1}{2}\arctan x - \dfrac{x+1}{2(x^2+1)}$

8. The partial fraction decomposition of the integrand has the form

   $$\dfrac{1}{(x-2)(x^2+1)} = \dfrac{A}{x-2} + \dfrac{Bx+C}{x^2+1}.$$

   To determine the constants, multiply both sides of the equation above by $(x-2)(x^2+1)$ to obtain

   $$1 = A(x^2+1) + (Bx+C)(x-2) = (A+B)x^2 + (C-2B)x + (A-2C).$$

   It follows that $A + B = 0$, $C - 2B = 0$, and $A - 2C = 1$. Inserting $x = 2$ into the equation displayed above, we find that $A = 1/5$ and, therefore, $B = -1/5$ and $C = -2/5$. Thus,

   $$\int_0^1 \dfrac{dx}{(x-2)(x^2+1)} = \dfrac{1}{5}\int_0^1 \dfrac{dx}{x-2} - \dfrac{1}{5}\int_0^1 \dfrac{x}{x^2+1}\,dx - \dfrac{2}{5}\int_0^1 \dfrac{dx}{x^2+1}$$

   $$= \left(\dfrac{1}{5}\ln|x-2| - \dfrac{1}{10}\ln\left|x^2+1\right| - \dfrac{2}{5}\arctan x\right)\Big]_0^1$$

   $$= \left(\dfrac{1}{10}\ln 2 - \dfrac{2}{5}\dfrac{\pi}{4}\right) - \dfrac{1}{5}\ln 2 = -\dfrac{3}{10}\ln 2 - \dfrac{\pi}{10}.$$

9. $\int \frac{dx}{x^3+1} = \int \frac{dx}{(x+1)(x^2-x+1)} = \frac{1}{3}\int \left(\frac{1}{x+1} + \frac{2-x}{x^2-x+1}\right) dx =$
$\frac{1}{3}\ln|x+1| - \frac{1}{6}\ln|x^2-x+1| + \frac{1}{2}\int \frac{dx}{x^2-x+1} =$
$\frac{1}{3}\ln|x+1| - \frac{1}{6}\ln|x^2-x+1| + \frac{1}{\sqrt{3}}\arctan\left(\frac{2x-1}{\sqrt{3}}\right)$. Thus, $\int_0^2 \frac{dx}{x^3+1} = \frac{1}{6}\left(\ln 3 + \sqrt{3}\pi\right)$.

10. $\int \frac{2x+1}{(x-2)(x+3)} dx = \int \left(\frac{1}{x-2} + \frac{1}{x+3}\right) dx = \ln(x-2) + \ln(x+3) + C$

11. $\int \frac{x+1}{(x-1)(x+2)} dx = \int \left(\frac{2}{3}\cdot\frac{1}{x-1} + \frac{1}{3}\cdot\frac{1}{x+2}\right) dx = \frac{2}{3}\ln|x-1| + \frac{1}{3}\ln|(x+2)| + C$

12. $\int \frac{x^2+x}{(x^2+4)^2} dx = \int \left(\frac{1}{x^2+4} + \frac{x-4}{(x^2+4)^2}\right) dx = \frac{1}{4}\arctan(x/2) - \frac{1}{2}\cdot\frac{x+1}{x^2+4} + C$

13. $\int \frac{5x^2+3x-2}{x^3+2x^2} dx = \int \left(-\frac{1}{x^2} + \frac{2}{x} + \frac{3}{x+2}\right) dx = \frac{1}{x} + 2\ln|x| + 3\ln|x+2| + C$

14. Let $I = \int \frac{4x^2-3x+2}{x(2x-1)^2} dx$. Use partial fractions: $\frac{4x^2-3x+2}{x(2x-1)^2} = \frac{A}{x} + \frac{B}{(2x-1)} + \frac{C}{(2x-1)^2}$. Solving gives $A = 2$, $B = -2$, and $C = 3$. Therefore, $I = 2\ln|x| - \ln|2x-1| - \frac{3}{2}\cdot\frac{1}{(2x-1)} + C$.

15. First complete the square: $I = \int \frac{x}{x^2+2x+6} dx = \int \frac{x}{(x+1)^2+5} dx$. Now let $u = x+1$, $x = u-1$, $dx = du$. Then,

$$I = \int \frac{u-1}{u^2+5} du = \int \frac{u}{u^2+5} du - \int \frac{du}{u^2+5}$$
$$= \frac{1}{2}\ln|u^2+5| - \frac{1}{\sqrt{5}}\arctan\left(\frac{u}{\sqrt{5}}\right) + C$$
$$= \frac{1}{2}\ln|x^2+2x+6| - \frac{1}{\sqrt{5}}\arctan\left(\frac{x+1}{\sqrt{5}}\right) + C$$

16. $\int \frac{x^4}{x^4-1} dx = \int \left(1 + \frac{1}{4}\cdot\frac{1}{x-1} - \frac{1}{4}\cdot\frac{1}{x+1} - \frac{1}{2}\cdot\frac{1}{1+x^2}\right) dx =$
$x + \frac{1}{4}\ln(x-1) - \frac{1}{4}\ln(x+1) - \frac{1}{2}\arctan x + C$

17. $\int \frac{x^3}{x^2+1} dx = \int \left(x - \frac{x}{x^2+1}\right) dx = \frac{1}{2}x^2 - \frac{1}{2}\ln(x^2+1) + C$

18. $\int \frac{x^3}{x^2-1} dx = \int \left(x + \frac{1}{2}\cdot\frac{1}{x-1} + \frac{1}{2}\cdot\frac{1}{x+1}\right) dx = \frac{1}{2}x^2 + \frac{1}{2}\ln(x-1) + \frac{1}{2}\ln(x+1) + C$

19. $\int \frac{3x^2-1}{(x-1)(x+2)} dx = \int \left(3 + \frac{2}{3}\cdot\frac{1}{x-1} - \frac{11}{3}\cdot\frac{1}{x+2}\right) dx = 3x + \frac{2}{3}\ln|x-1| - \frac{11}{3}\ln|x+2| + C$

20. $\int \frac{x^2}{(x^2+1)(x+1)^2} dx = \int \left(\frac{1}{2}\cdot\frac{1}{(1+x)^2} - \frac{1}{2}\cdot\frac{1}{x+1} + \frac{1}{2}\cdot\frac{x}{x^2+1}\right) dx =$
$-\frac{1}{2}\cdot\frac{1}{x+1} - \frac{1}{2}\ln(x+1) + \frac{1}{4}\ln(x^2+1)$

SECTION 9.2 PARTIAL FRACTIONS

21. $\int \dfrac{dx}{x\sqrt{x+1}} \to 2\int \dfrac{du}{u^2-1} = \int \left(\dfrac{1}{u-1} - \dfrac{1}{u+1}\right) du = \ln\left|\dfrac{u-1}{u+1}\right| + C \to \ln\left|\dfrac{\sqrt{x+1}-1}{\sqrt{x+1}+1}\right| + C$

22. $\int \dfrac{dx}{x\sqrt{x-1}} \to 2\int \dfrac{du}{u^2+1} = 2\arctan u + C \to 2\arctan(\sqrt{x-1}) + C$

23. $\int \dfrac{dx}{1+e^x} \to \int \dfrac{du}{u(u-1)} = \int \left(\dfrac{1}{u-1} - \dfrac{1}{u}\right) du = \ln\left|\dfrac{u-1}{u}\right| + C \to \ln\left(\dfrac{e^x}{1+e^x}\right) + C$

24. (a) Since p is a quadratic polynomial, $p(x)$ can be written as $p(x) = Ax^2 + Bx + C$, where A, B, and C are constants. The condition $p(0) = 1$ implies that $C = 1$. Since $p'(x) = 2Ax + B$, the condition $p'(0) = 0$ implies that $B = 0$. Thus, $p(x) = Ax^2 + 1$.

(b) A rational function is a function that can be written as the quotient of two polynomials. Any sum of rational functions is also a rational function. Now,

$$\int r(x)\, dx = B\ln x - Cx - \dfrac{D}{2x^2} + E\ln|x-1| - \dfrac{F}{x-1} + G$$

so $\int r(x)\, dx$ is not a rational function unless $B = E = 0$ (i.e., unless the two logarithm expressions are absent).

(c) Since $\int r(x)\, dx$ is a rational function, $\dfrac{q(x)}{x^3(x-1)^2} = \dfrac{C}{x^2} + \dfrac{D}{x^3} + \dfrac{F}{(x-1)^2}$

so $q(x) = Cx(x-1)^2 + D(x-1)^2 + Fx^3 = D + (C-2D)x + (D-2C)x^2 + (C+F)x^3$.

(d) Combining parts (a) and (c),

$$p(x) = Ax^2 + 1 = D + (C-2D)x + (D-2C)x^2 + (C+F)x^3.$$

Therefore, $D = 1$, $C = 2$, $F = -2$, and $A = D - 2C = -3$.

25. No. Since q is a quadratic polynomial, $\dfrac{q(x)}{(1-x)^2(3+x)} = \dfrac{P}{1-x} + \dfrac{Q}{(1-x)^2} + \dfrac{R}{3+x}$, where P, Q, and R are real numbers. Therefore, $\int \dfrac{q(x)}{(1-x)^2(3+x)}\, dx = -P\ln|1-x| + \dfrac{Q}{1-x} + R\ln(3+x) + S$, where P, Q, R, and S are real numbers. Since no choice of values for the constants P, Q, R, and S in the expression above leads to the expression $\dfrac{1}{1-x} + \arcsin x + \ln(3+x) + C$, we conclude that there is no quadratic polynomial q with the desired property.

26. No — the partial fraction decomposition of $\dfrac{q(x)}{(1-x)^2(x+3)} = \dfrac{A}{1-x} + \dfrac{B}{(1-x)^2} + \dfrac{C}{x+3}$, so $\int \dfrac{q(x)}{(1-x)^2(x+3)}\, dx = A\ln|1-x| + \dfrac{B}{1-x} + C\ln|x+3| + D$.

27. (a) $(\ln|x+a|)' = \dfrac{1}{x+a}$

(b) $\left(\dfrac{1}{1-n}\dfrac{1}{(x+a)^{n-1}}\right)' = \dfrac{1}{(x+a)^n}$

28. (a) $\left(\dfrac{1}{a}\arctan(x/a)\right)' = \dfrac{1}{a^2}\dfrac{1}{(x/a)^2+1} = \dfrac{1}{x^2+a^2}$

(b) Let $u = x^2 + a^2$. Then, $\int \dfrac{x}{x^2+a^2}\, dx \to \dfrac{1}{2}\int \dfrac{du}{u} = \dfrac{1}{2}\ln|u| + C \to \dfrac{1}{2}\ln(x^2+a^2) + C$.

29. (a) Integration by parts with $u = 1/(x^2+a^2)^n$ and $dv = dx$ leads to the desired identity.

(b) $\int \frac{x^2}{(x^2+a^2)^{n+1}} dx = \int \frac{(x^2+a^2)-a^2}{(x^2+a^2)^{n+1}} dx = \int \frac{dx}{(x^2+a^2)^n} - a^2 \int \frac{dx}{(x^2+a^2)^{n+1}}$

(c) Rearranging the terms in part (b) leads to the equation

$$2na^2 \int \frac{dx}{(x^2+a^2)^{n+1}} = \frac{x}{(x^2+a^2)^n} + (2n-1) \int \frac{dx}{(x^2+a^2)^n}.$$

The desired identity follows from dividing both sides of the equation above by $2na^2$.

30. $\begin{aligned} \int \frac{4x^2+2x+1}{(4x^2+5x+3)^2} dx &= \int \frac{dx}{4x^2+5x+3} - \int \frac{2+3x}{(4x^2+5x+3)^2} dx \\ &= \int \frac{dx}{4x^2+5x+3} - \frac{3}{8} \int \frac{8x+5}{(4x^2+5x+3)^2} dx - \frac{1}{8} \int \frac{dx}{(4x^2+5x+3)^2} \\ &= \frac{8-x}{23(4x^2+5x+3)} + \frac{44}{23\sqrt{23}} \arctan\left((8x+5)/\sqrt{23}\right) + C \end{aligned}$

9.3 Trigonometric Antiderivatives

1. No—the two answers are equal. To see this, use the identity $\cos^2 x = 1 - \sin^2 x$.

2. $\int \sin^2(3x)\, dx = \frac{1}{2}\int (1 - \cos(6x))\, dx = \frac{1}{2}x - \frac{1}{12}\sin(6x) + C$

3. $\int \cos^2(x/3)\, dx = \frac{1}{2}\int (1 + \cos(2x/3))\, dx = \frac{1}{2}x + \frac{3}{4}\sin(2x/3) + C$

4. Let $u = \sin x$. Then, $du = \cos x\, dx$ and

$$\int \sin^3 x \cos^3 x\, dx = \int \sin^3 x \cos^2 x \cos x\, dx = \int \sin^3 x(1 - \sin^2 x)\cos x\, dx$$
$$= \int \sin^3 x \cos x\, dx - \int \sin^5 x \cos x\, dx \to \int u^3\, du - \int u^5\, du$$
$$= \tfrac{1}{4}u^4 - \tfrac{1}{6}u^6 + C \to \tfrac{1}{4}\sin^4 x - \tfrac{1}{6}\sin^6 x + C$$

5. $\int \cos^2 x \sin^3 x\, dx = \int \left(\cos^2 x - \cos^4 x\right)\sin x\, dx = -\frac{1}{3}\cos^3 x + \frac{1}{5}\cos^5 x + C$

6. $\int \cos^3(2x)\sin^2(2x)\, dx = \int \cos(2x)\left(\sin^2(2x) - \sin^4(2x)\right)\, dx = \frac{1}{6}\sin^3(2x) - \frac{1}{10}\sin^5(2x) + C$

7. $\int \sin^2 x \cos^2 x\, dx = \int \left(\sin^2 x - \sin^4 x\right) = \frac{1}{8}x + \frac{1}{8}\cos x \sin x - \frac{1}{4}\cos^3 x \sin x + C$

8. $\int \cos^4 x \sin^2 x\, dx = \int \left(\cos^4 x - \cos^6 x\right) dx = \frac{1}{16}x + \frac{1}{16}\cos x \sin x + \frac{1}{24}\cos^3 x \sin x - \frac{1}{6}\sin x \cos^5 x + C$

9. $\int \frac{\sin^3 x}{\cos x}\, dx = \int \frac{(1 - \cos^2 x)\sin x}{\cos x}\, dx = \frac{1}{2}\cos^2 x - \ln|\cos x| + C$

10. $\int \sqrt{1 - x^2}\, dx \to \int \sqrt{1 - \sin^2 t}\, \cos t\, dt = \int \cos^2 t\, dt = \frac{1}{2}\cos t \sin t + \frac{1}{2}t + C$
$$\to \frac{1}{2}x\sqrt{1 - x^2} + \frac{1}{2}\arcsin x + C$$

11. $\int \frac{dx}{(x^2 + 4)^2} \to \frac{1}{8}\int \frac{\sec^2 t}{(1 + \tan^2 t)^2}\, dt = \frac{1}{8}\int \cos^2 t\, dt = \frac{t}{16} + \frac{\sin(2t)}{32} + C$
$$= \frac{t}{16} + \frac{\sin t \cos t}{16} + C \to \frac{1}{16}\arctan(x/2) + \frac{x}{8(4 + x^2)} + C$$
$[x = 2\tan t,\ dx = 2\sec^2 t\, dt,\ \sin t = x/\sqrt{4 + x^2},\ \cos t = 2/\sqrt{4 + x^2}]$

12. $\int x^2\sqrt{1 - x^2}\, dx \to \int \sin^2 t \cos^2 t\, dt = \int \left(\cos^2 t - \cos^4 t\right) dt$
$$= \frac{t}{8} + \frac{1}{16}\sin(2t) - \frac{1}{4}\cos^3 t \sin t + C$$
$$\to \frac{1}{8}\arcsin x + \frac{1}{8}x\sqrt{1 - x^2} - \frac{1}{4}x\left(1 - x^2\right)^{3/2} + C$$
$[x = \sin t,\ dx = \cos t\, dt,\ \cos t = \sqrt{1 - x^2}]$

13. $\displaystyle\int \frac{x^2}{\sqrt{9-x^2}}\,dx \;\to\; 9\int \sin^2 t\,dt = \frac{9t}{2} - \frac{9}{4}\sin(2t) + C = \frac{9t}{2} - \frac{9}{2}\sin t \cos t + C$
$\displaystyle\qquad\qquad\to\; \frac{9}{2}\arcsin(x/3) - \frac{1}{2}x\sqrt{9-x^2} + C$
$[x = 3\sin t,\, dx = 3\cos t\,dt,\, \cos t = \sqrt{1 - x^2/9}]$

14. $\displaystyle\int \frac{dx}{x^2\sqrt{x^2+1}} \;\to\; \int \frac{\cos t}{\sin^2 t}\,dt = -\frac{1}{\sin t} + C \;\to\; -\frac{\sqrt{x^2+1}}{x} + C$
$[x = \tan t,\, dx = \sec^2 t\,dt,\, \sin t = x/\sqrt{x^2+1}]$

15. $\displaystyle\int \frac{dx}{\sqrt{1+x^2}} \;\to\; \int \frac{\sec^2 t}{\sqrt{1+\tan^2 t}}\,dt = \int \sec t\,dt = \ln|\tan t + \sec t| + C \;\to\; \ln\left|x + \sqrt{1+x^2}\right| + C$

16. $\displaystyle\int \frac{\arctan x}{(1+x^2)^{3/2}} \;\to\; \int w\cos w\,dw = w\sin w + \cos w \;\to\; \frac{1 + x\arctan x}{\sqrt{1+x^2}} + C.$

17. $\displaystyle\int \tan^4 x\,dx = \frac{1}{3}\tan^3 x - \tan x + x + C$

18. $\displaystyle\int \sec^2 x \tan^2 x\,dx = \frac{1}{3}\tan^3 x + C$

19. $\displaystyle\int \sec^3 x \tan^2 x\,dx = \int \left(\sec^5 x - \sec^3 x\right) dx = \frac{1}{4}\sec^3 x \tan x - \frac{1}{8}\sec x \tan x - \frac{1}{8}\ln(\sec x + \tan x) + C$

20. $\displaystyle\int \sin(2x)\cos^2 x\,dx = \frac{1}{2}\int \sin(2x)(1 + \cos(2x))\,dx = -\frac{1}{4}\cos(2x) + \frac{1}{4}\cos^2(2x) + C$

21. $\displaystyle\int \sqrt{\cos x}\,\sin^5 x\,dx = \int \sqrt{\cos x}\left(1 - \cos^2 x\right)^2 \sin x\,dx = -\frac{2}{3}(\cos x)^{3/2} + \frac{4}{7}(\cos x)^{7/2} - \frac{2}{11}(\cos x)^{11/2} + C$

22. $\displaystyle\int \sqrt{1+\sin x}\,dx = \int \sqrt{1+\sin x}\cdot\frac{\sqrt{1-\sin x}}{\sqrt{1-\sin x}}\,dx = \int \frac{\cos x}{\sqrt{1-\sin x}}\,dx = -2\sqrt{1-\sin x} + C$

23. $\displaystyle\int \sqrt{1+x^2}\,dx \to \int \sec t \sec^2 t\,dt = \int \sec^3 t\,dt = \frac{1}{2}\tan t \sec t + \frac{1}{2}\ln|\sec t + \tan t| + C$
$\displaystyle\qquad\to\; \frac{1}{2}x\sqrt{1+x^2} + \frac{1}{2}\ln\left|\sqrt{1+x^2} + x\right| + C$
$[x = \tan t,\, dx = \sec^2 t,\, \sec t = \sqrt{1+x^2}]$

24. $\displaystyle\int \frac{dx}{x^2\sqrt{4-x^2}} \to \frac{1}{4}\int \csc^2 t\,dt = -\frac{1}{4}\cot^2 t + C = -\frac{1}{4}\frac{\cos t}{\sin t} + C \to -\frac{\sqrt{4-x^2}}{4x} + C$
$[x = 2\sin t,\, dx = 2\cos t\,dt,\, \cos t = \sqrt{1 - x^2/4}]$

25. $\displaystyle\int \frac{dx}{x^2\sqrt{x^2-4}} \to \frac{1}{4}\int \cos t\,dt = \frac{1}{4}\sin t + C \to \frac{\sqrt{x^2-4}}{4x} + C$
$[x = 2\sec t,\, dx = 2\sec t \tan t\,dt,\, \sin t = \sqrt{1 - 4/x^2}]$

26. $\displaystyle\int \frac{\sqrt{4-x^2}}{x^2}\,dx \;\to\; \int \frac{\cos^2 t}{\sin^2 t}\,dt = \int \frac{1-\sin^2 t}{\sin^2 t}\,dt = \int \left(\csc^2 t - 1\right) dt$
$\displaystyle\qquad = -\cot t - t + C = -\frac{\cos t}{\sin t} - t + C \to -\frac{\sqrt{4-x^2}}{x} - \arcsin(x/2) + C$
$[x = 2\sin t,\, dx = 2\cos t\,dt,\, \cos t = \sqrt{1 - x^2/4}]$

SECTION 9.3 TRIGONOMETRIC ANTIDERIVATIVES 75

27. First, note that $\int \dfrac{x+2}{x(x^2+1)}\,dx = \int \dfrac{dx}{x^2+1} + 2\int \dfrac{dx}{x(x^2+1)} = \arctan x + 2\int \dfrac{dx}{x(x^2+1)}$. Also,

$$\int \dfrac{dx}{x(x^2+1)} \to \int \dfrac{\cos t}{\sin t}\,dt = \ln|\sin t| + C \to \ln\left|\dfrac{x}{\sqrt{1+x^2}}\right| + C.$$

$[x = \tan t,\ dx = \sec^2 t,\ \sin t = x/\sqrt{1+x^2}]$

Therefore, $\int \dfrac{x+2}{x(x^2+1)}\,dx = \arctan x + 2\ln\left|\dfrac{x}{\sqrt{1+x^2}}\right| + C.$

28. Using integration by parts, $\int x \arcsin x\,dx = \dfrac{1}{2}x^2 \arcsin x - \dfrac{1}{2}\int \dfrac{x^2}{\sqrt{1-x^2}}\,dx.$ The integral on the right can be evaluated using the substitution $x = \sin t$: $\int \dfrac{x^2}{\sqrt{1-x^2}}\,dx = \dfrac{1}{2}\arcsin x - \dfrac{1}{2}x\sqrt{1-x^2} + C.$ Therefore,

$$\int x \arcsin x\,dx = \dfrac{1}{2}x^2 \arcsin x - \dfrac{1}{4}\arcsin x + \dfrac{1}{4}x\sqrt{1-x^2} + C.$$

29. (a) $\cos(x+y) + \cos(x-y) = (\cos x \cos y - \sin x \sin y) + (\cos x \cos(-y) - \sin x \sin(-y)) = 2\cos x \cos y$

   (b) $\int \cos(ax)\cos(bx)\,dx = \dfrac{1}{2}\int \big(\cos((a+b)x) + \cos((a-b)x)\big)\,dx = \dfrac{1}{2(a+b)}\sin((a+b)x) + \dfrac{1}{2(a-b)}\sin((a-b)x) + C$

   (c) $\int \cos(ax)\cos(ax)\,dx = \int \cos^2(ax)\,dx = \dfrac{x}{2} + \dfrac{1}{2a}\cos(ax)\sin(ax) + C$

30. (a) $\cos(x-y) - \cos(x+y) = (\cos x \cos(-y) - \sin x \sin(-y)) - (\cos x \cos y - \sin x \sin y) = 2\sin x \sin y$

   (b) $\int \sin(ax)\sin(bx)\,dx = \dfrac{1}{2}\int \big(\cos((a-b)x) - \cos((a+b)x)\big)\,dx = \dfrac{1}{2(a-b)}\sin((a-b)x) - \dfrac{1}{2(a+b)}\sin((a+b)x) + C$

   (c) $\int \sin(ax)\sin(ax)\,dx = \int \sin^2(ax)\,dx = x/2 - \dfrac{1}{2a}\cos(ax)\sin(ax) + C$

31. (a) $\sin(x+y) + \sin(x-y) = (\sin x \cos y + \cos x \sin y) + (\sin x \cos(-y) + \cos x \sin(-y)) = 2\sin x \cos y.$

   (b) $\int \sin(ax)\cos(bx)\,dx = \dfrac{1}{2}\int \big(\sin((a+b)x) + \sin((a-b)x)\big)\,dx = -\dfrac{1}{2(a+b)}\cos((a+b)x) - \dfrac{1}{2(a-b)}\cos((a-b)x) + C$

   (c) $\int \sin(ax)\cos(ax)\,dx = \dfrac{1}{2a}\sin^2(ax) + C$

32. (a) Let $x = \sec t$. When $x > 0$, $\sqrt{x^2-1} = \tan t$ (since $\tan t > 0$). Thus,

$$\int_1^2 \dfrac{\sqrt{x^2-1}}{x}\,dx = \int_0^{\pi/3} \tan^2 t\,dt = \tan t - t\Big]_0^{\pi/3} = \sqrt{3} - \pi/3.$$

   (b) Let $x = \sec t$. When $x < 0$, $\sqrt{x^2-1} = -\tan t$ (since $\tan t < 0$). Thus,

$$\int_{-2}^{-1} \dfrac{\sqrt{x^2-1}}{x}\,dx = -\int_{2\pi/3}^{\pi} \tan^2 t\,dt = t - \tan t\Big]_{2\pi/3}^{\pi} = \pi/3 - \sqrt{3}.$$

33. (a) $u = \sin^{n-1} x \implies du = (n-1)\sin^{n-2} x \cos x \, dx; dv = \sin x \, dx \implies v = -\cos x$. Thus,

$$\int \sin^n x \, dx = \int u \, dv = uv - \int v \, du = -\sin^{n-1} x \cos x + (n-1)\int \sin^{n-2} x \cos^2 x \, dx.$$

(b) $\int \sin^{n-2} x \cos^2 x \, dx = \int \sin^{n-2} x \left(1 - \sin^2 x\right) dx = \int \sin^{n-2} x \, dx - \int \sin^n x \, dx$. Thus,

$$\int \sin^n x \, dx = -\sin^{n-1} x \cos x + (n-1) \int \sin^{n-2} x \, dx - (n-1) \int \sin^n x \, dx$$

so $n \int \sin^n x \, dx = -\sin^{n-1} x \cos x + (n-1) \int \sin^{n-2} x \, dx$. Therefore, if $n \neq 0$,

$$\int \sin^n x \, dx = -\frac{\sin^{n-1} x \cos x}{n} + \frac{n-1}{n} \int \sin^{n-2} x \, dx.$$

34. Let $I_n = \int_0^{\pi/2} \sin^n x \, dx$. Then $I_1 = 1$, $I_2 = \pi/4$, and the reduction formula implies that $I_n = \frac{n-1}{n} I_{n-2}$.

   (a) If n is an odd integer such that $n > 1$,

   $$I_n = \frac{n-1}{n} I_{n-2} = \frac{n-1}{n} \frac{n-3}{n-2} I_{n-4} = \cdots = \frac{(n-1)}{n} \frac{(n-3)}{n-2} \cdots \frac{4}{5} \frac{2}{3}.$$

   (b) If n is an even integer such that $n > 2$,

   $$I_n = \frac{n-1}{n} I_{n-2} = \frac{n-1}{n} \frac{n-3}{n-2} I_{n-4} = \cdots = \frac{(n-1)}{n} \frac{(n-3)}{n-2} \cdots \frac{5}{6} \frac{3}{4} \frac{\pi}{4}.$$

35. The absolute value is required because $\sqrt{x^2 - 4} > 0$ when $x < -2$, but $2 \tan t < 0$ when $\pi/2 < t < \pi$.

36. Draw a right triangle with hypotenuse $\sqrt{1 + t^2}$ and sides of length 1 and t. If the angle opposite the side of length t is $x/2$, then $\sin(x/2) = t/\sqrt{1+t^2}$ and $\cos(x/2) = 1/\sqrt{1+t^2}$.

   (a) The desired result follows from the identity $\sin(2\theta) = 2\cos\theta\sin\theta$.

   (b) The desired result follows from the identity $\cos(2\theta) = 2\cos^2\theta - 1$.

   (c) Let $x = 2 \arctan t$. Then $dx = \frac{2}{1+t^2} \, dt$ and

   $$\int \frac{dx}{1 + \sin x + \cos x} \to \int \frac{1}{1 + \frac{2t}{1+t^2} + \frac{1-t^2}{1+t^2}} \cdot \frac{2}{1+t^2} \, dt$$

   $$= \int \frac{dt}{1+t} = \ln|1+t| + C \to \ln|1 + \tan(x/2)| + C.$$

37. $\int \frac{dx}{1 + \cos x} \to \int \frac{1}{1 + (1-t^2)/(1+t^2)} \frac{2}{1+t^2} \, dt = \int dt = t + C \to \tan(x/2) + C$

9.4 Miscellaneous Exercises

1. $\int \frac{\sin x}{(3 + \cos x)^2} dx = \frac{1}{3 + \cos x}$
   [substitution: $u = 3 + \cos x$]

2. $\int \frac{x^2}{x + 1} dx = \frac{1}{2}(x + 1)^2 - 2(x + 1) + \ln|x + 1|$
   [substitution: $u = x + 1$]

3. $\int x\left(3 + 4x^2\right)^5 dx = \frac{1}{48}\left(3 + 4x^2\right)^6$
   [subsitution: $u = 3 + 4x^2$]

4. $\int \frac{dx}{\sqrt{1 - x^2}} = \arcsin x$

5. $\int \frac{x}{\sqrt[3]{x^2 + 4}} dx = \frac{3}{4}\left(x^2 + 4\right)^{2/3}$
   [subsitution: $u = x^2 + 4$]

6. $\int \frac{(\ln x)^2}{x} dx = \frac{1}{3}(\ln|x|)^3$
   [substitution: $u = \ln x$]

7. $\int xe^x dx = e^x(x - 1)$
   [integration by parts: $u = x, dv = e^x dx$]

8. $\int e^x \sin x \, dx = \frac{1}{2}e^x(\sin x - \cos x)$
   [integration by parts (twice)]

9. $\int \frac{\ln x}{x} dx = \frac{1}{2}(\ln|x|)^2$
   [substituion: $u = \ln x$]

10. $\int x\sqrt{x + 2} \, dx = \frac{2}{5}(x + 2)^{5/2} - \frac{4}{3}(x + 2)^{3/2}$
    [substitution: $u = x + 2$]

11. $\int \sin^2(3x) \cos(3x) \, dx = \frac{1}{9}\sin^3(3x)$
    [subsitution: $u = \sin(3x)$]

12. $\int xe^{3x} dx = \frac{1}{3}e^{3x}\left(x - \frac{1}{3}\right)$
    [integration by parts: $u = x, dv = e^{3x} dx$]

13. $\int xe^{3x^2} dx = \frac{1}{6}e^{3x^2}$
    [substitution: $u = 3x^2$]

14. $\int \frac{dx}{1 + 4x^2} = \frac{1}{2}\arctan(2x)$
    [substitution: $u = 2x$]

15. $\int \dfrac{7-x}{(x+3)(x^2+1)}\,dx = \ln|x+3| + 2\arctan x - \dfrac{1}{2}\ln(x^2+1)$
    [partial fractions: $\dfrac{7-x}{(x+3)(x^2+1)} = \dfrac{1}{x+3} + \dfrac{2-x}{x^2+1}$]

16. $\int (2-3x)^{10}\,dx = -\dfrac{1}{33}(2-3x)^{11}$
    [substitution: $u = 2 - 3x$]

17. $\int \arctan x\,dx = x\arctan x - \dfrac{1}{2}\ln(1+x^2)$
    [integration by parts: $u = \arctan x, dv = dx$]

18. $\int \dfrac{\sec^2 x}{3+\tan x}\,dx = \ln|3+\tan x|$
    [subsitution: $u = 3 + \tan x$]

19. $\int x\sin x\,dx = \sin x - x\cos x$
    [integration by parts: $u = x, dv = \sin x\,dx$]

20. $\int \dfrac{dx}{(x-1)(x+2)} = \dfrac{1}{3}\ln|x-1| - \dfrac{1}{3}\ln|x+2| = \dfrac{1}{3}\ln\left|\dfrac{x-1}{x+2}\right|$
    [partial fractions: $\dfrac{1}{(x-1)(x+2)} = \dfrac{1}{3(x-1)} - \dfrac{1}{3(x+2)}$]

21. $\int x^2 \ln x\,dx = \dfrac{1}{3}x^3 \ln|x| - \dfrac{1}{9}x^3$
    [integration by parts: $u = \ln x, dv = x^2\,dx$]

22. $\int \dfrac{x+1}{x^2+1}\,dx = \int \dfrac{x}{x^2+1}\,dx + \int \dfrac{1}{x^2+1}\,dx = \dfrac{1}{2}\ln(x^2+1) + \arctan x$

23. $\int \dfrac{e^x}{\sqrt{1-e^{2x}}}\,dx = \arcsin(e^x)$
    [substitution: $u = e^x$]

24. $\int \dfrac{\sin x}{2+\cos x}\,dx = -\ln(2+\cos x)$
    [substitution: $u = 2 + \cos x$]

25. $\int \ln x\,dx = x(\ln x - 1)$
    [integration by parts: $u = \ln x, dv = dx$]

26. $\int x\cos(3x^2)\,dx = \dfrac{1}{6}\sin(3x^2)$
    [substitution: $u = 3x^2$]

27. $\int \arcsin x\,dx = x\arcsin x + \sqrt{1-x^2}$
    [integration by parts: $u = \arcsin x, dv = dx$]

28. $\int \dfrac{dx}{x^2+2x+3} = \dfrac{1}{\sqrt{2}}\arctan\left(\dfrac{x+1}{\sqrt{2}}\right)$
    [complete the square, substitution: $x^2 + 2x + 3 = (x+1)^2 + 2 = u^2 + 2$]

SECTION 9.4 MISCELLANEOUS EXERCISES

29. $\int \dfrac{x}{\sqrt{x-2}}\,dx = \dfrac{2}{3}(x-2)^{3/2} + 4\sqrt{x-2}$
 [substitution: $u = x - 2$]

30. $\int \dfrac{dx}{\sqrt{1-4x^2}} = \dfrac{1}{2}\arcsin(2x)$
 [substitution: $u = 2x$]

31. $\int \dfrac{x+6}{(x+1)(x^2+4)}\,dx = \ln|x+1| + \arctan(x/2) - \dfrac{1}{2}\ln(x^2+4)$
 [partial fractions: $\dfrac{x+6}{(x+1)(x^2+4)} = \dfrac{1}{x+1} + \dfrac{2-x}{x^2+4}$]

32. $\int x\sin^2 x \cos x\,dx = \dfrac{1}{3}x\sin^3 x + \dfrac{1}{3}\cos x - \dfrac{1}{9}\cos^3 x = \dfrac{1}{3}x\sin^3 x + \dfrac{1}{9}\sin^2 x \cos x + \dfrac{2}{9}\cos x$
 [integration by parts with $u = x$ and $dv = \sin^2 x \cos x\,dx$]

33. $\int \dfrac{x^3}{1+x^2}\,dx = \dfrac{1}{2}(1+x^2) - \dfrac{1}{2}\ln(1+x^2)$
 [substitution: $u = 1 + x^2$]

34. $\int \tan x\,dx = -\ln|\cos x|$
 [Write $\tan x = \sin x / \cos x$, then use the substitution $u = \cos x$.]

35. $\int \cos(2x)\,dx = \dfrac{1}{2}\sin(2x)$

36. $\int e^{2x}\sqrt{1+e^x}\,dx = \dfrac{2}{5}(1+e^x)^{5/2} - \dfrac{2}{3}(1+e^x)^{3/2}$
 [substitution: $u = 1 + e^x$]

37. $\int \dfrac{dx}{1+x^2} = \arctan x$

38. Let $u = (x+1)/2$. Then, $du = \tfrac{1}{2}dx$ and $\int \dfrac{dx}{\sqrt{3-2x-x^2}} \to \int \dfrac{du}{\sqrt{1-u^2}} = \arcsin u \to \arcsin\left(\tfrac{1}{2}(x+1)\right)$.

39. Let $u = \sin x$. Then, $du = \cos x\,dx$ and

$$\int \sin^3 x \cos^3 x\,dx = \int \sin^3 x \cos^2 x \cos x\,dx = \int \sin^3 x(1-\sin^2 x)\cos x\,dx$$

$$= \int \sin^3 x \cos x\,dx - \int \sin^5 x \cos x\,dx \to \int u^3\,du - \int u^5\,du$$

$$= \tfrac{1}{4}u^4 - \tfrac{1}{6}u^6 + C \to \tfrac{1}{4}\sin^4 x - \tfrac{1}{6}\sin^6 x + C$$

40. Using integration by parts with $u = \arcsin x$ and $dv = x^2\,dx$,

$$\int x^2 \arcsin x\,dx = \dfrac{1}{3}x^3 \arcsin x - \dfrac{1}{3}\int \dfrac{x^3}{\sqrt{1-x^2}}\,dx.$$

To find the remaining antiderivative, use the substituion $w = 1 - x^2$:

$$\int \dfrac{x^3}{\sqrt{1-x^2}}\,dx \to -\dfrac{1}{2}\int \dfrac{1-w}{\sqrt{w}}\,dw = -\dfrac{1}{2}\int \dfrac{dw}{\sqrt{w}} + \dfrac{1}{2}\int \sqrt{w}\,dw$$

$$= -\sqrt{w} + \dfrac{1}{3}w^{3/2} + C \to -\sqrt{1-x^2} + \dfrac{1}{3}(1+x^2)^{3/2} + C$$

Therefore, $\int x^2 \arcsin x \, dx = \frac{1}{3}x^3 \arcsin x + \frac{1}{3}\sqrt{1-x^2} - \frac{1}{9}(1+x^2)^{3/2} + C.$

41. Let $u = \ln x$. Then, $du = dx/x$ and $\int \dfrac{dx}{x(\ln x)^2} \to \int \dfrac{du}{u^2} = -\dfrac{1}{u} + C \to -\dfrac{1}{\ln|x|} + C.$

42. $\int x \arctan x \, dx = \dfrac{1}{2}(x^2 + 1)\arctan x - \dfrac{x}{2}$
    [integration by parts: $u = \arctan x$, $dv = x \, dx$]

43. $\int \dfrac{dx}{x^3 + x} = \ln|x| - \dfrac{1}{2}\ln(x^2 + 1)$
    [partial fractions: $\frac{1}{x^3+x} = \frac{1}{x} - \frac{x}{x^2+1}$]

44. $\int \tan^4 x \, dx = \dfrac{1}{3}\tan^3 x - \tan x + x$

45. $\int \dfrac{x+5}{x^2 + 3x - 4}\, dx = \dfrac{6}{5}\ln|x-1| - \dfrac{1}{5}\ln|x+4|$
    [partial fractions: $\frac{x+5}{x^2+3x-4} = \frac{6}{5(x-1)} - \frac{1}{5(x-4)}$]

46. $\int \dfrac{x^3}{\sqrt{4-x^2}} \, dx = -4\sqrt{4-x^2} + \dfrac{1}{3}\left(4-x^2\right)^{3/2} - x^2\sqrt{4-x^2} - \dfrac{2}{3}\left(4-x^2\right)^{3/2}$
    [substitution ($u = 4 - x^2$) or integration by parts ($u = x^2$, $dv = x/\sqrt{4-x^2}$)]

47. $\int \dfrac{dx}{\sqrt[3]{x-1}} = \dfrac{3}{2}(x-1)^{2/3}$

48. $\int \dfrac{x}{(x-1)(x+1)} \, dx = \dfrac{1}{2}\ln|x+1| + \dfrac{1}{2}\ln|x-1| = \dfrac{1}{2}\ln\left|x^2 - 1\right|$
    [partial fractions: $\frac{x}{(x-1)(x+1)} = \frac{1}{2(x+1)} + \frac{1}{2(x-1)}$]

49. $\int x^3 e^{x^2} \, dx = \dfrac{1}{2}e^{x^2}(x^2 - 1)$
    [substitution ($w = x^2$), then integration by parts ($u = w$, $dv = e^w \, dw$)]

50. $\int \dfrac{dx}{\sqrt{9+x^2}} = \ln\left|x + \sqrt{9+x^2}\right|$
    [trigonometric substitution: $x = 3 \tan t$]

51. $\int \dfrac{dx}{2x - x^2} = \dfrac{1}{2}\ln|x| - \dfrac{1}{2}\ln|x-2| = \dfrac{1}{2}\ln\left|\dfrac{x}{x-2}\right|$
    [partial fractions: $\frac{1}{2x-x^2} = \frac{1}{x(2-x)} = \frac{1}{2x} + \frac{1}{2(2-x)}$]

52. $\int \dfrac{x^2}{1-3x} \, dx = -\dfrac{1}{54}(1-3x)^2 + \dfrac{2}{27}(1-3x) - \dfrac{1}{27}\ln|1-3x| + C = -\dfrac{1}{6}x^2 - \dfrac{1}{9}x - \dfrac{1}{27}\ln|1-3x| + C$
    [substitution ($u = 1 - 3x$) or partial fractions]

53. $\int \left(x^2 + 2x + 3\right)^{3/2} dx = \dfrac{1}{4}(x+1)(x^2 + 2x + 3)^{3/2} + \dfrac{3}{4}(x+1)\sqrt{x^2+2x+3} +$
    $\dfrac{3}{2}\ln\left|\dfrac{\sqrt{x^2+2x+3}}{\sqrt{2}} + \dfrac{x+1}{\sqrt{2}}\right|$
    [Write $x^2 + 2x + 3 = (x+1)^2 + 2$, then use a trigonometric substitution ($x + 1 = \sqrt{2}\tan t$).]

Section 9.4 Miscellaneous Exercises

54. $\int \dfrac{x}{(x^2-1)^3}\, dx = -\dfrac{1}{4(x^2-1)^4}$

    [substitution: $u = x^2 - 1$]

55. $\int e^x e^{2x}\, dx = \int e^{3x}\, dx = \dfrac{1}{3} e^{3x}$

56. $\int \sqrt{4x-3}\, dx = \dfrac{1}{6}(4x-3)^{3/2}$

    [substitution: $u = 4x - 3$]

57. $\int \ln(1+x^2)\, dx = x\ln(1+x^2) - 2x + 2\arctan x$

    [integration by parts: $u = \ln(1+x^2),\ dv = dx$]

58. $\int \sin(\sqrt{x})\, dx = 2\sin(\sqrt{x}) - 2\sqrt{x}\cos(\sqrt{x})$

    [substitution ($w = \sqrt{x}$), then integration by parts ($u = w,\ dv = \sin w\, dw$)]

59. $\int \dfrac{x}{9+4x^4}\, dx = \dfrac{1}{12}\arctan\left(\dfrac{2x^2}{3}\right)$

60. $\int \dfrac{dx}{(4-x^2)^{3/2}} = \dfrac{x}{4\sqrt{4-x^2}}$

    [trigonometric substitution: $x = 2\sin t$]

61. $\int \sin(3x)\cos(5x)\, dx = \dfrac{1}{4}\cos(2x) - \dfrac{1}{16}\cos(8x)$

    [Write $\sin(3x)\cos(5x) = \tfrac{1}{2}\sin(8x) - \tfrac{1}{2}\sin(2x)$.]

62. $\int \ln\left(\sqrt{x^2+1}\right) dx = x\ln\left(\sqrt{1+x^2}\right) - x + \arctan x$

    [Write $\ln\left(\sqrt{1+x^2}\right) = \tfrac{1}{2}\ln(1+x^2)$, then use integration by parts ($u = \ln(1+x^2),\ dv = dx$).]

63. $\int x\sqrt{2x+1}\, dx = \dfrac{1}{10}(2x+1)^{5/2} - \dfrac{1}{6}(2x+1)^{3/2}$

    [substitution: $u = 2x + 1$]

64. $\int \dfrac{dx}{x(x+\sqrt[3]{x})} = -\dfrac{3}{\sqrt[3]{x}} - 3\arctan(\sqrt[3]{x})$

    [substitution ($u = x^{1/3}$), then partial fractions ($\dfrac{1}{u^4+u^2} = \dfrac{1}{u^2} - \dfrac{1}{u^2+1}$).]

65. $\int \dfrac{\tan x}{\sec^2 x}\, dx = -\dfrac{1}{2}\cos^2 x$

    [Write $\dfrac{\tan x}{\sec^2 x} = \sin x \cos x$, then use substitution ($u = \cos x$).]

66. $\int \dfrac{dx}{x^3+1} = \dfrac{1}{3}\ln|x+1| - \dfrac{1}{6}\ln\left|x^2-x+1\right| + \dfrac{1}{\sqrt{3}}\arctan\left(\dfrac{2x-1}{\sqrt{3}}\right)$

    [Write $x^3 + 1 = (x+1)(x^2-x+1) = (x+1)\cdot\left((2x-1)^2 + 3\right)/4$, then use partial fractions, etc.]

67. $\int \dfrac{x}{16+9x^2}\, dx = \dfrac{1}{18}\ln\left(16+9x^2\right)$

    [subsitution: $u = 16 + 9x^2$]

68. $\int \dfrac{dx}{e^x - 1} = \ln\left|1 - e^{-x}\right|$

[Write $\dfrac{1}{e^x-1} = \dfrac{e^{-x}}{1-e^{-x}}$, then use subsitution $u = 1 - e^{-x}$.]

69. $\int \dfrac{dx}{(e^x - e^{-x})^2} = \dfrac{1}{4}\left(\dfrac{1}{1+e^x} + \dfrac{1}{1-e^x} = \dfrac{1}{2 - 2e^{2x}}\right)$

[Write $(e^x - e^{-x})^{-2} = e^{2x}(e^{2x} - 1)^{-2} = (e^x)^2 (e^x - 1)^{-2} (e^x + 1)^{-2}$, then use substitution ($u = e^x$) and partial fractions.]

70. $\int \dfrac{dx}{\sqrt{2x - x^2}} = -\arcsin(1 - x)$

[Write $2x - x^2 = 1 - (1 - x)^2$, then substitute $u = 1 - x$.]

71. $\int x \tan^2 x \, dx = x \tan x - \dfrac{1}{2}x^2 - \ln|\sec x|$

[integration by parts: $u = x$, $dv = \tan^2 x \, dx$]

72. $\int \dfrac{dx}{1 + \sqrt{x}} = 2(1 + \sqrt{x}) - 2\ln(1 + \sqrt{x})$

[subsitution: $u = 1 + \sqrt{x}$]

73. $\int \dfrac{x^3}{(x^2 + 1)^2} = \dfrac{1}{2(x^2 + 1)} + \dfrac{1}{2}\ln(x^2 + 1)$

[substitution $u = x^2 + 1$]

74. $\int \cos^3 x \, dx = \sin x - \dfrac{1}{3}\sin^3 x$

[Write $\cos^3 x = \cos x \cos^2 x = \cos x - \cos x \sin^2 x$, then use the substitution $u = \sin x$.]

75. $\int \sin x \sin(2x) \, dx = \dfrac{1}{2}\sin x - \dfrac{1}{6}\sin(3x)$

[Write $\sin x \sin(2x) = \tfrac{1}{2}\cos x - \tfrac{1}{2}\cos(3x)$.]

76. $\int x^2 \ln(3x) \, dx = \dfrac{1}{3}x^3 \ln(3x) - \dfrac{1}{9}x^3$

[integration by parts: $u = \ln(3x)$, $dv = x^2 \, dx$]

77. $\int \dfrac{x}{1 + x^4} \, dx = \dfrac{1}{2}\arctan(x^2)$

[substitution: $u = x^2$]

78. $\int \sqrt{x} \ln x \, dx = \dfrac{2}{3}x^{3/2} \ln|x| - \dfrac{4}{9}x^{3/2}$

[integration by parts: $u = \ln x$, $dv = \sqrt{x} \, dx$]

79. $\int \sin^5 x \cos^2 x \, dx = \dfrac{1}{7}\sin^6 x \cos x - \dfrac{1}{35}\sin^4 x \cos x - \dfrac{4}{105}\sin^2 x \cos x - \dfrac{8}{105}\cos x$

[Write $\cos^2 x = 1 - \sin^2 x$, then use a reduction formula.]

80. $\int x \sec^2 x \, dx = x \tan x + \ln|\cos x|$

[integration by parts: $u = x$, $dv = \sec^2 x \, dx$]

10.1 When Is an Integral Improper?

1. The interval of integration is infinite.

2. The integrand is unbounded as $x \to 1^-$.

3. The integrand is unbounded as $x \to 1^+$.

4. The integrand is unbounded near $x = 2$.

5. The integrand is unbounded as $x \to \pi/2^-$

6. The integrand is unbounded near $x = \pi$.

7. $\int_0^\infty \frac{dx}{x^2} = \int_0^1 \frac{dx}{x^2} + \int_1^\infty \frac{dx}{x^2}$. Since $\int_0^1 \frac{dx}{x^2} = \infty$, the original improper integral diverges.

8. $\lim_{x \to \pi/2} \frac{\cos x}{\sqrt{1 - \sin^2 x}} = 1$ so the integrand is bounded over the interval of integration.

9. $\int_0^\infty e^{-x} dx = \lim_{t \to \infty} \int_0^t = \lim_{t \to \infty} -e^{-x}\Big]_0^t = \lim_{t \to \infty} (1 - e^{-t}) = 1$

10. $\int_e^\infty \frac{dx}{x(\ln x)^2} = \lim_{t \to \infty} \frac{-1}{\ln x}\Big]_e^t = \lim_{t \to \infty} \left(1 - \frac{1}{\ln t}\right) = 1$

11. $\int_1^\infty \frac{dx}{x(1+x)} = \lim_{t \to \infty} \ln\left(\frac{x}{1+x}\right)\Big]_1^t = \lim_{t \to \infty} \ln\left(\frac{t}{1+t}\right) - \ln \frac{1}{2} = \ln 2$

12. $\int_0^4 \frac{dx}{\sqrt{x}} = \lim_{t \to 0^+} \int_t^4 \frac{dx}{\sqrt{x}} = \lim_{t \to 0^+} 2\sqrt{x}\Big]_t^4 = \lim_{t \to 0^+} (4 - 2\sqrt{t}) = 4$

13. $\int_{-2}^2 \frac{2x+1}{\sqrt[3]{x^2+x-6}} dx = \lim_{t \to 2^-} \int_{-2}^t \frac{2x+1}{\sqrt[3]{x^2+x-6}} dx = \lim_{t \to 2^-} \frac{3}{2}(x^2+x-6)^{2/3}\Big]_{-2}^t$
$= \lim_{t \to 2^-} \frac{3}{2}\left((t^2+t-6)^{2/3} - (-4)^{2/3}\right) = \frac{3}{2}\sqrt[3]{16}$

14. $\int_\pi^\infty e^{-x} \sin x \, dx = \lim_{t \to \infty} -\frac{1}{2} e^{-x}(\sin x + \cos x)\Big]_\pi^t = \frac{1}{2} e^{-\pi}$

15. $\int_0^\infty \frac{\arctan x}{(1+x^2)^{3/2}} \to \int_0^{\pi/2} w \cos w \, dw = w \sin w + \cos w\Big]_0^{\pi/2} = \frac{\pi}{2} - 1$.

    NOTE: $\int \frac{\arctan x}{(1+x^2)^{3/2}} \to \int w \cos w \, dw = w \sin w + \cos w \to \frac{1 + x \arctan x}{\sqrt{1+x^2}}$.

16. $\int_2^4 \frac{x}{\sqrt{|x^2-9|}} dx = \int_2^3 \frac{x}{\sqrt{9-x^2}} dx + \int_3^4 \frac{x}{\sqrt{x^2-9}} dx$
$= \lim_{s \to 3^-} -\sqrt{9-x^2}\Big]_2^s + \lim_{t \to 3^+} \sqrt{x^2-9}\Big]_t^4 = \sqrt{5} + \sqrt{7}$

17. $\int_a^\infty e^{-x} dx = e^{-a} \leq 10^{-5}$ if $a \geq 11.6$.

18. $\int_a^\infty \dfrac{dx}{x^2+1} = \dfrac{\pi}{2} - \arctan a \leq 10^{-5}$ if $a \geq 100{,}000$.

19. $\int_a^\infty \dfrac{dx}{x(\ln x)^3} = \dfrac{1}{2(\ln a)^2} \leq 10^{-5}$ if $a \geq 5 \times 10^{97}$.

20. (a) Let $I = \int_1^\infty f(x)\,dx$. Then, if $a > 1$, $I - \int_1^a f(x)\,dx = \int_a^\infty f(x)\,dx$. Since $\lim_{a\to\infty} \int_1^a f(x)\,dx = I$, $\lim_{a\to\infty} \int_a^\infty f(x)\,dx = 0$. This implies that if a is large enough, $\int_a^\infty f(x)\,dx \leq 10^{-10}$.

   (b) Since $g(x) \geq 0$ for all $x \geq 1$, $\int_1^a g(x)\,dx$ is an increasing function of a. Since $\lim_{t\to\infty} \int_1^t g(x)\,dx = \infty$, one can make $\int_1^a g(x)\,dx$ as large as desired by choosing a large enough.

21. $I = \int_0^\infty f(x)\,dx = \int_0^a f(x)\,dx + \int_a^\infty f(x)\,dx \implies \left| I - \int_0^a f(x)\,dx \right| = \left| \int_a^\infty f(x)\,dx \right| \leq 0.0001$.

22. (a) $\lim_{a\to\infty} \int_{-a}^a x\,dx = \lim_{a\to\infty} \tfrac{1}{2}x^2 \Big]_{-a}^a = \lim_{a\to\infty} \left(\tfrac{1}{2}(a)^2 - \tfrac{1}{2}(-a)^2 \right) = \lim_{a\to\infty} (0) = 0$

   (b) $\int_{-\infty}^\infty x\,dx = \lim_{s\to-\infty} \int_s^0 x\,dx + \lim_{t\to\infty} \int_0^t x\,dx$. Since neither $\lim_{s\to-\infty} \int_s^0 x\,dx$ nor $\lim_{t\to\infty} \int_0^t x\,dx$ exists, the original improper integral diverges.

23. (a) The integral is improper because the integrand is unbounded near $x = 0$.

   (b) No — $\int_{-1}^1 x^{-3}\,dx = \int_{-1}^0 x^{-3}\,dx + \int_0^1 x^{-3}\,dx$ and both of the latter improper integrals diverge.

24. Diverges. $\int_0^\infty \dfrac{x}{\sqrt{1+x^2}}\,dx = \lim_{t\to\infty} \left(\sqrt{1+t^2} - 1 \right) = \infty$

25. Converges. $\int_0^\infty \dfrac{\arctan x}{1+x^2}\,dx = \lim_{t\to\infty} \tfrac{1}{2}(\arctan t)^2 = \pi^2/8$

26. Diverges. $\int_e^\infty \dfrac{dx}{x \ln x} = \lim_{t\to\infty} \ln(\ln t) = \infty$

27. Converges. $\int_3^\infty \dfrac{x}{(x^2-4)^3}\,dx = \lim_{t\to\infty} \dfrac{1}{4}\left(\dfrac{1}{25} - \dfrac{1}{t^2-4} \right) = \dfrac{1}{100}$

28. Converges. $\int_{-\infty}^1 e^x\,dx = \lim_{t\to-\infty} \left(e^1 - e^t \right) = e$

29. Converges. $\int_0^8 \dfrac{dx}{\sqrt[3]{x}} = \lim_{t\to 0^+} \dfrac{3}{2}\left(8^{2/3} - t^{2/3} \right) = 6$

30. Converges. $\int_1^3 \dfrac{dx}{\sqrt[3]{x-2}} = \lim_{s\to 2^-} \int_1^s \dfrac{dx}{\sqrt[3]{x-2}} + \lim_{t\to 2^+} \int_t^3 \dfrac{dx}{\sqrt[3]{x-2}} = \lim_{s\to 2^-} \dfrac{3}{2}(x-2)^{2/3} \Big|_1^s + \lim_{t\to 2^+} \dfrac{3}{2}(x-2)^{2/3} \Big|_t^3 = -\dfrac{3}{2} + \dfrac{3}{2} = 0$

31. Converges. $\int_2^3 \dfrac{x}{\sqrt{3-x}}\,dx = \lim_{t\to 3^-} \left(\dfrac{2}{3}(3-t)^{3/2} - 6\sqrt{3-t} + \dfrac{16}{3} \right) = \dfrac{16}{3}$

SECTION 10.1 WHEN IS AN INTEGRAL IMPROPER? 85

32. Converges. $\int_0^2 \frac{dx}{\sqrt{4-x^2}} = \lim_{t \to 2^-} \arcsin(t/2) = \pi/2$

33. Diverges. $\int_1^\infty \frac{dx}{x(\ln x)^2} = \int_1^2 \frac{dx}{x(\ln x)^2} + \int_2^\infty \frac{dx}{x(\ln x)^2} = \lim_{s \to 1} \left(\frac{1}{\ln s} - \frac{1}{\ln 2}\right) + \lim_{t \to \infty} \left(\frac{1}{\ln 2} - \frac{1}{\ln t}\right) = \infty$

34. Diverges. $\int_0^\infty \frac{dx}{(x-1)^2} = \int_0^1 \frac{dx}{(x-1)^2} + \int_1^\infty \frac{dx}{(x-1)^2} = \infty$

35. Diverges. $\int_0^\infty \frac{dx}{e^x - 1} = \int_0^1 \frac{dx}{e^x - 1} + \int_1^\infty \frac{dx}{e^x - 1}$

36. Diverges. $\int_{-\infty}^\infty e^{-x}\, dx = \int_{-\infty}^0 e^{-x}\, dx + \int_0^\infty e^{-x}\, dx$

37. Converges. $\int_{-\infty}^\infty \frac{dx}{e^x + e^{-x}} = \int_{-\infty}^\infty \frac{e^x}{e^{2x}+1}\, dx = \frac{\pi}{2}$

38. Converges. $\int_0^1 \frac{x}{\sqrt{1-x^2}}\, dx = 1$

39. Converges. $\int_0^1 \frac{e^{-\sqrt{x}}}{\sqrt{x}}\, dx = 2 - 2/e$

40. Converges. $\int_0^{\pi/2} \frac{\cos x}{\sqrt{\sin x}}\, dx = 2$

41. (a) $\int_1^\infty \frac{dx}{x} = \lim_{t \to \infty} \int_1^t \frac{dx}{x} = \lim_{t \to \infty} \ln t = \infty$

 (b) If $p > 1$, $\int_1^\infty \frac{dx}{x^p} = \lim_{t \to \infty} \frac{1 - t^{1-p}}{p-1} = \frac{1}{p-1}.$

 (c) If $p < 1$, $\int_1^\infty \frac{dx}{x^p} = \lim_{t \to \infty} \frac{1 - t^{1-p}}{p-1} = \infty.$

42. $\int_0^1 \frac{dx}{x^p} = \lim_{t \to 0^+} \int_t^1 \frac{dx}{x^p} = \lim_{t \to 0^+} \frac{1 - t^{1-p}}{p-1}$. This limit is a finite number only if $p < 1$. Thus, $\int_0^1 \frac{dx}{x^p}$ converges if $p < 1$ and diverges if $p \geq 1$.

43. (a) $p < 1$
 (b) $p > 1$

44. (a) $\int_0^t f(x)\, dx \geq \int_0^t c\, dx = ct$ implies that the area between the x-axis and the curve $y = f(x)$ is greater than the area of a rectangle with height c and length t. As $t \to \infty$, the area of this rectangle becomes infinite so $\int_0^\infty f(x)\, dx$ diverges.

 (b) The area between the x-axis and the curve $y = x^{-2}$ from $x = 1$ to $x = t$ is $1 - t^{-1}$. Since $g(x) \leq x^{-2}$ for all $x \geq 1$, the area between the x-axis and the curve $y = g(x)$ from $x = 1$ to $x = t$ must be less than $1 - t^{-1}$. Thus, $\int_1^\infty g(x)\, dx \leq \lim_{t \to \infty} (1 - t^{-1}) = 1$.

 (c) The area between the x-axis and the curve $y = h(x)$ from $x = 1$ to $x = t$ is greater than $\ln t$. Since $\lim_{t \to \infty} \ln t = \infty$, $\int_1^\infty h(x)\, dx$ must diverge.

45. $\int \left(\frac{2x}{x^2+1} - \frac{C}{2x+1} \right) dx = \ln(x^2+1) - \frac{C}{2}\ln(2x+1) = \ln\left(\frac{x^2+1}{\sqrt{(2x+1)^C}}\right)$. Thus, the improper integral converges to $-2\ln 2$ if $C = 4$ and diverges for all other values of C.

46. $\int \left(\frac{C}{x+1} - \frac{3x}{2x^2+C} \right) dx = C\ln(x+1) - \frac{3}{4}\ln(2x^2+C)$ Thus, the improper integral converges to $-\frac{3}{4}\ln 2$ if $C = \frac{3}{2}$ and diverges for all other values of C.

47. $\int \left(\frac{Cx^2}{x^3+1} - \frac{1}{3x+1} \right) dx = \frac{C}{3}\ln(x^3+1) - \frac{1}{3}\ln(3x+1)$. Thus, the improper integral converges to $-\frac{1}{3}\ln 3$ if $C = \frac{1}{3}$ and diverges for all other values of C.

48. $\int_0^\infty \left(\frac{1}{\sqrt{x^2+4}} - \frac{C}{x+2} \right) dx = \ln\left(x + \sqrt{x^2+4}\right) - C\ln(x+2)$. Thus, the improper integral converges to $\ln 2$ if $C = 1$ and diverges for all other values of C.

49. $\int_0^\infty \left(\frac{x}{x^2+1} - \frac{C}{3x+1} \right) dx = \frac{1}{2}\ln\left(x^2+1\right) - \frac{C}{3}\ln(3x+1)$. Thus, the improper integral converges to $-\ln 3$ if $C = 3$ and diverges for all other values of C.

50. $\int_1^\infty \left(\frac{Cx}{x^2+1} - \frac{1}{2x} \right) dx = \frac{C}{2}\ln\left(x^2+1\right) - \frac{1}{2}\ln x$. Thus, the improper integral converges to $-(\ln 2)/4$ if $C = 1/2$ and diverges for all other values of C.

51. $\int_1^\infty \frac{x}{x^3+1} dx = \lim_{t \to \infty} \int_1^t \frac{x}{x^3+1} dx = \lim_{t \to \infty} -\int_1^{1/t} \frac{du}{1+u^3} = \int_0^1 \frac{du}{1+u^3}$

52. $\int_0^{\pi/2} \frac{\cos x}{\sqrt{\pi - 2x}} dx = \lim_{t \to \frac{\pi}{2}^-} \int_0^t \frac{\cos x}{\sqrt{\pi - 2x}} dx = \lim_{t \to \frac{\pi}{2}^-} -\int_{\sqrt{\pi}}^{\sqrt{\pi-2t}} \cos\left(\frac{1}{2}\left(\pi - u^2\right)\right) du = \int_0^{\sqrt{\pi}} \cos\left(\frac{1}{2}\left(\pi - u^2\right)\right) du$

53. Use the substitution $u = 1/x$:

$$\int_0^\infty \frac{dx}{1+x^4} = \int_0^1 \frac{dx}{1+x^4} + \int_1^\infty \frac{dx}{1+x^4} = \lim_{s \to 0^+} \int_s^1 \frac{dx}{1+x^4} + \lim_{t \to \infty} \int_1^t \frac{dx}{1+x^4}$$

$$= \lim_{s \to 0^+} -\int_{1/s}^1 \frac{u^2}{u^4+1} du + \lim_{t \to \infty} -\int_1^{1/t} \frac{u^2}{u^4+1} du$$

$$= \int_1^\infty \frac{u^2}{u^4+1} du + \int_0^1 \frac{u^2}{u^4+1} du = \int_0^\infty \frac{u^2}{u^4+1} du$$

54. Let $u = e^{-x}$. Then $du = -e^{-x} dx$ and

$$\int_0^\infty x^3 e^{-x} dx = \lim_{t \to \infty} \int_0^t x^3 e^{-x} dx = \lim_{t \to 0^+} -\int_1^t (-\ln u)^3 du = \int_0^1 (-\ln x)^3 dx.$$

55. $\int_0^\infty \frac{x \ln x}{1+x^4} dx = 0$. Using the substitution $u = 1/x$: $\int_1^\infty \frac{x \ln x}{1+x^4} dx = -\int_0^1 \frac{u \ln u}{u^4+1} dx.$

56. $\tan x = u^{-2} \implies x = \arctan(u^{-2})$ so $dx = -\dfrac{2u^{-3}}{1+(u^{-2})^2}\,du$. Thus,

$$\int \sqrt{1+\tan x}\,dx = \int \sqrt{1+u^{-2}}\left(-\frac{2u^{-3}}{1+u^{-4}}\,du\right) = -2\int \frac{\sqrt{u^2+1}}{u^4+1}\,du.$$

Since, $\tan(\pi/4) = 1 \implies u = 1$ and $x \to \pi/2^- \implies u \to 0^+$,

$$\int_{\pi/4}^{\pi/2} \sqrt{1+\tan x}\,dx = -2\int_1^0 \frac{\sqrt{u^2+1}}{1+u^4}\,du = 2\int_0^1 \frac{\sqrt{u^2+1}}{1+u^4}\,du.$$

10.2 Detecting Convergence, Estimating Limits

1. (a) For every $x \in \mathbb{R}$, $-1 \leq \sin x \leq 1$

   (b) $\int_2^\infty \dfrac{dx}{x + \sin x} \geq \int_2^\infty \dfrac{dx}{x+1} = \infty$

2. (a) $0 \leq \sqrt{x} \leq x^2$ for all $x \geq 1$, so $x^2 \leq x^2 + \sqrt{x} \leq x^2 + x^2 = 2x^2$ for all $x \geq 1$.

   (b) $0 \leq \int_1^\infty \dfrac{dx}{x^2 + \sqrt{x}} \leq \int_1^\infty \dfrac{dx}{x^2} = 1$

   (c) $0 \leq x^2 \leq \sqrt{x}$ if $0 \leq x \leq 1$, so $\sqrt{x} \leq x^2 + \sqrt{x} \leq 2\sqrt{x}$ if $0 \leq x \leq 1$.

   (d) The improper integral $\int_0^1 \dfrac{dx}{x^2 + \sqrt{x}}$ **converges** because $0 \leq \int_0^1 \dfrac{dx}{x^2 + \sqrt{x}} \leq \int_0^1 \dfrac{dx}{\sqrt{x}} = 2$.

   (e) Since $\int_0^\infty \dfrac{dx}{x^2 + \sqrt{x}} = \int_0^1 \dfrac{dx}{x^2 + \sqrt{x}} + \int_1^\infty \dfrac{dx}{x^2 + \sqrt{x}}$ and both of the improper integrals on the right converge, the improper integral on the left converges.

3. (a) $0 \leq \sqrt{x} \leq x^2/2$ for all $x \geq 2$, so $x^2/2 = x^2 - x^2/2 \leq x^2 - \sqrt{x} \leq x^2$ for all $x \geq 2$.

   (b) $0 \leq \int_3^\infty \dfrac{dx}{x^2 - \sqrt{x}} \leq 2 \int_3^\infty \dfrac{dx}{x^2} = \dfrac{1}{3}$

4. (a) The inequality $\dfrac{e^x}{1 + e^x} < 1$ holds for all x because $e^x < 1 + e^x$ for all x.

   Let $f(x) = \dfrac{e^x}{1 + e^x}$. The racetrack principle can be used to show that $\dfrac{1}{2} \leq f(x)$ all $x \geq 0$: $f(0) = 1/2$ and $f'(x) = e^x/(1 + e^x)^2 > 0$ for all x.

   (b) The improper integral **diverges** because $\int_0^\infty \dfrac{1}{2}\, dx \leq \int_0^\infty \dfrac{e^x}{1 + e^x}\, dx$ and the improper integral on the left diverges.

5. (a) When $x \geq 1$, $\sqrt{x^3} \leq \sqrt{1 + x^3} \leq \sqrt{2x^3}$. Therefore,

   $$\dfrac{1}{\sqrt{2x}} = \dfrac{x}{\sqrt{2x^3}} \leq \dfrac{x}{\sqrt{1 + x^3}} \leq \dfrac{x}{\sqrt{x^3}} = \dfrac{1}{\sqrt{x}}$$

   when $x \geq 1$.

   (b) The improper integral **diverges** since

   $$0 \leq \int_0^\infty \dfrac{x}{\sqrt{1 + x^3}}\, dx = \int_0^1 \dfrac{x}{\sqrt{1 + x^3}}\, dx + \int_1^\infty \dfrac{x}{\sqrt{1 + x^3}}\, dx$$
   $$\geq \int_0^1 \dfrac{x}{\sqrt{1 + x^3}}\, dx + \int_1^\infty \dfrac{dx}{\sqrt{2x}}$$

6. (a) Since $1 > 0$, $\sqrt{x} < \sqrt{x} + 1$. Also, $\sqrt{x} \geq 1 > 0$ for all $x \geq 1$ so $1 + \sqrt{x} \leq \sqrt{x} + \sqrt{x} = 2\sqrt{x}$.

   (b) **Diverges**: $\int_0^\infty \dfrac{dx}{1 + \sqrt{x}} \geq \int_0^1 \dfrac{dx}{1 + \sqrt{x}} + \int_1^\infty \dfrac{dx}{2\sqrt{x}} = \infty$

7. (a) When $x \geq 1$, $\sqrt{x} \geq 1$. Therefore, $\sqrt{x} \cdot x^4 = \sqrt{x} \cdot \sqrt{x^8} = \sqrt{x^9} < \sqrt{1 + x^9}$.
   When $x \geq 1$, $x^9 \geq 1$. Therefore, $\sqrt{1 + x^9} \leq \sqrt{x^9 + x^9} = \sqrt{2}\, x^{9/2}$.

   (b) **Converges**. $0 \leq \int_0^\infty \dfrac{dx}{\sqrt{1 + x^9}} \leq \int_0^1 \dfrac{dx}{\sqrt{1 + x^9}} + \int_1^\infty \dfrac{dx}{x^{9/2}} < \infty$

Section 10.2 Detecting Convergence, Estimating Limits

8. $0 < \int_a^\infty \dfrac{dx}{x^2+e^x} < \int_a^\infty e^{-x}\,dx = e^{-a} \le 10^{-5}$ if $a \ge \ln(100000) \approx 11.513$. Thus, $\int_0^{12} \dfrac{dx}{x^2+e^x}$ approximates $\int_0^\infty \dfrac{dx}{x^2+e^x}$ within 10^{-5}.

9. $0 < \int_a^\infty \dfrac{dx}{x^4\sqrt{2x^3+1}} < \dfrac{1}{\sqrt{2}}\int_a^\infty \dfrac{dx}{x^{11/2}} = \dfrac{\sqrt{2}}{9} a^{-9/2} \le 10^{-5}$ if $a \ge \left(\dfrac{2\times 10^{10}}{81}\right)^{1/9} \approx 8.5606$. Thus, $\int_1^9 \dfrac{dx}{x^4\sqrt{2x^3+1}}$ approximates $\int_1^\infty \dfrac{dx}{x^4\sqrt{2x^3+1}}$ within 10^{-5}.

10. $0 < \int_a^\infty \dfrac{\arctan x}{(1+x^2)^3}\,dx < \dfrac{\pi}{2}\int_a^\infty \dfrac{dx}{x^6} = \dfrac{\pi}{10a^5} \le 10^{-5}$ if $a \ge (10^4\pi)^{1/5} \approx 7.9329$. Thus, $\int_0^8 \dfrac{\arctan x}{(1+x^2)^3}\,dx$ approximates $\int_0^\infty \dfrac{\arctan x}{(1+x^2)^3}\,dx$ within 10^{-5}.

11. $0 < \int_a^\infty \dfrac{e^{-x}}{2+\cos x}\,dx < \int_a^\infty e^{-x}\,dx = e^{-a} \le 10^{-5}$ if $a \ge \ln(100000) \approx 11.513$. Thus, $\int_0^{12} \dfrac{e^{-x}}{2+\cos x}\,dx$ approximates $\int_0^\infty \dfrac{e^{-x}}{2+\cos x}\,dx$ within 10^{-5}.

12. (a) The interval of integration is unbounded.

(b) Let $f(x) = x - \ln x$. Then $f(e) = e - 1 > 0$ and $f'(x) = 1 - 1/x > 0$ for all $x \ge e$. Thus, $1 \le \ln x \le x$ for all $x \ge e$. Finally, since $u^2 \ge u$ if $u \ge 1$, $1 \le (\ln x)^2 \le x^2$ for all $x \ge e$.

(c) Using the inequalities in part (b): $\int_e^\infty \dfrac{dx}{x^2} \le I \le \int_e^\infty dx$. The improper integral on the left converges and the improper integral on the right diverges. Thus, all that can be concluded is that I is larger than $1/e$.

(d) Let $g(x) = \sqrt{x} - \ln x$. Then $g(1) = 1 > 0$ and $g'(x) = 1/\sqrt{x} - 1/x = (\sqrt{x}-1)/x > 0$ for all $x \ge 1$. Therefore, since $\ln e = 1$, $1 \le \ln x \le \sqrt{x}$ for all $x \ge e$.

(e) The inequality in part (d) implies that $(\ln x)^2 \le x$ for all $x \ge 1$. Therefore, $I = \int_e^\infty \dfrac{dx}{(\ln x)^2}\,dx \ge \int_e^\infty \dfrac{dx}{x}$. Because the improper integral on the right diverges, I also diverges.

13. (a) Let $f(x) = \sin x - x/2$. Then $f(0) = 0$ and $f'(x) = \cos x - 1/2 > 0$ if $0 \le x \le 1$. Therefore, $f(x) > 0$ if $0 \le x \le 1$.

(b) $0 \le \int_0^1 \dfrac{dx}{\sqrt{\sin x}} \le \sqrt{2}\int_0^1 \dfrac{dx}{\sqrt{x}} = 2\sqrt{2}$.

14. (a) The integral I is doubly improper. The interval of integration is unbounded and the integrand is unbounded as $x \to 0^+$.

(b) Let $I_1 = \int_0^1 \dfrac{dx}{\sqrt{x+x^4}}$ and let $I_2 = \int_1^\infty \dfrac{dx}{\sqrt{x+x^4}}$. Then, since $0 \le 1/\sqrt{x+x^4} \le 1/\sqrt{2x}$ when $0 < x \le 1$, $0 \le I_1 \le \int_0^1 dx/\sqrt{2x} = \sqrt{2}$. Furthermore, since $0 \le 1/\sqrt{x+x^4} \le 1/\sqrt{2x^4}$ for all $x \ge 1$, $0 \le I_2 \le \int_1^\infty dx/\sqrt{2x^4} = \sqrt{2}/2$. Therefore, $0 \le I = I_1 + I_2 \le 3\sqrt{2}/2 \approx 2.12132 < 3$.

15. Converges. $0 \le \int_1^\infty \dfrac{dx}{x^4+1}\,dx \le \int_1^\infty \dfrac{dx}{x^4} = \dfrac{1}{3}$

16. Diverges. Since $x^4 \le x$ if $0 \le x \le 1$, $\int_0^\infty \dfrac{dx}{x^4+x}\,dx = \int_0^1 \dfrac{dx}{x^4+x}\,dx + \int_1^\infty \dfrac{dx}{x^4+x}\,dx > \int_0^1 \dfrac{dx}{2x}\,dx + \int_0^1 \dfrac{dx}{x^4+x}\,dx$. Since $\int_0^1 \dfrac{dx}{2x}\,dx$ diverges, the original improper integral also diverges.

17. Diverges. $\int_2^\infty \frac{dx}{\sqrt{x-1}} dx > \int_2^\infty \frac{dx}{\sqrt{x}} = \infty$.

18. Diverges because $\lim_{x\to\infty} e^{\sin x}$ does not exist.

19. Converges. $0 \leq \int_1^\infty \frac{dx}{x\sqrt{1+x}} \leq \int_1^\infty \frac{dx}{x^{3/2}} = 2$.

20. Converges. $0 \leq \int_0^\infty \frac{dx}{x+e^x} \leq \int_0^\infty e^{-x}\, dx = 1$.

21. Diverges. $\int_0^\infty \frac{dx}{x+e^{-x}} = \int_0^1 \frac{dx}{x+e^{-x}} + \int_1^\infty \frac{dx}{x+e^{-x}} > \int_0^1 \frac{dx}{x+e^{-x}} + \int_1^\infty \frac{dx}{2x}$. Since the improper integral on the right diverges, the original improper integral diverges.

22. Converges. $0 \leq \int_1^\infty \frac{dx}{\sqrt[3]{x^6+x}} \leq \int_1^\infty \frac{dx}{x^2} = 1$.

23. Diverges. $\int_1^\infty \frac{\sqrt{x}}{x+1}\, dx \geq \frac{1}{2}\int_1^\infty \frac{dx}{\sqrt{x}} = \infty$.

24. Converges. Let $I = \int_2^\infty \frac{\sin x}{x^2\sqrt{x-1}}\, dx$. Then $|I| \leq \int_2^\infty \frac{dx}{x^2}$.

25. Diverges. $\int_1^\infty \frac{dx}{1+\sqrt{x}} \geq \int_1^\infty \frac{dx}{2\sqrt{x}} = \infty$.

26. Diverges. Since $x > \ln x$ for all $x \geq 1$, $\int_3^\infty \frac{x}{\ln x}\, dx > \int_3^\infty dx = \infty$.

27. Converges. $0 \leq \int_0^\infty \frac{dx}{\sqrt{x}(1+x)} = \int_0^1 \frac{dx}{\sqrt{x}(1+x)} + \int_1^\infty \frac{dx}{\sqrt{x}(1+x)} \leq \int_0^1 \frac{dx}{\sqrt{x}} + \int_1^\infty \frac{dx}{x^{3/2}} = 2+2 = 4$.

28. Converges. $0 \leq \int_0^\infty \frac{\sin^2 x}{(1+x)^2}\, dx \leq \int_0^\infty \frac{dx}{(1+x)^2} = 1$.

29. Let $u = 1/x$. Then, $x = u^{-1}$ and $-u^{-2}\, du = dx$ so

$$\int_0^1 \frac{dx}{\sqrt{x}(1+x)} = \lim_{t\to 0^+} \int_t^1 \frac{dx}{\sqrt{x}(1+x)} \to \lim_{t\to 0^+} -\int_{1/t}^1 \frac{du}{u^2\sqrt{u^{-1}+u^{-3}}}$$

$$= \lim_{t\to 0^+} \int_1^{1/t} \frac{du}{\sqrt{u^3+u}}$$

$$= \int_1^\infty \frac{du}{\sqrt{u^3+u}} = \int_1^\infty \frac{dx}{\sqrt{x+x^3}}$$

30. (a) $0 < \int_0^\infty \frac{e^{-x}}{\sqrt{x}}\, dx = \int_0^1 \frac{e^{-x}}{\sqrt{x}}\, dx + \int_1^\infty \frac{e^{-x}}{\sqrt{x}}\, dx \leq \int_0^1 \frac{dx}{\sqrt{x}} + \int_1^\infty e^{-x}\, dx = 2 + e^{-1}$.
Therefore, the improper integral converges.

(b) Let $f(x) = e^{-x}/\sqrt{x}$. Since $\int_0^\infty f(x)\, dx = \int_0^\alpha f(x)\, dx + \int_\alpha^\beta f(x)\, dx + \int_\beta^\infty f(x)\, dx$, $\int_\alpha^\beta f(x)\, dx$ approximates $\int_0^\infty f(x)\, dx$ within 0.005 if $\int_0^\alpha f(x)\, dx + \int_\beta^\infty f(x)\, dx \leq 0.005$.
Now, $0 \leq \int_0^\alpha f(x)\, dx \leq \int_0^\alpha x^{-1/2}\, dx = 2\sqrt{\alpha} \leq 0.0025$ if $\alpha \leq 1.5625 \times 10^{-6}$. Also,
$0 \leq \int_\beta^\infty f(x)\, dx \leq \int_\beta^\infty e^{-x}\, dx = e^{-\beta} \leq 0.0025$ if $\beta \geq -\ln(0.0025) \approx 6$.

SECTION 10.2 DETECTING CONVERGENCE, ESTIMATING LIMITS

31. (a) f is an increasing function for all $x \geq 0$ since $f'(x) = \sqrt{x}e^{-x} > 0$ for these values of x. Thus, the limit exists if and only if f is bounded above. Since $f(x) = \int_3^x \sqrt{t}e^{-t}\,dt < \int_3^x te^{-t}\,dt = 4e^{-3} - (x+1)e^{-x}$, $\lim_{x \to \infty} f(x) < 4e^{-3}$. Therefore, the limit exists.

    (b) $\int_a^\infty \sqrt{t}e^{-t}\,dt \leq \int_a^\infty te^{-t}\,dt = (a+1)e^{-a}$ for all $a \geq 1$. Since $(a+1)e^{-a} < 0.001$ for all $a \geq 9.2335$, $f(a)$ approximates $\int_3^\infty \sqrt{x}e^{-x}\,dx$ within 0.001 when $a \geq 9.2335$.

32. (a) I is an improper integral because the interval of integration is unbounded (i.e., infinitely long).

    (b) Since $I = \int_0^\infty f(x)\,dx = \int_0^1 f(x)\,dx + \int_1^\infty f(x)\,dx$, I converges if and only if $\int_1^\infty f(x)\,dx$ converges. The comparison test can be used to show that this integral converges: since $0 \leq e^{-x^3}\cos^2 x \leq e^{-x^3} \leq x^2 e^{-x^3}$ if $x \geq 1$, it follows that $0 \leq \int_1^\infty f(x)\,dx \leq \int_1^\infty x^2 e^{-x^3}\,dx = \dfrac{1}{3e}$ and, therefore, $\int_1^\infty f(x)\,dx$ converges.

    (c) $I = \int_0^a f(x)\,dx + \int_a^\infty f(x)\,dx \leq \int_0^a f(x)\,dx + \int_a^\infty x^2 e^{-x^3}\,dx = \int_0^a f(x)\,dx + \tfrac{1}{3}e^{-a^3}$ for any $a \geq 1$. Now, $\tfrac{1}{3}e^{-a^3} \leq 0.0025$ if $a \geq \sqrt[3]{-\ln 0.0075} \approx 1.6977$ so $\int_0^2 f(x)\,dx$ approximates I within 0.0025 (i.e., $0 < I - \int_0^2 f(x)\,dx < 0.0025$). Thus, an estimate of $\int_0^2 f(x)\,dx$ that is in error by no more than 0.0025 will be an estimate of I that is guaranteed to be correct within $0.0025 + 0.0025 = 0.005$.

    Using the derivative bounds given, it can be shown that $\int_0^2 f(x)\,dx$ is approximated within 0.0025 by M_n if $n \geq 20$ and by S_n if $n \geq 8$. $M_{20} \approx 0.64425$; $S_8 \approx 0.64441$. Thus, $|I - 0.64425| \leq 0.005$.

33. $\int_a^\infty \dfrac{dx}{x^4+\sqrt{x}} \leq \int_a^\infty \dfrac{dx}{x^4} = \tfrac{1}{3}a^{-3} \leq 0.0025$ if $a \geq \sqrt[3]{400/3} \approx 5.1087$. Thus, $\int_1^6 \dfrac{dx}{x^4+\sqrt{x}}$ approximates $\int_1^\infty \dfrac{dx}{x^4+\sqrt{x}}$ with an error no greater than 0.0025.

    Using $K_2 = 2.5$ and $K_4 = 21$, we find that a midpoint rule estimate computed with $n \geq 73$ (or a Simpson's rule estimate computed with $n \geq 20$) approximates $\int_1^6 \dfrac{dx}{x^4+\sqrt{x}}$ with an error no greater than 0.0025. Therefore, $M_{73} \approx 0.23603$ and $S_{20} \approx 0.23629$ are estimates of $\int_a^\infty \dfrac{dx}{x^4+\sqrt{x}}$ guaranteed to be accurate within 0.005.

34. $\left|\int_a^\infty e^{-x^2}\sin x\,dx\right| \leq \int_a^\infty \left|e^{-x^2}\sin x\right|\,dx \leq \int_a^\infty e^{-x^2}\,dx \leq \int_a^\infty xe^{-x^2}\,dx = \tfrac{1}{2}e^{-a^2} \leq 0.0025$ if $a \geq \sqrt{\ln 200} \approx 2.3018$. Thus, $\int_0^3 e^{-x^2}\sin x\,dx$ approximates $\int_0^\infty e^{-x^2}\sin x\,dx$ with an error no greater than 0.0025.

    Using $K_2 = 2.25$ and $K_4 = 21$, we find that a midpoint rule estimate computed with $n \geq 32$ (or a Simpson's rule estimate computed with $n \geq 12$) approximates $\int_0^3 e^{-x^2}\sin x\,dx$ with an error no greater than 0.0025. Therefore, $M_{32} \approx 0.42480$ and $S_{12} \approx 0.42460$ are estimates of $\int_0^\infty e^{-x^2}\sin x\,dx$ guaranteed to be accurate within 0.005.

35. $\int_a^\infty \dfrac{dx}{e^{x^2}+x} \leq \int_a^\infty e^{-x^2}\,dx \leq \tfrac{1}{2}e^{-a^2} \leq 0.0025$ if $a \geq \sqrt{\ln 200} \approx 2.3018$. Thus, $\int_0^3 \dfrac{dx}{e^{x^2}+x}$ approximates $\int_0^\infty \dfrac{dx}{e^{x^2}+x}$ with an error no greater than 0.0025.

Using $K_2 = 0.7$ and $K_4 = 36$, we find that the midpoint rule with $n \geq 18$ and Simpson's rule with $n \geq 12$ approximate $\int_0^3 \dfrac{dx}{e^{x^2}+x}$ with an error no greater than 0.0025. Therefore, $M_{18} \approx 0.69764$ and $S_{12} \approx 0.69870$ are estimates of $\int_0^\infty \dfrac{dx}{e^{x^2}+x}$ guaranteed to be accurate within 0.005.

36. $\int_a^\infty \dfrac{\arctan x}{(1+x^2)^4}\,dx \leq \dfrac{\pi}{2}\int_a^\infty x^{-8}\,dx = \dfrac{\pi}{14a^7} \leq 0.0025$ if $a \geq 1.9011$. Thus, $\int_1^2 \dfrac{\arctan x}{(1+x^2)^4}\,dx$ approximates $\int_1^\infty \dfrac{\arctan x}{(1+x^2)^4}\,dx$ with an error no greater than 0.0025.

Using $K_2 = 0.6$ and $K_4 = 2.25$, we find that the midpoint rule with $n \geq 4$ and Simpson's rule with $n \geq 2$ approximate $\int_1^2 \dfrac{\arctan x}{(1+x^2)^4}\,dx$ with an error no greater than 0.0025. Therefore, $M_4 \approx 0.013483$ and $S_2 \approx 0.014349$ are estimates of $\int_1^\infty \dfrac{\arctan x}{(1+x^2)^4}\,dx$ guaranteed to be accurate with 0.005.

37. (a) Since $\left|\dfrac{\cos x}{x^2}\right| \leq \dfrac{1}{x^2}$ and $\int_1^\infty x^{-2}\,dx$ converges, Theorem 2 asserts that $\int_1^\infty \dfrac{\cos x}{x^2}\,dx$ converges.

(b) Let $u = x^{-1}$ and $dv = \sin x\,dx$. Then, $\int_1^\infty \dfrac{\sin x}{x}\,dx = -\dfrac{\cos x}{x}\Big]_1^\infty + \int_1^\infty \dfrac{\cos x}{x^2}\,dx = \cos 1 + \int_1^\infty \dfrac{\cos x}{x^2}\,dx$. Thus, $\int_1^\infty \dfrac{\sin x}{x}\,dx$ can be written as the sum of two numbers (i.e., it converges).

(c) Let $w = e^x$. Then, $\int_0^\infty \sin(e^x)\,dx = \int_1^\infty \dfrac{\sin w}{w}\,dw$.

10.3 Improper Integrals and Probability

1. Graphs appear below; for reference, the the standard normal graph appears, too:

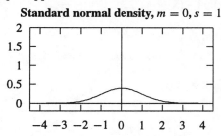
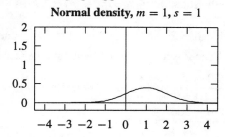
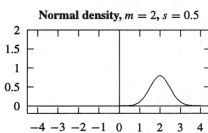
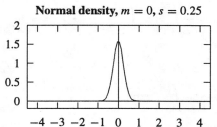

   Notice the similarities among all the graphs—m locates the center; s determines the "spread."

2. (a) The probability density function for X is $f(x) = \dfrac{1}{\sqrt{2\pi}} e^{-x^2/2}$. The area under this function over the interval $[-3, 3]$ is the fraction of X-values which lie no more than 3 standard deviations below the mean and no more than 3 standard deviations above the mean.

   (b) $\dfrac{1}{2\cdot\sqrt{2\pi}} \displaystyle\int_{-6}^{6} \exp\left(-\dfrac{x^2}{8}\right) dx$

   (c) If $u = x/2$, $du = \tfrac{1}{2}dx$. Therefore, $\dfrac{1}{2\cdot\sqrt{2\pi}} \displaystyle\int_{-6}^{6} \exp\left(-\dfrac{x^2}{8}\right) dx = \dfrac{1}{\sqrt{2\pi}} \displaystyle\int_{-3}^{3} e^{-u^2/2} du$.

   (d) Using Simpson's rule with $n = 20$, $\dfrac{1}{\sqrt{2\pi}} \displaystyle\int_{-3}^{3} e^{-x^2/2} dx \approx 0.9973$.

3. When $Z = \dfrac{x-500}{100}$, $dZ = \dfrac{dx}{100}$. Furthermore, when $x = 500$, $Z = 0$; when $x = 700$, $Z = 2$. Thus,

$$I_1 = \dfrac{1}{100\sqrt{2\pi}} \int_{500}^{700} \exp\left(-\dfrac{(x-500)^2}{2\cdot 100^2}\right) dx = \dfrac{1}{100\sqrt{2\pi}} \int_{0}^{2} \exp\left(-\dfrac{Z^2}{2}\right) 100\, dZ = \dfrac{1}{\sqrt{2\pi}} \int_{0}^{2} e^{-Z^2/2}\, dZ$$

4. Since $z = (x-m)/s$, $dz = dx/s$ and

$$\dfrac{1}{\sqrt{2\pi}\, s} \int \exp\left(-\dfrac{(x-m)^2}{2s^2}\right) dx = \dfrac{1}{\sqrt{2\pi}} \int e^{-z^2/2}\, dz.$$

   Using the rules for substitution in a definite integral, the limits of integration change from x_1 and x_2 to z_1 and z_2, respectively.

5. (a) Since the integrand is an even function $\displaystyle\int_{-\infty}^{\infty} e^{-x^2} dx = 2\int_{0}^{\infty} e^{-x^2} dx = 2\cdot \dfrac{\sqrt{\pi}}{2} = \sqrt{\pi}$.

(b) Using the substitution $u = x/\sqrt{2}$, $\dfrac{1}{\sqrt{2\pi}} \displaystyle\int_{-\infty}^{\infty} \exp\left(-x^2/2\right) = \dfrac{1}{\sqrt{\pi}} \displaystyle\int_{-\infty}^{\infty} e^{-u^2}\, du = 1.$

(c) Using the substitution $u = (x-m)/s$, $\dfrac{1}{\sqrt{2\pi}\, s} \displaystyle\int_{-\infty}^{\infty} \exp\left(-(x-m)^2/2s^2\right) = \dfrac{1}{2\sqrt{\pi}} \displaystyle\int_{-\infty}^{\infty} e^{-u^2/2}\, du = 1.$

6. (a) No. Negative Z-scores correspond to data values that are smaller than the mean since $z = (x-m)/s$.

   (b) 1.3 — The Z-score measures how far an observed value is from the mean in units of standard deviations.

7. (a) $z = \dfrac{600 - 500}{100} = 1$

   (b) $z = \dfrac{450 - 500}{100} = -0.5$

   (c) $\dfrac{1}{\sqrt{2\pi}\cdot 100} \displaystyle\int_{450}^{\infty} \exp\left(-\dfrac{x-500}{2\cdot 100^2}\right) dx = \dfrac{1}{\sqrt{2\pi}} \displaystyle\int_{-0.5}^{\infty} e^{-x^2/2}\, dx$

   (d) $\dfrac{1}{\sqrt{2\pi}\cdot 100} \displaystyle\int_{-\infty}^{600} \exp\left(-\dfrac{x-500}{2\cdot 100^2}\right) dx = \dfrac{1}{\sqrt{2\pi}} \displaystyle\int_{-\infty}^{1} e^{-x^2/2}\, dx$

   (e) $\dfrac{1}{\sqrt{2\pi}\cdot 100} \displaystyle\int_{450}^{600} \exp\left(-\dfrac{x-500}{2\cdot 100^2}\right) dx = \dfrac{1}{\sqrt{2\pi}} \displaystyle\int_{-0.5}^{1} e^{-x^2/2}\, dx$

8. $A(z_2) - A(z_1) = \displaystyle\int_{-\infty}^{z_2} n(t)\, dt - \int_{-\infty}^{z_1} n(t)\, dt = \int_{z_1}^{z_2} n(t)\, dt$

9. (a) Since $n(-t) = n(t)$, $A(-z) = \displaystyle\int_{-\infty}^{-z} n(t)\, dt = \int_{z}^{\infty} n(t)\, dt = \int_{-\infty}^{\infty} n(t)\, dt - \int_{-\infty}^{z} n(t)\, dt = 1 - A(z)$

   (b) $A(-1.2) = 1 - A(1.2) \approx 1 - 0.8849 = 0.1151$

10. Raw scores of 350, 500, and 600 correspond to Z-scores of -1.5, 0, and 1, respectively. Looking at the table (and the graph above it) shows:

    (a) Probability that $Z \geq 1$ is $1 - A(1) \approx 1 - 0.8413 = 0.1587$. Thus about 16% of scores are above 600.

    (b) Probability that $-1.5 \leq Z \leq 0$ is the same as $A(1.5) - A(0) \approx 0.9333 - 0.5 = 0.4333$. Thus about 43% of scores fall in this range.

11. In both parts $m = 10$ and $s = 5$.

    (a) $z = (14-10)/5 = 0.8$ and $A(0.8) \approx 0.7881$. Thus, $\dfrac{1}{5\sqrt{2\pi}} \displaystyle\int_{-\infty}^{14} \exp\left(-\dfrac{(x-10)^2}{50}\right) \approx 0.7881$.

    (b) $z = (4-10)/5 = -1.2$ and $A(-1.2) = 1 - A(1.2) \approx 0.1151$. Thus, $\dfrac{1}{5\sqrt{2\pi}} \displaystyle\int_{-\infty}^{4} \exp\left(-\dfrac{(x-10)^2}{50}\right) \approx 0.1151$.

12. We want $I_1 = \displaystyle\int_{6.5}^{8.5} f(x)\, dx$ and

    $I_2 = \displaystyle\int_{10}^{\infty} f(x)\, dx = 0.5 - \int_{7.5}^{10} f(x)\, dx$, where (given that $m = 7.5$, $s = 1$)

    $$f(x) = \dfrac{1}{\sqrt{2\pi}} \exp\left(-\dfrac{(x-7.5)^2}{2}\right).$$

    Using the midpoint rule, M_{20} gave (approximate) answers of 0.6829 and 0.4938, respectively. Applying the trapezoid rule with 20 subdivisions to I_1 and I_2 respectively gives $I_1 \approx T_{20} \approx 0.6823$; $I_2 \approx T_{20} \approx 0.4937$.

SECTION 10.3 IMPROPER INTEGRALS AND PROBABILITY

These numerical results are consistent with those from the midpoint rule. Bounding the errors is done in the usual way, using the second derivatives. A look at the graph of f'' shows that $K_2 = 0.5$ is OK; thus the errors, respectively are:

$$I_1 \text{ error} \leq \frac{0.5 \cdot 2^3}{12 \cdot 20^2} \approx 0.0008;$$

$$I_2 \text{ error} \leq \frac{0.5 \cdot 2.5^3}{12 \cdot 20^2} \approx 0.0016.$$

Since the errors are so small, the answers are properly consistent with those from the midpoint rule.

13. Looking at the table shows that the top 10% starts at about 1.3 standard deviations above the mean, i.e., at the raw score 630. Similarly, the top 5% starts at about $Z = 1.7$, i.e., at a raw score of 670.

14. The Z-scores that corresponds to differing from the average by more than 1 inch are $Z = \pm 1/0.6 \approx \pm 1.67$. Thus, the probability is (approximately) $\int_{-\infty}^{-1.67} n(x)\,dx + \int_{1.67}^{\infty} n(x)\,dx \approx 0.095581.$

15. (a) The nearest edge of the net should be placed 135 feet from the cannon (i.e., so that the center of the net is 150 feet from the cannon). This position maximizes the probability that the performer will land in the net.

    (b) Missing the net corresponds to a Z-score of magnitude greater than 1.5. Thus, the probability that the performer will miss the landing net is $\int_{-\infty}^{-1.5} n(x)\,dx + \int_{1.5}^{\infty} n(x)\,dx \approx 0.13361.$

16. Let D be the "nominal" diameter at which the machine is set. Then the Z-score of a can top with diameter greater than 3 inches is greater than $Z_0 = (3-D)/0.01$. No more than 10% of the can tops produced will have diameters greater than 3 inches if $\int_{Z_0}^{\infty} n(x)\,dx = 0.1$, that is if $Z_0 = 1.2816$. This corresponds to $D = 2.9872$ inches.

17. (a) Since $A''(x) = -\frac{x}{\sqrt{2\pi}} e^{-x^2/2}$, the graph of A is concave down over the interval $[0.7, 0.8]$. Therefore, the graph lies *above* the secant line joining $(0.7, A(0.7))$ and $(0.8, A(0.8))$. In other words, $A(0.75) > 0.77305$.

    (b) $A(0.75) \approx A(0.7) + 0.05 \cdot A'(0.7) = 0.7580 + 0.05 \cdot 0.31225 \approx 0.7736$

    (c) Since A is concave down over the interval $[0.7, 0.8]$, the line tangent to the graph $y = A(x)$ at $x = 0.7$ lies above the graph at $x = 0.75$. Thus, the estimate in part (b) overestimates $A(0.75)$.

18. (a) If $\beta < 0$, then g is not a positive function. If $\beta = 0$, $\int_{-\infty}^{\infty} g(x)\,dx = 0 \neq 1$. Therefore, $\beta > 0$ must be true for g to be a probability distribution.

    (b) If $u = (\ln x - \alpha)/\beta$, then $du = (1/(\beta x))\,dx$ so

$$\int_a^b g(x)\,dx \rightarrow \int_{(\ln a - \alpha)/\beta}^{(\ln b - \alpha)/\beta} \frac{1}{\sqrt{2\pi}} e^{-u^2/2}\,du$$

$$= \int_{-\infty}^{(\ln b - \alpha)/\beta} \frac{1}{\sqrt{2\pi}} e^{-u^2/2}\,du - \int_{-\infty}^{(\ln a - \alpha)/\beta} \frac{1}{\sqrt{2\pi}} e^{-u^2/2}\,du$$

$$= A\left(\frac{\ln b - \alpha}{\beta}\right) - A\left(\frac{\ln a - \alpha}{\beta}\right).$$

19. The idea is to estimate various areas under the graph, perhaps using the fact that each grid rectangle has area 0.05. Reasonable answers are below; they correspond to probabilities that an observation falls in the given range:

(a) area about 0.2.
(b) area about 0.4.
(c) area about 0.875.
(d) area about 0.125.

20. (a) $\displaystyle\int_{-\infty}^{\infty} f(x)\,dx = \int_{1}^{\infty} \frac{3}{x^4}\,dx = -\frac{1}{x^3}\Big]_{1}^{\infty} = 1$

(b) $\displaystyle\int_{2}^{\infty} f(x)\,dx = \int_{2}^{\infty} \frac{3}{x^4}\,dx = -\frac{1}{x^3}\Big]_{2}^{\infty} = \frac{1}{8}$

(c) $\displaystyle\int_{-\infty}^{3/2} f(x)\,dx = \int_{1}^{3/2} \frac{3}{x^4}\,dx = -\frac{1}{x^3}\Big]_{1}^{3/2} = \frac{19}{27}$

(d) $\displaystyle\int_{-\infty}^{\infty} xf(x)\,dx = \int_{1}^{\infty} \frac{3}{x^3}\,dx = -\frac{3}{2x^2}\Big]_{1}^{\infty} = \frac{3}{2}$

(e) We wish to determine m such that $\displaystyle\int_{-\infty}^{m} f(x)\,dx = 0.5$. Since $\displaystyle\int_{-\infty}^{m} f(x)\,dx = 1 - 1/m^3$, $m = 2^{1/3} \approx 1.26$ minutes.

21. Since $\displaystyle\int_{-\infty}^{\infty} f(x)\,dx = k\pi$, f is a probability density function if $k = 1/\pi$.

22. Since $\displaystyle\int_{-\infty}^{\infty} f(x)\,dx = \int_{0}^{1} k(1-x^2)\,dx = 2k/3$, f is a probability density function if $k = 3/2$.

23. (a) f is a probability density function because it is a positive function and
$$\int_{-\infty}^{\infty} f(x)\,dx = \int_{0}^{\infty} \lambda e^{-\lambda x}\,dx = -e^{-\lambda x}\Big]_{0}^{\infty} = 1.$$

(b) $\displaystyle\int_{-\infty}^{\infty} xf(x)\,dx = \lambda \int_{0}^{\infty} xe^{-\lambda x}\,dx = -\frac{1+\lambda x}{\lambda}e^{-\lambda x}\Big]_{0}^{\infty} = \frac{1}{\lambda}$

24. (a) The probability that a device fails between time $t = 0$ and $t = T$ months is $\int_{0}^{T} f(x)\,dx$. Thus, for this device, $0.1 = \int_{0}^{6} f(x)\,dx = 1 - e^{-6\lambda}$ so $\lambda = -\ln(0.9)/6 \approx 0.01756$.

(b) The probability that a device is still functioning after T months is $\int_{T}^{\infty} f(x)\,dx$. Using the value of λ from part (a), the probability is 0.729.

(c) The mean lifetime of the device is $\int_{-\infty}^{\infty} xf(x)\,dx = -6/\ln(0.9) \approx 56.95$ months.

10.4 l'Hôpital's Rule: Comparing Rates

1. (a)

x	1	10	100	1000
x^2	1	100	10,000	1×10^6
2^x	2	1024	1.268×10^{30}	1.07×10^{301}
$(3x+10)^2$	169	1600	96,100	9.0601×10^6

   (b) One might guess $\lim_{x \to \infty} \left(\dfrac{x^2}{2^x}\right) = 0$; $\lim_{x \to \infty} \left(\dfrac{x^2}{(3x+10)^2}\right) = \dfrac{1}{9}$; $\lim_{x \to \infty} \left(\dfrac{2^x}{(3x+10)^2}\right) = \infty$

   (c) Using l'Hôpital's rule:

   $$\lim_{x \to \infty} \left(\frac{x^2}{2^x}\right) = \lim_{x \to \infty} \left(\frac{2x}{2^x \ln 2}\right) = \lim_{x \to \infty} \left(\frac{2}{2^x (\ln 2)^2}\right) = 0$$

   $$\lim_{x \to \infty} \left(\frac{x^2}{(3x+10)^2}\right) = \lim_{x \to \infty} \left(\frac{2x}{2(3x+10) \cdot 3}\right) = \lim_{x \to \infty} \frac{2}{18} = \frac{1}{9}$$

   $$\lim_{x \to \infty} \left(\frac{2^x}{(3x+10)^2}\right) = \lim_{x \to \infty} \left(\frac{2^x \ln 2}{2(3x+10) \cdot 3}\right) = \lim_{x \to \infty} \left(\frac{2^x (\ln 2)^2}{18}\right) = \infty$$

2. $\lim_{x \to 1} \dfrac{x^3 + x - 2}{x^2 - 3x + 2} = \lim_{x \to 1} \dfrac{3x^2 + 1}{2x - 3} = -4$

3. $\lim_{x \to \infty} \left(\dfrac{x^2 + 1}{2x^2 + 3}\right) = \lim_{x \to \infty} \left(\dfrac{2x}{4x}\right) = \dfrac{1}{2}$

4. $\lim_{x \to \infty} \dfrac{\ln x}{x^{2/3}} = \lim_{x \to \infty} \dfrac{x^{-1}}{2x^{-1/3}/3} = \lim_{x \to \infty} \dfrac{3}{2x^{2/3}} = 0.$

5. $\lim_{x \to 0} \left(\dfrac{5x - \sin x}{x}\right) = \lim_{x \to 0} \left(\dfrac{5 - \cos x}{1}\right) = 4$

6. $\lim_{x \to 0} \dfrac{1 - \cos x}{\sin(2x)} = \lim_{x \to 0} \dfrac{\sin x}{2 \cos(2x)} = 0$

7. $\lim_{x \to 0} \left(\dfrac{1 - \cos(5x)}{4x + 3x^2}\right) = \lim_{x \to 0} \left(\dfrac{\sin(5x) \cdot 5}{4 + 6x}\right) = 0$

8. $\lim_{x \to \infty} \dfrac{e^x}{x^2 + x} = \lim_{x \to \infty} \dfrac{e^x}{2x + 1} = \lim_{x \to \infty} \dfrac{e^x}{2} = \infty$

9. $\lim_{x \to 0} \dfrac{e^x - x - 1}{x^2} = \lim_{x \to 0} \dfrac{e^x - 1}{2x} = \lim_{x \to 0} \dfrac{e^x}{2} = \dfrac{1}{2}$

10. No. Since $\lim_{x \to \pi/2^-} \tan x = \infty$ and $\lim_{x \to \pi/2^+} \tan x = -\infty$, $\lim_{x \to \pi/2} \tan x$ does not exist. Therefore, l'Hôpital's rule cannot be used. Moreover, the limits $\lim_{x \to \pi/2^\pm} \dfrac{\tan x}{x - \pi/2}$ have the form $\mp\infty/0$ so l'Hôpital's rule cannot be used to evaluate these limits either.
    [NOTE: L'Hôpital's rule can be used only for limits of the form 0/0 or ∞/∞.]

11. (a) Since $\lim_{x \to 1} f(x) = \lim_{x \to 1}(x^2 - 1) = 0$, $\lim_{x \to 1} \dfrac{f(x)}{x^2 - 1} = \lim_{x \to 1} \dfrac{f'(x)}{2x} = \dfrac{f'(1)}{2} = -1.5$.

(b) Since $\lim_{x \to 2} f(x) = -1.6$, $\lim_{x \to 2^-}(x^2 - 4)^{-1} = -\infty$, and $\lim_{x \to 2^+}(x^2 - 4)^{-1} = \infty$, $\lim_{x \to 2} \dfrac{f(x)}{x^2 - 4}$ does not exist.

(c) Since $\lim_{x \to 1} f(x) = 0$ and $\lim_{x \to 1} f(-2x) = 0$, $\lim_{x \to 1} \dfrac{f(x)}{f(-2x)} = \lim_{x \to 1} \dfrac{f'(x)}{-2f'(-2x)} = \dfrac{f'(1)}{-2f'(-2)} \approx 0.3125$.

(d) Since $\lim_{x \to 1} f(x - 3) = 0$ and $\lim_{x \to 1} f(x + 3) = 3.6$, $\lim_{x \to 1} \dfrac{f(x - 3)}{f(x + 3)} = 0$.

12. $\lim_{x \to \infty} e^{-x} \ln x = \lim_{x \to \infty} \left(\dfrac{\ln x}{e^x} \right) = \lim_{x \to \infty} \left(\dfrac{\frac{1}{x}}{e^x} \right) = \lim_{x \to \infty} \left(\dfrac{1}{xe^x} \right) = 0$ [Denominator blows up!]

13. $\lim_{x \to 0} x \cot x = \lim_{x \to 0} \left(\dfrac{x}{\tan x} \right) = \lim_{x \to 0} \left(\dfrac{1}{\sec^2 x} \right) = 1$

14. $\lim_{x \to 8} \left(\dfrac{x - 8}{\sqrt[3]{x} - 2} \right) = \lim_{x \to 8} \left(\dfrac{1}{\frac{1}{3} x^{-2/3}} \right) = \dfrac{1}{\frac{1}{3} \cdot 8^{-2/3}} = 3 \cdot 8^{2/3} = 3 \cdot 4 = 12$

15. $\lim_{x \to 0^+} \left(\dfrac{\sin x}{x + \sqrt{x}} \right) = \lim_{x \to 0^+} \left(\dfrac{\cos x}{1 + \frac{1}{2\sqrt{x}}} \right) \to \dfrac{1}{\infty} = 0$

16. $\lim_{x \to 0} \left(\dfrac{\sin x}{x - \sin x} \right) = \lim_{x \to 0} \left(\dfrac{\cos x}{1 - \cos x} \right) = \infty$ [NOTE: numerator $\to 1$ and denominator $\to 0$ from *above*.]

17. $\lim_{x \to 0} \dfrac{\tan(3x)}{\ln(1 + x)} = \lim_{x \to 0} \dfrac{3 \sec^2(3x)}{(1 + x)^{-1}} = 3$

18. $\lim_{x \to 0} \dfrac{e^x - 1}{x} = \lim_{x \to 0} \dfrac{e^x}{1} = 1$

19. $\lim_{x \to 0} \dfrac{\arctan(2x)}{\sin(3x)} = \lim_{x \to 0} \dfrac{2/(1 + x^2)}{3 \cos(3x)} = \dfrac{2}{3}$

20. $\lim_{x \to 0} \dfrac{e^x - e^{-x}}{x} = \lim_{x \to 0} (e^x + e^{-x}) = 2$

21. $\lim_{x \to 0} \dfrac{\sin^2 x}{\cos(3x) - 1} = \lim_{x \to 0} \dfrac{2 \sin x \cos x}{-3 \sin(3x)} = \lim_{x \to 0} \dfrac{2 \cos^2 x - 2 \sin^2 x}{-9 \cos(3x)} = -\dfrac{2}{9}$

22. $\lim_{x \to 0} \dfrac{1 - \cos^2 x}{x^2} = \lim_{x \to 0} \dfrac{2 \cos x \sin x}{2x} = \lim_{x \to 0} \dfrac{2 \cos^2 x - 2 \sin^2 x}{2} = 1$

23. $\lim_{x \to 0} \dfrac{1 - x - e^{-x}}{1 - \cos x} = \lim_{x \to 0} \dfrac{e^{-x} - 1}{\sin x} = \lim_{x \to 0} \dfrac{-e^{-x}}{\cos x} = -1$

24. $\lim_{x \to 1} \dfrac{\ln x}{x^2 - x} = \lim_{x \to 1} \dfrac{1}{2x^2 - x} = 1$

25. $\lim_{x \to \infty} x^2 e^{-x^2/2} = \lim_{x \to \infty} \dfrac{x^2}{e^{x^2/2}} = \lim_{x \to \infty} \dfrac{2}{e^{x^2/2}} = 0$

26. $\lim_{x \to 1} \dfrac{\sin(\pi x)}{x^2 - 1} = \lim_{x \to 1} \dfrac{\pi \cos(\pi x)}{2x} = -\dfrac{\pi}{2}$

27. $\lim\limits_{x\to 0} \dfrac{1-x^2-e^{-x^2}}{x^4} = \lim\limits_{x\to 0} \dfrac{e^{-x^2}-1}{2x^2} = \lim\limits_{x\to 0} \dfrac{-e^{-x^2}}{2} = -\dfrac{1}{2}$

28. $\lim\limits_{x\to 1} \dfrac{\cos^3(\pi x/2)}{\sin(\pi x)} = \lim\limits_{x\to 1} \dfrac{-3\cos^2(\pi x/2)\sin(\pi x/2)}{2\cos(\pi x)} = 0$

29. $\lim\limits_{x\to \pi/2} \dfrac{\ln(\sin x)}{(x-\pi/2)^2} = \lim\limits_{x\to \pi/2} \dfrac{\cos x}{2(x-\pi/2)\sin x} = \lim\limits_{x\to \pi/2} \dfrac{-\sin x}{2\sin x + 2(x-\pi/2)\cos x} = -\dfrac{1}{2}$

30. $\lim\limits_{x\to 0^+} x^2 \ln x = \lim\limits_{x\to 0^+} \dfrac{\ln x}{x^{-2}} = \lim\limits_{x\to 0^+} -\dfrac{x^2}{2} = 0$

31. $\lim\limits_{x\to \infty} x\sin(1/x) = \lim\limits_{x\to \infty} \dfrac{\sin(1/x)}{1/x} = \lim\limits_{x\to \infty} \cos(1/x) = 1$

32. $\lim\limits_{x\to 0} x^2 \ln(\cos x) = 0$ since $\lim\limits_{x\to 0} \ln(\cos x) = 0$.

33. $\lim\limits_{x\to 0^+} \sin x \ln(\sin x) = \lim\limits_{x\to 0^+} \dfrac{\ln(\sin x)}{1/\sin x} = \lim\limits_{x\to 0^+} -\sin x = 0$

34. $\lim\limits_{x\to \infty} x\left(\dfrac{\pi}{2} - \arctan x\right) = \lim\limits_{x\to \infty} \dfrac{\pi/2 - \arctan x}{x^{-1}} = \lim\limits_{x\to \infty} \dfrac{x^2}{1+x^2} = 1$

35. $\lim\limits_{x\to \pi/2} \left(\dfrac{\pi}{2} - x\right) \tan x = \lim\limits_{x\to \pi/2} \dfrac{\pi/2 - x}{\cot x} = \lim\limits_{x\to \pi/2} \sin^2 x = 1$

36. $\lim\limits_{x\to 0} \left(\dfrac{1}{\sin x} - \dfrac{1}{x}\right) = \lim\limits_{x\to 0} \dfrac{x-\sin x}{x\sin x} = \lim\limits_{x\to 0} \dfrac{1-\cos x}{\sin x + x\cos x} = \lim\limits_{x\to 0} \dfrac{\sin x}{2\cos x - x\sin x} = 0$

37. $\lim\limits_{x\to \infty} \dfrac{\int_1^x \sqrt{1+e^{-3t}}\,dt}{x} = \lim\limits_{x\to \infty} \dfrac{\sqrt{1+e^{-3x}}}{1} = 1$
[NOTE: The improper integral in the numerator diverges to ∞ because $\sqrt{1+e^{-3t}} > 1$ for all $t \geq 1$.]

38. $\lim\limits_{x\to 0} \dfrac{\int_0^x \sin(t^2)\,dt}{x^3} = \lim\limits_{x\to 0} \dfrac{\sin(x^2)}{3x^2} = \lim\limits_{x\to 0} \dfrac{\cos(x^2)}{3} = \dfrac{1}{3}$

39. $\lim\limits_{x\to \infty} e^{x^2} \int_0^x e^{-t^2}\,dt = \infty$ since $\lim\limits_{x\to \infty} \int_0^x e^{-t^2}\,dt = \sqrt{\pi}/2$ and $\lim\limits_{x\to \infty} e^{x^2} = \infty$.

40. $\lim\limits_{x\to \infty} \dfrac{\int_0^x e^{t^2}\,dt}{e^{x^2}} = \lim\limits_{x\to \infty} \dfrac{1}{2x} = 0$

41. $f(k) = \begin{cases} \dfrac{3k-6}{k-2} & \text{when } k \neq 2 \\ -1 & \text{when } k = 2 \end{cases}$

42. $\lim\limits_{x\to 1} \dfrac{(f(x))^2 - 1}{x^2 - 1} = \lim\limits_{x\to 1} \dfrac{f(x)f'(x)}{x} = 2$

43. $\lim\limits_{x\to 0} \dfrac{f(x)}{\sin(2x)} = \lim\limits_{x\to 0} \dfrac{f'(x)}{2\cos(2x)} = \dfrac{f'(0)}{2} = 5$ so $f'(0) = 10$.

44. $\lim\limits_{x\to 0} \dfrac{xf(x)}{(e^x-1)g(x)} = \lim\limits_{x\to 0} \dfrac{f(x)+xf'(x)}{e^x g(x) + (e^x-1)g'(x)} = \dfrac{f(0)}{g(0)}$

45. $\lim\limits_{x\to \infty} \dfrac{g(1/x)}{f(1/x)} = \lim\limits_{u\to 0^+} \dfrac{g(u)}{f(u)} = -\dfrac{1}{2}$

46. $I = \int_0^\infty xe^{-x}dx = \lim_{t\to\infty}\int_0^t xe^{-x}dx = \lim_{t\to\infty} -xe^{-x} - e^{-x}\Big]_0^t = \lim_{t\to\infty}(1 - te^{-t} - e^{-t}) = 1 - \lim_{t\to\infty}(te^{-t})$.

The remaining limit can be evaluated using l'Hôpital's rule: $\lim_{t\to\infty} te^{-t} = \lim_{t\to\infty}\left(\dfrac{t}{e^t}\right) = \lim_{t\to\infty}\left(\dfrac{1}{e^t}\right) = 0$. Now we're done: $I = 1$.

47. (a) $\lim_{x\to 0} \sin x = \lim_{x\to 0} x = 0$ but $\lim_{x\to 0} \cos x = 1$.

   (b) $\lim_{x\to 0^+} \dfrac{\cos x}{x} = \infty$

   (c) **Diverges** since $\cos x > 1/2$ when $0 \leq x \leq 1 \Longrightarrow \int_0^1 \dfrac{\cos x}{x}\,dx > \dfrac{1}{2}\int_0^1 \dfrac{dx}{x} = \infty$.

   (d) $\lim_{x\to 0^+} \dfrac{\sin x}{x} = 1$ so the integrand is bounded over the entire interval of integration.

   (e) $\int_0^\infty \sin(e^{-x})\,dx = \lim_{t\to\infty}\int_0^t \sin(e^{-x})\,dx = \lim_{t\to\infty} -\int_1^{e^{-t}} \sin u\,\dfrac{du}{u} = \int_0^1 \dfrac{\sin u}{u}\,du$. The desired inequalities follow from the fact that $0.8 < \dfrac{\sin x}{x} < 1$ when $0 < x < 1$. [$u = e^{-x}$]

48. No. Since $\lim_{x\to 0^+} x\ln x = 0$, the integrand is bounded over the entire interval of integration (i.e., there is no impropriety).

49. No. Using L'Hôpital's rule, $\lim_{x\to 0} \dfrac{\arctan x}{\sqrt[3]{x}} = \lim_{x\to 0} \dfrac{(1+x^2)^{-1}}{\frac{1}{3}x^{-2/3}} = \lim_{x\to 0} \dfrac{3x^{2/3}}{1+x^2} = 0$. Therefore, since the integrand is finite throughout the interval of integration, the integral is not improper.

50. Since the interval of integration is finite, the integral is improper only if the integrand is unbounded on this interval. The integrand is well-behaved (i.e., bounded) throughout the interval of integration except, possibly, at $x = 0$. Using L'Hôpital's rule, $\lim_{x\to 0^+} \dfrac{\tan x}{\sqrt{x}} = \lim_{x\to 0^+} \dfrac{\sec^2 x}{\frac{1}{2}x^{-1/2}} = \lim_{x\to 0^+} 2x^{1/2}\sec^2 x = 0$. Therefore, the integrand is finite throughout the interval of integration and the integral is not improper.

51. (a) $\left|\int_0^\infty \dfrac{\sin x}{x^{3/2}}\,dx\right| = \left|\int_0^1 \dfrac{\sin x}{x^{3/2}}\,dx + \int_1^\infty \dfrac{\sin x}{x^{3/2}}\,dx\right| \leq \left|\int_0^1 \dfrac{\sin x}{x^{3/2}}\,dx\right| + \left|\int_1^\infty \dfrac{\sin x}{x^{3/2}}\,dx\right|$
$\leq \int_0^1 \dfrac{x}{x^{3/2}}\,dx + \int_1^\infty \dfrac{1}{x^{3/2}}\,dx = \int_0^1 \dfrac{1}{x^{1/2}}\,dx + \int_1^\infty \dfrac{1}{x^{3/2}}\,dx = 2 + 2$

Therefore, Theorem 2 (p. 195) asserts that the improper integral $\int_0^\infty f(x)\,dx$ converges.

   (b) $\int_0^\alpha \dfrac{\sin x}{x^{3/2}}\,dx \leq \int_0^\alpha \dfrac{dx}{\sqrt{x}} = 2\sqrt{\alpha} \leq 0.0005$ when $\alpha \leq 6.25\times 10^{-8}$.

$\left|\int_\beta^\infty \dfrac{\sin x}{x^{3/2}}\,dx\right| \leq \int_\beta^\infty \dfrac{dx}{x^{3/2}} = \dfrac{2}{\sqrt{\beta}} \leq 0.0005$ when $\beta \geq 1.6\times 10^7$.

Therefore, $\int_{6.25\times 10^{-8}}^{1.6\times 10^7} \dfrac{\sin x}{x^{3/2}}\,dx$ approximates $\int_0^\infty \dfrac{\sin x}{x^{3/2}}\,dx$ within 0.001.

52. (a) $\lim_{x\to a} \ln(f(x)) = \ln\left(\lim_{x\to a} f(x)\right) = \ln A$ since $\ln x$ is a continuous function.

   (b) From part (a) it follows that $B = \ln\left(\lim_{x\to a} f(x)\right)$. Therefore, $\lim_{x\to a} f(x) = e^B$.

SECTION 10.4 L'HÔPITAL'S RULE: COMPARING RATES

53. $\lim\limits_{x \to 0} \dfrac{\tan x}{x} = \lim\limits_{x \to 0} \sec^2 x = 1$ so $\lim\limits_{x \to 0} \cos\left(\dfrac{\tan x}{x}\right) = \cos 1$

54. $\lim\limits_{x \to 0^+} x \ln x = 0$ so $\lim\limits_{x \to 0^+} \arctan(x \ln x) = \arctan 0 = 0$

55. $\lim\limits_{x \to 0^+} 2x \ln x = 0$ so $\lim\limits_{x \to 0^+} x^{2x} = e^0 = 1$

56. $\lim\limits_{x \to \infty} \dfrac{\ln(1+x)}{x} = \lim\limits_{x \to \infty} \dfrac{1}{1+x} = 0$ so $\lim\limits_{x \to \infty} (1+x)^{1/x} = e^0 = 1$.

57. $\lim\limits_{x \to 0} \dfrac{\ln(1+x)}{x} = \lim\limits_{x \to 0} \dfrac{1}{1+x} = 1$ so $\lim\limits_{x \to \infty} (1+x)^{1/x} = e^1 = e$.

58. $\lim\limits_{x \to 0} \dfrac{\ln(e^x + x)}{x} = \lim\limits_{x \to 0} \dfrac{e^x + 1}{e^x + x} = 2 \implies \lim\limits_{x \to 0} (e^x + x)^{1/x} = e^2$

59. $\lim\limits_{x \to 1} (\ln x)^{\sin x} = 0$ since $\lim\limits_{x \to 1} \ln x = 0$ and $\lim\limits_{x \to 1} \sin x = \sin 1 \approx 0.84147$.

60. $\lim\limits_{x \to (\pi/2)^-} \tan x \ln x = \infty$ so $\lim\limits_{x \to (\pi/2)^-} \sin\left(x^{\tan x}\right)$ does not exist.

61. $\displaystyle\int_1^\infty f'(x)\, dx = \lim\limits_{t \to \infty} f(t) - f(1)$. Thus, the improper integral converges if (and only if) $\lim\limits_{t \to \infty} f(t)$ is a finite number.

    Since $|f(x)| \le e^{-x} \ln x$, $\lim\limits_{t \to \infty} |f(t)| \le \lim\limits_{t \to \infty} e^{-t} \ln t = 0$ (see problem 3d above). Thus, $\lim\limits_{t \to \infty} f(t) = 0$.

    Finally, therefore, $\displaystyle\int_1^\infty f'(x)\, dx = -f(1)$.

62. $\lim\limits_{x \to 0} \dfrac{f(x)}{x} = \lim\limits_{x \to 0} \dfrac{f(x) - f(0)}{x - 0}$. The limit on the right is the definition of $f'(0)$.

63. (a) $\lim\limits_{x \to a} \dfrac{f(x) - P_2(x)}{x - a} = \lim\limits_{x \to a} (f'(x) - P_2'(x)) = f'(a) - P_2'(a) = 0$

    (b) $\lim\limits_{x \to a} \dfrac{f(x) - P_2(x)}{(x-a)^2} = \lim\limits_{x \to a} \dfrac{f'(x) - P_2'(x)}{2(x-a)} = \lim\limits_{x \to a} \dfrac{f''(x) - P_2''(x)}{2} = 0$

    (c) $\lim\limits_{x \to a} \dfrac{f(x) - q(x)}{(x-a)^2} = \lim\limits_{x \to a} \dfrac{f'(x) - q'(x)}{2(x-a)} = \lim\limits_{x \to a} \dfrac{f''(x) - q''(x)}{2} \ne 0$ since $q''(a) \ne f''(a)$. (If $q''(a) = f''(a)$, then $q(x) = P_2(x)$ would be true.)

    (d) The graph of P_2 is more like the graph of f near $x = a$ than the graph of any other quadratic polynomial.

    (e) Since $f(a) = P_n(a)$ and the first n derivatives of f and P_n are equal at $x = a$, l'Hôpital's rule implies that $\lim\limits_{x \to a} \dfrac{f(x) - P_n(x)}{(x-a)^n} = 0$. No other n^{th}-order polynomial has this property since any n^{th}-order polynomial that is not P_n will not have the same derivative values at $x = a$ as f does.

64. (a) $\Gamma(1) = \displaystyle\int_0^\infty e^{-t}\, dt = \lim\limits_{a \to \infty} \int_0^a e^{-t}\, dt = \lim\limits_{a \to \infty} (1 - e^{-a}) = 1$

    (b) If $x > 1$, an integration by parts (with $u = t^{x-1}$ and $dv = e^{-t}\, dt$) shows that
    $$\Gamma(x) = \int_0^\infty t^{x-1} e^{-t}\, dt = -t^{x-1} e^{-t} \Big]_0^\infty + (x-1) \int_0^\infty t^{x-2} e^{-t}\, dt = 0 + (x-1)\Gamma(x-1).$$

    (c) $\Gamma\left(\dfrac{2}{3}\right) = \displaystyle\int_0^\infty \dfrac{dt}{\sqrt[3]{t} e^t} = \int_0^1 \dfrac{dt}{\sqrt[3]{t} e^t} + \int_1^\infty \dfrac{dt}{\sqrt[3]{t} e^t} < \int_0^1 \dfrac{dt}{\sqrt[3]{t}} + \int_1^\infty \dfrac{dt}{e^t} = 3/2 + e^{-1} \approx 1.8679 < 2$. Finally, $\Gamma\left(\dfrac{2}{3}\right) > 0$ since the integrand $(t^{-1/3} e^{-t})$ is positive throughout the interval of integration.

(d) The substitution $u = \sqrt{t}$ shows that $\Gamma\left(\frac{1}{2}\right) = \int_0^\infty \dfrac{dt}{\sqrt{t}e^t} = 2\int_0^\infty e^{-u^2}\,du = \sqrt{\pi}$.

65. Let $t = \ln(1/x)$. Then $e^{-t} = x$ and $dt = -(1/x)\,dx$ so

$$\int_0^\infty t^z e^{-t}\,dt = \lim_{s\to\infty}\int_0^s t^z e^{-t}\,dt = \lim_{w\to 0^+} -\int_1^w \big(\ln(1/x)\big)^z\,dx = \int_0^1 \big(\ln(1/x)\big)^z\,dx.$$

66. First make the subsitution $x = e^{-u}$, then make the substitution $w = (m+1)u$:

$$\int_0^1 x^m(\ln x)^n\,dx \to -\int_\infty^0 e^{-(m+1)u}(-u)^n\,du = (-1)^n\int_0^\infty u^n e^{-(m+1)u}\,du$$

$$\to \dfrac{(-1)^n}{(m+1)^{n+1}}\int_0^\infty w^n e^{-w}\,dw = \dfrac{(-1)^n\,n!}{(m+1)^{n+1}}.$$

11.1 Sequences and Their Limits

1. $a_k = (-1/3)^k$ for $k = 0, 1, 2, \ldots$
2. $a_j = 3j$ for $j = 1, 2, 3, \ldots$
3. $a_k = k/2^k$ for $k = 1, 2, 3 \ldots$
4. $a_n = 1/(n^2 + 1)$ for $n = 1, 2, 3, \ldots$
5. $\lim_{k \to \infty} a_k$ does not exist
6. $\lim_{k \to \infty} a_k = 0$
7. $\lim_{k \to \infty} a_k = \infty$
8. $\lim_{k \to \infty} a_k = 0$
9. $\lim_{k \to \infty} a_k = \infty$
10. $\lim_{k \to \infty} a_k = 0$
11. $\lim_{k \to \infty} a_k = \infty$
12. $\lim_{k \to \infty} a_k = 0$
13. $\lim_{k \to \infty} a_k$ does not exist
14. $\lim_{k \to \infty} a_k = \pi/2$
15. $\lim_{k \to \infty} a_k = 1$
16. $\lim_{k \to \infty} a_k = 0$
17. $\lim_{k \to \infty} a_k$ does not exist
18. $\lim_{k \to \infty} a_k = 1$
19. $\lim_{k \to \infty} a_k = \sqrt{2}$
20. $\lim_{k \to \infty} a_k = \infty$
21. $\lim_{n \to \infty} a_n = \lim_{n \to \infty} \dfrac{n+2}{n^3+4} = \lim_{n \to \infty} \dfrac{1}{3n^2} = 0$
22. $\lim_{m \to \infty} a_m = \lim_{m \to \infty} \dfrac{m^2}{e^m} = \lim_{m \to \infty} \dfrac{2m}{e^m} = \lim_{m \to \infty} \dfrac{2}{e^m} = 0$
23. $\lim_{j \to \infty} b_j = \lim_{j \to \infty} \dfrac{\ln j}{\sqrt[3]{j}} = \lim_{j \to \infty} \dfrac{3j^{2/3}}{j} = \lim_{j \to \infty} \dfrac{3}{\sqrt[3]{j}} = 0$
24. $\lim_{k \to \infty} c_k = \lim_{k \to \infty} \dfrac{\ln(1+k^2)}{\ln(4+k)} = \lim_{k \to \infty} \dfrac{2k(4+k)}{1+k^2} = 2$
25. $\lim_{n \to \infty} d_n = \lim_{n \to \infty} n \sin(1/n) = \lim_{n \to \infty} \dfrac{\sin(n^{-1})}{n^{-1}} = \lim_{n \to \infty} \cos(n^{-1}) = 1$

26. $\lim_{k\to\infty} \dfrac{\ln(2^k + 3^k)}{k} = \ln 3 \implies \lim_{k\to\infty} (2^k + 3^k)^{1/k} = e^{\ln 3} = 3$

27. $\lim_{n\to\infty} \dfrac{\ln x}{n} = 0 \implies \lim_{n\to\infty} x^{1/n} = 1$ for all $x > 0$.

28. (a) $\lim_{k\to\infty} \ln(a_k) = \lim_{k\to\infty} \ln(1 + x/k)^k = \lim_{k\to\infty} k \ln(1 + x/k) = \lim_{k\to\infty} \dfrac{\ln(1 + x/k)}{k^{-1}} = \lim_{k\to\infty} \dfrac{x}{1 + x/k} = x$.

   (b) $\lim_{k\to\infty} a_k = e^x$.

29. Let $a_n = \left(1 - \dfrac{1}{2n}\right)^n$. Then $\ln(a_n) = n \ln\left(1 - \dfrac{1}{2n}\right) = \dfrac{\ln\left(1 - \frac{1}{2n}\right)}{\frac{1}{n}}$ so, using l'Hôpital's rule, $\lim_{n\to\infty} \ln(a_n) = -1/2$. It follows that $\lim_{n\to\infty} a_n = e^{-1/2}$.

30. $\lim_{k\to\infty} a_k = 0$

31. $\lim_{k\to\infty} a_k = 0$

32. $\lim_{k\to\infty} a_k = 1$

33. $\lim_{k\to\infty} a_k = 0$

34. $\lim_{k\to\infty} a_k = 0$

35. $\lim_{k\to\infty} a_k = 0$

36. $\lim_{k\to\infty} a_k = 0$

37. $\lim_{k\to\infty} a_k = 1/2$

38. $\lim_{k\to\infty} a_k = 1$

39. (a) $a_k = (-1)^k/k$. $\lim_{k\to\infty} a_k = 0$, but $\{a_k\}$ is neither an increasing nor a decreasing sequence.

   (b) $a_k = \cos k$. $|a_k| \le 1$ for all integers $k \ge 1$, but $\lim_{k\to\infty} a_k$ does not exist.

   (c) $a_k = k$. $a_k < a_{k+1}$ for all integers $k \ge 1$, but $\lim_{k\to\infty} a_k = \infty$.

   (d) $a_k = -k$. $a_{k+1} < a_k$ for all integers $k \ge 1$, and $\lim_{k\to\infty} a_k = -\infty$.

   (e) $a_k = e^{-k}$. $a_{k+1} < a_k$ for all integers $k \ge 1$, and $\lim_{k\to\infty} a_k = 0$.

   (f) $a_k = (-1)^k k$. $a_k < a_{k+1}$ when k is an odd integer, but $a_{k+1} < a_k$ when k is an even integer. Thus, $\{a_k\}$ is not a monotone sequence. Also, $\lim_{k\to\infty} |a_k| = \infty$ so the sequence is not bounded.

40. (b) $a_n = R_n$, the right Riemann sum approximation to $\displaystyle\int_0^1 x\,dx$ with n subintervals.

41. (a) $a_{n+1} = \displaystyle\sum_{k=1}^{n+1} \dfrac{1}{(n+1)+k} = \sum_{k=2}^{n+2} \dfrac{1}{n+k} = \sum_{k=2}^{n} \dfrac{1}{n+k} + \dfrac{1}{2n+1} + \dfrac{1}{2n+2}$

   $> \displaystyle\sum_{k=2}^{n} \dfrac{1}{n+k} + \dfrac{2}{2n+2} = \sum_{k=1}^{n} \dfrac{1}{n+k} = a_n$

   (b) $a_n = \displaystyle\sum_{k=1}^{n} \dfrac{1}{n+k} \le \sum_{k=1}^{n} \dfrac{1}{n+1} = \dfrac{n}{n+1} < 1$ when $n \ge 1$.

Section 11.1 Sequences and Their Limits

(c) Parts (a) and (b) imply that the sequence $\{a_n\}$ is monotonically increasing and bounded above. Therefore, $\lim_{n\to\infty} a_n$ exists.

(d) Part (a) implies that the sequence $\{a_n\}$ is montonically increasing. Therefore, $\lim_{n\to\infty} a_n > a_k$ for any integer $k \geq 1$. Since $a_1 = 1/2$, $\lim_{n\to\infty} a_n > 1/2$.

(e) $\lim_{n\to\infty} a_n = \int_0^1 \dfrac{dx}{1+x} = \ln 2$

42. $\lim_{n\to\infty} a_n = \int_0^1 x^2\, dx = \dfrac{1}{3}$

43. (a) First, note that $\ln(\sqrt[n]{n!}) = \ln(n!)^{1/n} = \dfrac{1}{n}\ln(n!) = \dfrac{1}{n}\sum_{k=1}^n \ln k$. Also, $n = (n^n)^{1/n}$ so

$\ln n = \dfrac{1}{n}\ln(n^n) = \dfrac{1}{n}\sum_{k=1}^n \ln n$. Therefore, $\ln a_n = \dfrac{\sqrt[n]{n!}}{n} = \ln(\sqrt[n]{n!}) - \ln n = \dfrac{1}{n}\sum_{k=1}^n \ln k - \dfrac{1}{n}\sum_{k=1}^n \ln n$.

(b) The right sum approximation to $\int_0^1 \ln x\, dx$ is

$$R_n = \dfrac{1}{n}\sum_{k=1}^n \ln\left(\dfrac{k}{n}\right) = \dfrac{1}{n}\sum_{k=1}^n \ln k - \dfrac{1}{n}\sum_{k=1}^n \ln n.$$

(c) The result in part (b) implies that $\lim_{n\to\infty} \ln a_n = \int_0^1 \ln x\, dx = x\ln x - x\Big]_0^1 = -1$. Therefore, $\lim_{n\to\infty} a_n = e^{-1}$.

44. (a) $a_{n+1} \leq a_n$ for all $n \geq 4$ (i.e., $N = 3$).

(b) The sequence $\{a_n\}$ is monotonically decreasing for all $n \geq 3$ and bounded below (by zero). Therefore, the sequence must converge.

(c) Observe that $0 < a_n \leq (4/5)^{n-4}(32/3)$ for all $n \geq 4$. Since $\lim_{n\to\infty}(4/5)^{n-4} = 0$, $\lim_{n\to\infty} a_n = 0$ (by the "squeeze" theorem).

45. Observe that $\dfrac{n+3}{2n+1} \leq \dfrac{6}{7}$ for all $n \geq 3$. Now,

$$\dfrac{a_{n+1}}{a_3} = \dfrac{a_{n+1}}{a_n} \cdot \dfrac{a_n}{a_{n-1}} \cdots \dfrac{a_4}{a_3} \leq \left(\dfrac{6}{7}\right)^{n-2} \implies a_{n+1} \leq (6/7)^{n-2} a_3 \implies \lim_{n\to\infty} a_n = 0.$$

46. (a) $\lim_{n\to\infty} x^n = 0$ when $|x| < 1$.

(b) x^n *diverges* when $x > 1$ or $x \leq -1$.

(c) x^n *converges* to 1 when $x = 1$.

47. $\lim_{k\to\infty} a_k$ exists only when $x \leq 0$ because $e^x > 1$ when $x > 0$. $\lim_{k\to\infty} e^k x = \infty$. When $x < 0$, $\lim_{k\to\infty} a_k = 0$. When $x = 0$, $\lim_{k\to\infty} a_k = 1$.

48. $\lim_{k\to\infty} a_k = 0$ when $1/e < x < e$ and $\lim_{k\to\infty} a_k = 1$ when $x = e$.

49. $\lim_{k\to\infty} a_k = 0$ when $-\sin 1 < x < \sin 1$ and $\lim_{k\to\infty} a_k = 1$ when $x = \sin 1$.

50. $\lim_{k\to\infty} a_k = 0$ when $-\infty < x < \infty$

51. (a) $a_1 \approx 0.5403, a_2 \approx -0.2248, a_3 \approx 0.2226, a_4 \approx -0.1455$
    (b) Yes. $|a_n| \leq 1$ since $|\cos x| \leq 1$ for all x.
    (c) No. The terms of the sequence change sign.
    (d) Yes. $|a_n| \leq 1$.
    (e) Yes. $|a_{n+1}| < a_n$

52. The sequence is bounded above (e.g, by 0.8) and is monotone increasing.

53. (a) $a_1 = 1/2; a_2 = 3/8; a_5 = 945/3840 = 63/256$
    (b) Since $0 < a_{n+1} < a_n$, the sequence is bounded below and monotonically decreasing. Therefore, $\lim_{n \to \infty} a_n$ exists.

54. Let $a_n = \sin\left(\pi/n^2\right)$. Then, the inequality $0 < \sin x < x$ when $0 < x < 1$ implies that $0 < a_n < \pi/2^n$. Since $\lim_{n \to \infty} \pi/2^n = 0$, Theorem 2 implies the result.

55. No. The terms of the sequence are 1 and 0 alternately.

56. Yes. The inequalities $0 < a_{n+1} = a_n/2 < a_n$ imply that the sequence is monotonically decreasing and bounded below.

57. (a) Since $0 < a_{n+1} < a_n$, the sequence is bounded below and monotone decreasing. Therefore, it must have a limit.
    (b) $\lim_{n \to \infty} a_n = \lim_{n \to \infty} 1/n = 0$

58. $\lim_{k \to \infty} b_k = \lim_{k \to \infty} (L - a_k) = \lim_{k \to \infty} L - \lim_{k \to \infty} a_k = L - \lim_{k \to \infty} a_k = L - L = 0$

59. $a_{n+1} = x^{1/2^n}$ so $\lim_{n \to \infty} a_n = 1$ for all $x > 0$ by the previous problem. When $x = 0$, $\lim_{n \to \infty} a_n = 0$. Thus, $\lim_{n \to \infty} a_n$ exists for all $x \geq 0$.

60. (a) Using l'Hôpital's rule, $\lim_{n \to \infty} nf(1/n) = \lim_{n \to \infty} \dfrac{f(n^{-1})}{n^{-1}} = \lim_{n \to \infty} \dfrac{-n^{-2} \cdot f'(n^{-1})}{-n^{-2}} = \lim_{n \to \infty} f'(n^{-1}) = f'(0)$
    (b) $\lim_{n \to \infty} n \arctan(1/n) = \lim_{n \to \infty} \left(1 + (1/n)^2\right)^{-1} = 1$

61. (a) Dividing both sides of the equation $F_{n+1} = F_n + F_{n-1}$ by F_n leads to the equation
$$\frac{F_{n+1}}{F_n} = 1 + \frac{F_{n-1}}{F_n} = 1 + \frac{1}{F_n/F_{n-1}}.$$
Now, taking the limit as $n \to \infty$, we obtain the equation $L = 1 + 1/L$ which is equivalent to $L^2 = L + 1$.
    (b) Since $x = xF_1 + F_0 = x + 1$, the equation holds when $n = 1$. Therefore, assume that the equation is true when $n = N$. We must show that the equation holds when $n = N + 1$:
$$\begin{aligned} x^{N+1} &= x\, x^N = x\,(xF_N + F_{N-1}) \\ &= (x+1)F_N + xF_{N-1} = x\,(F_N + F_{N-1}) + F_N \\ &= xF_{N+1} + F_N \end{aligned}$$
    (c) The numbers r_1 and r_2 are solutions of the equation $x^2 = x + 1$. Therefore, part (b) implies that $r_1^n - r_2^n = (r_1 F_n + F_{n-1}) - (r_2 F_n + F_{n-1}) = (r_1 - r_2) F_n = \sqrt{5} F_n$.
    (d) Since $r_2 < r_1$,
$$\lim_{n \to \infty} \frac{F_{n+1}}{F_n} = \lim_{n \to \infty} \frac{r_1^{n+1} - r_2^{n+1}}{r_1^n - r_2^n} = \lim_{n \to \infty} \frac{r_1 - \left(\frac{r_2}{r_1}\right)^n r_2}{1 - \left(\frac{r_2}{r_1}\right)^n} = r_1 = \frac{1 + \sqrt{5}}{2}.$$

11.2 Infinite Series, Convergence, and Divergence

1. (a) $S_5 = 63$; yes

   (b) $S_5 = \dfrac{101010101010}{100000000000}$; yes

   (c) When $r = 1$, $S_n = (n+1)a$ but the right-hand side of the expression is undefined (because of the zero in the denominator).

   (d) $S_n - rS_n = (a + ar + ar^2 + \cdots + ar^n) - r(a + ar + ar^2 + \cdots + ar^n) = a - ar^{n+1} = a(1 - r^{n+1})$

   (e) $S_n - rS_n = (1 - r)S_n = a(1 - r^{n+1})$. The desired result is now obtained by dividing through by $1 - r$.

   (f) The sum is a geometric series with $a = 3$, $r = 2$, and $n = 10$. Thus, $3 + 6 + 12 + \cdots + 3072 = 6141$.

2. (a) $a_1 = 1/1! = 1$, $a_2 = 1/2! = 1/2$, $a_5 = 1/5! = 1/120 \approx 0.0083333$, $a_{10} = 1/10! = 1/3628800 \approx 2.7557 \times 10^{-7}$

   (b) $S_1 = 2$; $S_2 = 2.5$; $S_5 = 163/60 \approx 2.71667$; $S_{10} = 9864101/3628800 \approx 2.71828$

   (c) The terms of the sequence are positive.

   (d) $R_1 = e - S_1 \approx 0.71828$, $R_2 = e - S_2 \approx 0.21828$, $R_5 = e - S_5 \approx 0.0016152$, $R_{10} = e - S_{10} \approx 2.7313 \times 10^{-8}$

   (e) $R_n = \sum_{k=n+1}^{\infty} 1/k!$. Since each term in the sum is positive, $R_n > 0$.

   (f) Since $R_n = e - S_n$ and S_n is a monotonically increasing sequence, $R_{n+1} < R_n$ for all $n \geq 0$.

   (g) Any $n \geq 6$ will do.

   (h) Any $n \geq 8$ will do.

   (i) $R_{50} < R_8 < 10^{-5}$

   (j) $\lim_{n \to \infty} R_n = \lim_{n \to \infty} (e - S_n) = e - \lim_{n \to \infty} S_n = e - e = 0$

3. (a) $a_1 = 1$; $a_2 = 1/4$; $a_5 = 1/25$; $a_{10} = 1/100$

   (b) $S_1 = 1$; $S_2 = 5/4$; $S_5 = 5269/3600 \approx 1.46361$; $S_{10} = 1968329/1270080 \approx 1.54977$

   (c) Yes. S_n is an increasing sequence because the terms of the series are all positive.

   (d) $R_1 = \pi^2/6 - S_1 \approx 0.64493$; $R_2 = \pi^2/6 - S_2 \approx 0.39493$; $R_5 = \pi^2/6 - S_5 \approx 0.18132$; $R_{10} = \pi^2/6 - S_{10} \approx 0.095166$

   (e) Yes because $R_n = \pi^2/6 - S_n$ and $S_{n+1} > S_n$.

   (f) $R_{20} \approx 0.048771$. Since $0 < R_{n+1} < R_n$ for all $n \geq 1$, $0 < R_n < 0.05$ for all $n \geq 20$.

   (g) $\lim_{n \to \infty} R_n = 0$

4. (a) $a_1 = 1/5$, $a_2 = 1/25$, $a_5 = 1/5^5 = 1/3125$, $a_{10} = 1/9765625$

   (b) $S_n = \sum_{k=0}^{n} \dfrac{1}{5^k} = \dfrac{1 - (1/5)^{n+1}}{1 - (1/5)} = \dfrac{5 - (1/5)^n}{4}$. Thus, $S_1 = 6/5 = 1.2$, $S_2 = 31/25 = 1.24$, $S_5 = 3906/3125 = 1.24992$, and $S_{10} = 12207031/9765625 = 1.2499999744$.

   (c) $S_{n+1} = S_n + 1/5^{n+1} > S_n$, so the sequence is increasing. Since $S_n = \dfrac{5}{4} - \dfrac{1}{4}\left(\dfrac{1}{5}\right)^n$, $S_n < 5/4$ for all $n \geq 0$. Because the sequence of partial sums is increasing and bounded above, it must converge.

   (d) $S_n = \dfrac{5}{4} - \dfrac{1}{4}\left(\dfrac{1}{5}\right)^n \implies \lim_{n \to \infty} S_n = \dfrac{5}{4}$

   (e) $R_1 = 1/20$, $R_2 = 1/100$, $R_5 = 1/12500$, $R_{10} = 1/39062500$.

   (f) $R_n = \sum_{k=n+1}^{\infty} 5^{-k} = \dfrac{1}{4}\left(\dfrac{1}{5}\right)^n$. Thus, $0 < R_{n+1} < R_n$ for all $n \geq 0$.

(g) $\lim_{n\to\infty} R_n = 0$

5. (a) $a_1 = -4/5 = -0.8$; $a_2 = 16/25 = 0.64$; $a_5 = 1024/3125 = -0.32768$; $a_{10} = 1048576/9765625 = 0.1073741824$

    (b) $S_1 = 1/5 = 0.2$; $S_2 = 21/25 = 0.84$; $S_5 = 1281/3125 = 0.40992$; $S_{10} = 5891381/9765625 \approx 0.60328$

    (c) $\sum_{k=0}^{\infty} (-0.8)^k = \dfrac{1}{1-(-0.8)} = \dfrac{5}{9} \approx 0.5555555555$

    (d) $R_1 = 16/45 \approx 0.35556$; $R_2 = -64/225 \approx -0.28444$; $R_5 = 4096/28125 \approx 0.14564$; $R_{10} = -4194304/87890625 \approx -0.047722$

    (e) No — $S_{2m+1} < S_{2m} < S_{2m-1}$ for $m = 1, 2, 3, \ldots$ (the partial sums are go up and down in value)

    (f) $R_n = \sum_{k=n+1}^{\infty} (-0.8)^k = \dfrac{5}{9}(-0.8)^{n+1}$ so the terms of the sequence are alternately positive and negative.

    (g) $|R_{n+1}| = \left|\sum_{k=n+2}^{\infty} (-0.8)^k\right| = \left|-0.8 \sum_{k=n+1}^{\infty} (-0.8)^k\right| = 0.8|R_n| < |R_n|$ (i.e., the sequence is decreasing).

    (h) $\lim_{n\to\infty} R_n = 0$.

6. (a) $S_1 = 4/3 \approx 1.33333$; $S_2 = 3/2 = 1.5$; $S_5 = 13577/8140 \approx 1.66794$; $S_{10} = 127807216183/75344540040 \approx 1.69630$

    (b) $S_{n+1} = S_n + \dfrac{1}{n+1+2^{n+1}} > S_n$ for all $n \geq 0$

    (c) Since $k + 2^k \geq 2^k$ for all $k \geq 0$, $a_k = 1/(k+2^k) \leq 1/2^k = 2^{-k}$ for all $k \geq 0$.

    (d) $S_n = \sum_{k=0}^{n} a_k \leq \sum_{k=0}^{n} 2^{-k} = 2 - 2^{-n} < 2$ for all $n \geq 1$.

    (e) From parts (b) and (d) we know that the sequence of partial sums is increasing and bounded above. Therefore, the sequence of partial sums converges (i.e., the series converges).

    (f) $0 \leq R_n = \sum_{k=n+1}^{\infty} a_k \leq \sum_{k=n+1}^{\infty} 2^{-k} = 2^{-n}$ for all $n \geq 0$. Thus, $0 \leq \lim_{n\to\infty} R_n \leq \lim_{n\to\infty} 2^{-n} = 0 \implies \lim_{n\to\infty} R_n = 0$.

7. (a) $S_1 = 8/15 \approx 0.53333$; $S_2 = 103/165 \approx 0.62424$; $S_5 = 2626616/3892119 \approx 0.67486$; $S_{10} = 292378957513217078064425 3/4319382292184626122988455 \approx 0.67690$

    (b) The sequence of partial sums is increasing because each term of the series is a positive number (i.e., $S_{n+1} = S_n + a_{n+1} > S_n$ because $a_{n+1} > 0$).

    Since $a_j < 3^{-j}$ for all $j \geq 0$, $S_n = \sum_{j=0}^{n} a_j < \sum_{j=0}^{n} 3^{-j} = \dfrac{1}{2}(3 - 3^{-n}) \leq \dfrac{3}{2}$ for all $n \geq 0$. Thus, each term in the sequence of partial sums is bounded above by $3/2$.

    (c) Since the sequence of partial sums is increasing and bounded above, it must converge. This implies that the infinite series converges.

8. (a) $\sum_{i=0}^{\infty} \dfrac{1}{(i+1)^4} = \sum_{j=1}^{\infty} \dfrac{1}{j^4} = \pi^4/90$

    (b) $\sum_{k=3}^{\infty} \dfrac{1}{k^4} = \sum_{j=1}^{\infty} \dfrac{1}{j^4} - 1 - 1/2^4 = \pi^4/90 - 17/16 \approx 0.019823$

9. $\dfrac{1}{16} + \dfrac{1}{32} + \dfrac{1}{64} + \dfrac{1}{128} + \cdots + \dfrac{1}{2^{i+4}} + \cdots = \dfrac{1}{2^4} \sum_{i=0}^{\infty} \dfrac{1}{2^i} = \dfrac{1}{2^4} \cdot 2 = \dfrac{1}{8}$

Section 11.2 Infinite Series, Convergence, and Divergence

10. $2 - 5 + 9 + \frac{1}{3} + \frac{1}{9} + \frac{1}{27} + \frac{1}{81} + \cdots + \frac{1}{3^n} + \cdots = 6 + \frac{1}{3} \sum_{j=0}^{\infty} \left(\frac{1}{3}\right)^j = 6 + 1/2 = 6.5$

11. $\sum_{n=0}^{\infty} e^{-n} = \frac{e}{e-1}$

12. $\sum_{k=3}^{\infty} \left(\frac{e}{\pi}\right)^k = \sum_{k=0}^{\infty} \left(\frac{e}{\pi}\right)^k - \sum_{k=0}^{2} \left(\frac{e}{\pi}\right)^k = \frac{1}{1-e/\pi} - 1 - \frac{e}{\pi} - \left(\frac{e}{\pi}\right)^2 = \frac{e^3}{\pi^2(\pi-e)} \approx 4.8076.$

13. $\sum_{m=1}^{\infty} (\arctan 1)^m = \sum_{m=1}^{\infty} (\pi/4)^m = \sum_{m=0}^{\infty} (\pi/4)^m - 1 = \frac{1}{1-\pi/4} - 1 = \frac{\pi}{4-\pi}$

14. $\sum_{i=10}^{\infty} \left(\frac{2}{3}\right)^i = \sum_{i=0}^{\infty} \left(\frac{2}{3}\right)^i - \sum_{i=0}^{9} \left(\frac{2}{3}\right)^i = \frac{1}{1-(2/3)} - \frac{1-(2/3)^{10}}{1-(2/3)} = \frac{1024}{19683} \approx 0.052025$

15. $\sum_{j=5}^{\infty} \left(-\frac{1}{2}\right)^j = \sum_{j=0}^{\infty} \left(-\frac{1}{2}\right)^j - \sum_{j=0}^{4} \left(-\frac{1}{2}\right)^j = \frac{1}{1-(-1/2)} - \frac{1-(-1/2)^5}{1-(-1/2)} = -\frac{1}{48} \approx -0.020833$

16. $\sum_{j=0}^{\infty} \frac{3^j + 4^j}{5^j} = \sum_{j=0}^{\infty} \frac{3^j}{5^j} + \sum_{j=0}^{\infty} \frac{4^j}{5^j} = 5/2 + 5 = 15/2$

17. Since $1/(2 + \sin k) \geq 1$ for all $k \geq 1$, the series diverges by the nth-term test.

18. A series diverges if the sequence of its partial sums does not have a limit. Since $S_{2m} = 1$ and $S_{2m+1} = 0$ for $m = 0, 1, 2, 3, \ldots$, $\lim_{n\to\infty} S_n$ does not exist and, therefore, the series diverges.

19. $S_n = \arctan(n+1) - \arctan(0) = \arctan(n+1)$. Since $\lim_{n\to\infty} S_n = \frac{\pi}{2}$, the series converges to $\frac{\pi}{2}$.

20. $S_n = 1 - 1/n \implies \lim_{n\to\infty} S_n = 1$. Thus, the series converges to 1.

21. $\sum_{k=1}^{\infty} \frac{2}{k^2+k} = \sum_{k=1}^{\infty} \frac{2}{k} - \frac{2}{k+1}$ so $S_n = 2 - \frac{2}{n+1}$. Since $\lim_{n\to\infty} S_n = 2$, the series converges to 2.

22. $\sum_{n=1}^{\infty} \frac{n}{(n+1)!} = \sum_{n=1}^{\infty} \left(\frac{1}{n!} - \frac{1}{(n+1)!}\right)$ so $S_m = 1 - \frac{1}{(m+1)!}$. Since $\lim_{m\to\infty} S_m = 1$, the series converges to 1.

23. $S_n = 1 + \frac{1}{\sqrt{2}} - \frac{1}{\sqrt{n+1}} - \frac{1}{\sqrt{n+2}}$ when $n \geq 1$. Since $\lim_{n\to\infty} S_n = 1 + 1/\sqrt{2}$, the series converges to $1 + 1/\sqrt{2}$.

24. $S_n = \ln(n+1)$. Since $\lim_{n\to\infty} S_n = \infty$, the series diverges.

25. $S_n = \left(1 + (-1)^n\right)/2$. Since $\lim_{n\to\infty} S_n$ doesn't exist, the series diverges.

26. The series diverges by the nth-term test: $\lim_{j\to\infty} (-1)^j j$ does not exist.

27. (a) $\frac{1}{4} + \frac{1}{16} + \frac{1}{36} + \frac{1}{100} + \cdots = \frac{1}{4}\left(1 + \frac{1}{4} + \frac{1}{9} + \cdots\right) = \frac{\pi^2}{24}.$

    (b) $\sum_{k=0}^{\infty} \frac{1}{(2k+1)^2} = \sum_{k=1}^{\infty} \frac{1}{k^2} - \frac{1}{4}\sum_{k=1}^{\infty} \frac{1}{k^2} = \frac{\pi^2}{6} - \frac{\pi^2}{24} = \frac{\pi^2}{8}.$

(c) $\sum_{m=1}^{\infty} \frac{(-1)^{m+1}}{m^2} = \sum_{m=1}^{\infty} \frac{1}{m^2} - \frac{1}{2}\sum_{m=1}^{\infty} \frac{1}{m^2} = \frac{\pi^2}{12}$.

Alternatively, $\sum_{m=1}^{\infty} \frac{(-1)^{m+1}}{m^2} = \sum_{k=0}^{\infty} \frac{1}{(2k+1)^2} - \sum_{j=1}^{\infty} \frac{1}{(2j)^2} = \frac{\pi^2}{8} - \frac{\pi^2}{24} = \frac{\pi^2}{12}$.

28. $\sum_{m=3}^{\infty} \frac{2^{m+4}}{5^m} = \frac{2^7}{5^3} \sum_{m=0}^{\infty} \left(\frac{2}{5}\right)^m = \frac{2^7}{5^3} \cdot \frac{5}{3} = \frac{128}{75}$

29. The series converges for all values of x such that $-1 < x < 1$. $\sum_{k=0}^{\infty} x^k = \frac{1}{1-x}$.

30. The series converges when $-1 < x/5 < 1$. Thus, the series converges for all values of x such that $-5 < x < 5$.
$\sum_{m=2}^{\infty} \left(\frac{x}{5}\right)^m = \sum_{m=0}^{\infty} \left(\frac{x}{5}\right)^m - \sum_{m=0}^{1} \left(\frac{x}{5}\right)^m = \frac{1}{1-(x/5)} - 1 - \frac{x}{5} = \frac{x^2}{5(5-x)}$.

31. Since $\sum_{j=5}^{\infty} x^{2j} = x^{10} \sum_{j=0}^{\infty} \left(x^2\right)^j$, the series converges when $-1 < x^2 < 1$. Thus, the series converges to $\frac{x^{10}}{1-x^2}$ for all values of x such that $-1 < x < 1$.

32. Since $\sum_{k=1}^{\infty} x^{-k} = \sum_{k=1}^{\infty} \left(x^{-1}\right)^k$, the series converges when $\left|x^{-1}\right| < 1$. Therefore, the series converges for all values of x such that $x < -1$ or $x > 1$. $\sum_{k=1}^{\infty} x^{-k} = \frac{1}{1-x^{-1}} - 1 = \frac{1}{x-1}$.

33. The series converges when $|1+x| < 1$. Thus, the series converges for all values of x such that $-2 < x < 0$.
$\sum_{n=3}^{\infty} (1+x)^n = (1+x)^3 \sum_{n=0}^{\infty} x^n = -\frac{(1+x)^3}{x}$.

34. The series converges when $\left|\frac{1}{1-x}\right| < 1$. Thus, the series converges for all values of x such that $x < 0$ or $x > 2$.
$\sum_{j=4}^{\infty} \frac{1}{(1-x)^j} = \frac{x-1}{x(1-x)^4}$.

35. Since $|\sin x|/2 < 1$ for all x, the series converges to $\frac{2}{2-\sin x}$ for all x.

36. The series converges when $|\ln x| < 1$. Thus, the series converges when $e^{-1} < x < e$. For these values of x,
$\sum_{m=2}^{\infty} (\ln x)^m = \frac{(\ln x)^2}{1 - \ln x}$.

37. The series converges when $|\arctan x| < 1$. Thus, the series converges when $|x| < \pi/4$. For these values of x,
$\sum_{n=0}^{\infty} (\arctan x)^n = \frac{1}{1-\arctan x}$.

38. $S_{n+1} = \sum_{k=0}^{n} (1/3)^k \implies \lim_{n \to \infty} S_n = 3/2$

39. $a_{n+1} = 4 - \sum_{k=1}^{n} (1/2)^k \implies \lim_{n \to \infty} a_n = 3$

40. Diverges by the nth term test.

SECTION 11.2 INFINITE SERIES, CONVERGENCE, AND DIVERGENCE

41. The series $\sum_{k=0}^{\infty} \frac{3}{10} \left(-\frac{1}{2}\right)^k$ converges to 1/5.

42. Converges to 1. The partial sums are $S_{2n-1} = 1$ and $S_{2n} = 1 - 1/n$ $(n = 1, 2, 3, \ldots)$ so $\lim_{k \to \infty} S_k = 1$.

43. Diverges by the nth term test.

44. $\sum_{k=5}^{\infty} \frac{4}{7^{2k}} = \sum_{k=0}^{\infty} \frac{4}{49^k} - \sum_{k=0}^{4} \frac{4}{49^k} = \frac{4}{1 - (1/49)} - \frac{4\left(1 - (1/49)^5\right)}{1 - (1/49)} = \frac{49}{12} - \frac{23539604}{5764801} = \frac{1}{69177612}$

45. Diverges by the nth term test: $\lim_{n \to \infty} \frac{n+1}{2n+1} = \frac{1}{2} \neq 0$.

46. $\sum_{j=0}^{\infty} (\ln 2)^j = \frac{1}{1 - \ln 2} \approx 3.2589$

47. The partial sums of the series are $S_N = \sum_{n=2}^{N} \frac{2}{n^2 - 1} = \sum_{n=2}^{N} \left(\frac{1}{n-1} - \frac{1}{n+1}\right) = 1 + \frac{1}{2} - \frac{1}{N} - \frac{1}{N+1}$. Therefore, the series converges to 3/2.

48. Diverges by the nth term test. (Since $0 < \pi - e$, $\lim_{k \to \infty} k^{\pi - e} = \infty$.)

49. $\sum_{m=2}^{\infty} \frac{1}{(\ln 3)^m} = \frac{1}{(\ln 3)^2} \frac{1}{1 - (1/\ln 3)} = \frac{1}{(\ln 3)(\ln 3 - 1)}$

50. $\sum_{j=0}^{\infty} \left(\frac{1}{2^j} + \frac{1}{3^j}\right)^2 = \sum_{j=0}^{\infty} \left(\frac{1}{2^{2j}} + \frac{2}{6^j} + \frac{1}{3^{2j}}\right) = \sum_{j=0}^{\infty} \left(\frac{1}{4^j} + \frac{2}{6^j} + \frac{1}{9^j}\right) = \frac{4}{3} + \frac{12}{5} + \frac{9}{8} = \frac{583}{120}$.

51. $\sum_{k=1}^{\infty} \left(\int_k^{k+1} \frac{dx}{x^2}\right) = \sum_{k=1}^{\infty} \frac{1}{k(k+1)} = \lim_{n \to \infty} \left(1 - \frac{1}{n+1}\right) = 1$

52. Diverges by the nth term test: $\lim_{m \to \infty} \int_0^m e^{-x^2} dx = \sqrt{\pi}/2 \neq 0$.

53. Diverges by the nth term test: $\lim_{n \to \infty} \left(1 + \frac{1}{n}\right)^n = e \neq 0$.

54. Diverges by the nth term test: $\lim_{n \to \infty} \sqrt[n]{\pi} = 1 \neq 0$,

55. Diverges by the nth term test: $\lim_{n \to \infty} \frac{\ln n}{\ln(3 + n^2)} = \frac{1}{2} \neq 0$.

56. After the first bounce, the ball rebounds to a height of $4 \cdot (2/3)$ feet. The ball then falls from this height and rebounds to $4 \cdot (2/3)^2$ feet, etc. Thus, the total distance the ball travels is

$$4 + 2 \cdot 4 \cdot \frac{2}{3} + 2 \cdot 4 \cdot \left(\frac{2}{3}\right)^2 + 2 \cdot 4 \cdot \left(\frac{2}{3}\right)^3 + \cdots = 4 + 2 \cdot \frac{2}{3} \cdot \frac{1}{1 - (2/3)} = 8 \text{ feet.}$$

57. (a) $\lim_{k \to \infty} \frac{1}{\sqrt{k}} = 0$

(b) Since $1/\sqrt{n} \leq 1/\sqrt{k}$ when $1 \leq k \leq n$, $\sum_{k=1}^{n} \frac{1}{\sqrt{k}} \geq \sum_{k=1}^{n} \frac{1}{\sqrt{n}} = \frac{n}{\sqrt{n}} = \sqrt{n}$.

(c) It follows from part (a) that $\lim_{n \to \infty} S_n = \infty$.

58. (a) Yes. The terms of the sequence are increasing and bounded above, so Theorem 3 (p. 219) implies that the sequence has a limit.

    (b) Since the terms of the sequence are increasing, $a_k \geq a_1 > 0$ when $k \geq 1$. Therefore, $S_n = \sum_{k=1}^{n} a_k \geq \sum_{k=1}^{n} a_1 = na_1 > 0$ so $\lim_{n \to \infty} S_n = \infty$. This implies that the series diverges.

59. The series $\sum_{k=1}^{\infty} a_k$ converges because its partial sums are bounded above (by 100) and increasing (the terms of the series are positive). Therefore, Theorem 5 (p. 230) implies that $\lim_{k \to \infty} a_k = 0$.

60. (a) $S_{100} = 5 - \dfrac{3}{100} = 4.97$

    (b) $\sum_{k=1}^{\infty} a_k = \lim_{N \to \infty} \sum_{k=1}^{N} a_k = \lim_{N \to \infty} S_N = \lim_{N \to \infty} \left(5 - \dfrac{3}{N}\right) = 5$

    (c) $\lim_{k \to \infty} a_k = 0$ since $\sum_{k=1}^{\infty} a_k$ converges (Theorem 5).

61. (a) $\lim_{n \to \infty} S_n = \ln 2$.

    (b) Yes. Since the sequence of partial sums has a limit, the series converges.

    (c) For any $n \geq 1$,
    $$b_n = S_n - S_{n-1} = \ln\left(\dfrac{2n+3}{n+1}\right) - \ln\left(\dfrac{2n+1}{n}\right)$$
    $$= \ln\left(\dfrac{2n+3}{2n+1} \dfrac{n}{n+1}\right) = \ln\left(\dfrac{2n^2+3n}{2n^2+3n+1}\right) < 0.$$

62. (a) $\sum_{k=1}^{\infty} a_k = \lim_{n \to \infty} S_n = 3$ (by the Squeeze Theorem).

    (b) Since the series converges, $\lim_{k \to \infty} a_k = 0$.

63. (a) $a_k = (-1)^{k+1}$

    (b) By definition, the infinite series $\sum_{k=1}^{\infty} a_k$ converges if the sequence of its partial sums has a limit—that is, the series converges if $\lim_{n \to \infty} S_n$ exists. Since $a_k \geq 0$, S_n is an increasing sequence bounded above by 100 so Theorem 3 implies that this sequence has a limit and, therefore, that the infinite series converges.

    (c) $\sum_{k=1}^{\infty} a_k = \sum_{k=1}^{10^6-1} a_k + \sum_{k=10^6}^{\infty} a_k$. $\sum_{k=1}^{10^6-1} a_k$ is a finite sum, so it is a real number. The infinite series $\sum_{k=10^6}^{\infty} a_k$ converges because the sequence of its partial sums is increasing and bounded above. Thus, the original infinite series converges.

64. (a) Since $a_k \geq 0$, the partial sums of the series $\sum_{k=1}^{\infty} a_k$ form a monotonically increasing sequence. Since the series diverges, the partial sums must increase without bound.

    (b) If $a_k = (-1)^k$, $S_n = -1$ when n is odd and $S_n = 0$ when n is even. Thus, $\lim_{n \to \infty} S_n$ doesn't exist.

65. The series defining S doesn't converge.

Section 11.2 Infinite Series, Convergence, and Divergence

66. (a) Using the trigonometric identity given,

$$\frac{1}{2^k} \tan\left(\frac{x}{2^k}\right) = \frac{1}{2^k} \cot\left(\frac{x}{2^k}\right) - \frac{1}{2^{k-1}} \cot\left(\frac{x}{2^{k-1}}\right).$$

Thus, $\sum_{k=1}^{n} \frac{1}{2^k} \tan\left(\frac{x}{2^k}\right)$ can be written as a telescoping sum and

$$\sum_{k=1}^{n} \frac{1}{2^k} \tan\left(\frac{x}{2^k}\right) = \frac{1}{2^n} \cot\left(\frac{x}{2^n}\right) - \cot x.$$

(b) Using l'Hôpital's Rule,

$$\lim_{n \to \infty} \sum_{k=1}^{n} \frac{1}{2^k} \tan\left(\frac{x}{2^k}\right) = \lim_{n \to \infty} \frac{1}{2^n} \cot\left(\frac{x}{2^n}\right) - \cot x = \frac{1}{x} - \cot x.$$

68. (a) If the interval $[1, n+1]$ is divided into n equal subintervals, each subinterval has length one and the left endpoint of the jth subinterval is $1/j$.

(b) Since $1/x$ is a decreasing function on the interval $[1, \infty)$, $L_n > I_n$ for all $n \geq 1$. Now, $I_n = \ln(n+1) \implies \lim_{n \to \infty} L_n = \infty$. Thus, since $H_n = L_n$, the sequence of partial sums of the harmonic series does not have a finite limit (i.e., the harmonic series diverges).

69. (c) $a_k = \sum_{j=1}^{2^{k-1}} \frac{1}{2^{k-1}+j} \geq \sum_{j=1}^{2^{k-1}} \frac{1}{2^{k-1}+2^{k-1}} = \frac{1}{2}$

(d) $H_{2^n} = 1 + \sum_{k=1}^{n} a_k \geq 1 + \sum_{k=1}^{n} \frac{1}{2} = 1 + n/2 \implies \lim_{n \to \infty} H_{2^n} = \infty$

70. (a) When the interval $[1, 2]$ is divided into n equal subintervals, each subinterval has length $1/n$ and the left endpoint of the kth subinterval is $1 + (k-1)/n = (n+k-1)/n$. Thus,

$$L_n = \frac{1}{n} \sum_{k=1}^{n} \frac{n}{n+k-1} = \sum_{k=1}^{n} \frac{1}{n+k-1} = \sum_{k=0}^{n-1} \frac{1}{n+k}.$$

(b) Since $1/x$ is a decreasing function on the interval $[1, 2]$, $L_n > I = \ln 2$ for all $n \geq 1$.

(c) $H_{2n} - H_n = \sum_{k=1}^{2n} \frac{1}{k} - \sum_{k=1}^{n} \frac{1}{k} = \sum_{k=n+1}^{2n} \frac{1}{k}$

$$= \frac{1}{n+1} + \frac{1}{n+2} + \cdots + \frac{1}{2n-1} + \frac{1}{2n} = \left(L_n - \frac{1}{n}\right) + \frac{1}{2n}$$

$$= L_n - \frac{1}{2n}$$

(d) $\lim_{n \to \infty} (H_{2n} - H_n) = \lim_{n \to \infty} \left(L_n - \frac{1}{2n}\right) = \ln 2 - 0 = \ln 2$

(e) The result in part (d) implies that the partial sums of the harmonic series do not have a finite limit.

(f) Since $1/(2n) \leq 1/2$ for all $n \geq 1$, $H_{2n} - H_n = L_n - 1/(2n) \geq L_n - 1/2$ for all $n \geq 1$. Since, by part (b), $L_n > \ln 2$ for all $n \geq 1$, $H_{2n} - H_n > \ln 2 - 1/2$ for all $n \geq 1$.

(g) $H_2 = H_1 + L_1 - \frac{1}{2} = 1 + L_1 - \frac{1}{2} > 1 + \ln 2 - \frac{1}{2}$

(h) By part (g), the formula holds when $m = 1$. To show that the formula holds for all $m \geq 1$, we use induction: Assume that the formula holds for $m - 1$. That is, assume that $H_{2^{m-1}} > 1 + (m-1)\left(\ln 2 - \frac{1}{2}\right)$. Then, by part (f), $H_{2^m} - H_{2^{m-1}} > \ln 2 - \frac{1}{2}$ so $H_{2^m} > H_{2^{m-1}} + \left(\ln 2 - \frac{1}{2}\right) > 1 + (m-1)\left(\ln 2 - \frac{1}{2}\right) + \left(\ln 2 - \frac{1}{2}\right) = 1 + m\left(\ln 2 - \frac{1}{2}\right)$. Therefore, the formula holds for all $m \geq 1$.

71. (a) Since $\sin 0 = \tan 0 = 0$, and $\cos x \leq 1 \leq \sec^2 x$ when $0 \leq x < \pi/2$ and the racetrack principle implies that $\sin x \leq x \leq \tan x$ when $0 \leq x < \pi/2$.

(b) When $0 < x < \pi/2$, $0 < \sin x \leq x \leq \tan x$ so $0 < \sin^2 x \leq x^2 \leq \tan^2 x$. Therefore, $\csc^2 x \geq 1/x^2 \geq \cot^2 x$ when $0 < x < \pi/2$. The desired result follows from the trigonometric identity $\csc^2 x = 1 + \cot^2 x$.

(c) If k is an integer such that $1 \leq k \leq n$, then $x_k = k\pi/(2n+1)$ satisfies the inequality $0 < x_k < \pi/2$. Therefore, part (b) implies
$$\sum_{k=1}^{n} \cot^2 x_k \leq \sum \frac{1}{x_k^2} \leq \sum_{k=1}^{n}\left(1 + \cot^2 x_k\right).$$
Observing that $\sum_{k=1}^{n} 1 = n$ and inserting the definition of x_k, we obtain the desired result:
$$\sum_{k=1}^{n} \cot^2\left(\frac{k\pi}{2n+1}\right) \leq \frac{(2n+1)^2}{\pi^2}\sum_{k=1}^{n}\frac{1}{k^2} \leq n + \sum_{k=1}^{n}\cot^2\left(\frac{k\pi}{2n+1}\right).$$

(d) Let $S_n = \sum_{k=1}^{n}\frac{1}{k^2}$. After multiplication by $\pi^2/(2n+1)^2$, the given inequality becomes
$$\frac{\pi^2 n(2n-1)}{3(2n+1)^2} \leq S_n \leq \frac{\pi^2 n}{(2n+1)^2} + \frac{\pi^2 n(2n-1)}{3(2n+1)^2}.$$
Evaluating the limit of each term of this inequality as $n \to \infty$ leads to the result
$$\frac{\pi^2}{6} \leq \sum_{k=1}^{\infty}\frac{1}{k^2} \leq \frac{\pi^2}{6}.$$
This establishes the desired result.

72. (a) Since $0 \leq \tan x \leq 1$ when $0 \leq x \leq \pi/4$, $I_{n+1} = \int_0^{\pi/4} \tan^{n+1} x \, dx \leq \int_0^{\pi/4} \tan^n x \, dx = I_n$ for any integer $n \geq 0$.

(b)
$$I_n + I_{n-2} = \int_0^{\pi/4} \tan^n x \, dx + \int_0^{\pi/4} \tan^{n-2} x \, dx$$
$$= \int_0^{\pi/4} (1 + \tan^2 x)\tan^{n-2} x \, dx = \int_0^{\pi/4} \sec^2 x \tan^{n-2} x \, dx$$
$$= \int_0^1 u^{n-2} \, du = \frac{1}{n-1}.$$

(c) Since $\{I_n\}$ is a decreasing sequence, $\frac{1}{n-1} = I_n + I_{n-2} \geq 2I_n$ so $\frac{1}{2(n-1)} \geq I_n$.
Similarly, $\frac{1}{n+1} = I_{n+2} + I_n \leq 2I_n$ so $\frac{1}{2(n+1)} \leq I_n$.

(d) Part (b) may be rewritten $I_n = \frac{1}{n-1} - I_{n-2} = \frac{1}{n-1} - \int_0^{\pi/4} \tan^{n-2} x \, dx.$

(e) First, note that $\int_0^{\pi/4} \tan^0 x \, dx = \dfrac{\pi}{4}$. Then, part (d) can be used repeatedly to show that

$$I_{2n} = \frac{1}{2n-1} - I_{2n-2} = \frac{1}{2n-1} - \left(\frac{1}{2n-3} - I_{2n-4}\right)$$

$$= \frac{1}{2n-1} - \frac{1}{2n-3} + \left(\frac{1}{2n-5} - I_{2n-6}\right) = \cdots = (-1)^n \left(\frac{\pi}{4} + \sum_{k=1}^{n} \frac{(-1)^k}{2k-1}\right)$$

for all integers $n \geq 1$.

(f) It follows from part (c) that $\lim_{n \to \infty} I_n = 0$. Therefore,

$$-\frac{\pi}{4} = \lim_{n \to \infty} \sum_{k=1}^{n} \frac{(-1)^k}{2k-1} = \lim_{n \to \infty} \sum_{k=0}^{n-1} \frac{(-1)^{k+1}}{2k+1}$$

$$= \sum_{k=0}^{\infty} \frac{(-1)^{k+1}}{2k+1} = -\sum_{k=0}^{\infty} \frac{(-1)^k}{2k+1}.$$

(g) First, note that $\int_0^{\pi/4} \tan^1 x \, dx = \tfrac{1}{2} \ln 2$. Then, part (d) can be used repeatedly to show that

$$I_{2n+1} = \frac{1}{2n} - I_{2n-1} = \frac{1}{2n} - \left(\frac{1}{2n-2} - I_{2n-3}\right) = \frac{1}{2n} - \frac{1}{2n-2} + \left(\frac{1}{2n-4} - I_{2n-5}\right)$$

$$= \cdots = (-1)^n \left(\tfrac{1}{2} \ln 2 + \sum_{k=1}^{n} \frac{(-1)^k}{2k}\right)$$

for all integers $n \geq 1$.

(h) It follows from part (c) that $\lim_{n \to \infty} I_n = 0$. Therefore,

$$-\frac{\ln 2}{2} = \lim_{n \to \infty} \sum_{k=1}^{n} \frac{(-1)^k}{2k} = \lim_{n \to \infty} \frac{1}{2} \sum_{k=1}^{n} \frac{(-1)^k}{k} = \frac{1}{2} \sum_{k=1}^{\infty} \frac{(-1)^k}{k} = -\frac{1}{2} \sum_{k=0}^{\infty} \frac{(-1)^{k+1}}{k}.$$

73. (a) $S_{n+1} = S_n + \dfrac{1}{(n+1)^p} > S_n$ for all $n \geq 1$ since $\dfrac{1}{(n+1)^p} > 0$ for all $n \geq 1$.

(b) $S_{2m+1} = \displaystyle\sum_{k=1}^{2m+1} \frac{1}{k^p} = 1 + \sum_{k=2}^{2m+1} \frac{1}{k^p} = 1 + \sum_{k=1}^{m} \frac{1}{(2k)^p} + \sum_{k=1}^{m} \frac{1}{(2k+1)^p}$

(c) $S_{2m+1} = 1 + \displaystyle\sum_{k=1}^{m} \frac{1}{(2k)^p} + \sum_{k=1}^{m} \frac{1}{(2k+1)^p} < 1 + \sum_{k=1}^{m} \frac{1}{(2k)^p} + \sum_{k=1}^{m} \frac{1}{(2k)^p} = 1 + 2 \sum_{k=1}^{m} \frac{1}{(2k)^p}$

since $\dfrac{1}{(2k+1)^p} < \dfrac{1}{(2k)^p}$ for all $k \geq 1$ (since $p > 1$).

(d) $S_{2m+1} < 1 + 2 \displaystyle\sum_{k=1}^{m} \frac{1}{(2k)^p} = 1 + 2^{1-p} \sum_{k=1}^{m} \frac{1}{k^p} = 1 + 2^{1-p} S_m < 1 + 2^{1-p} S_{2m+1}$. (The last inequality holds because the partial sums are strictly increasing.)

(e) By part (d), $S_{2m+1} < 1 + 2^{1-p} S_{2m+1}$. Therefore, $\left(1 - 2^{1-p}\right) S_{2m+1} < 1$ so $S_{2m+1} < \dfrac{1}{1 - 2^{1-p}}$ for every $m \geq 1$. Finally, since any $n \geq 1$ can be written as either $n = 2m$ or $n = 2m+1$ where $m \geq 1$ is an integer, and since $S_{2m} < S_{2m+1}$ for all $m \geq 1$, it follows that sequence of partial sums is bounded above. Since the sequence of partial sums of the series is increasing and bounded above, this sequence has a finite limit. By definition, this implies that the series converges.

11.3 Testing for Convergence; Estimating Limits

1. (a) When $k \geq 0$, $k + 2^k \geq 2^k \implies a_k \leq 1/2^k = 2^{-k}$. Since $\sum_{k=0}^{\infty} 2^{-k}$ converges, the comparison test implies that $\sum_{k=0}^{\infty} a_k$ converges.

   (b) $R_{10} = \sum_{k=11}^{\infty} a_k < \sum_{k=11}^{\infty} 2^{-k} = 2^{-11} \sum_{k=0}^{\infty} 2^{-k} = 2^{-10}$.

   (c) Since $R_{10} < 2^{-10} \approx 0.00097656$, S_n has the desired accuracy if $n \geq 10$.
   $$S_{10} = \sum_{k=0}^{10} a_k = \frac{127807216183}{75344540040} \approx 1.6963.$$

   (d) Since the terms of the series are all positive, the estimate in part (c) *underestimates* the limit.

2. (a) The series converges by the comparison test: $\frac{1}{2+3^j} < 3^{-j}$ for all $j \geq 0$ and $\sum_{j=0}^{\infty} 3^{-j}$ converges.

   (b) $R_n = \sum_{j=n+1}^{\infty} \frac{1}{2+3^j} < \sum_{j=n+1}^{\infty} 3^{-j} = \frac{1}{2\cdot 3^n}$. Since $R_n < 0.01$ for all $n \geq 4$, the partial sum $S_4 = 266401/397155 \approx 0.67077$ approximates the limit with the desired accuracy.

   (c) The estimate in part (b) *underestimates* the limit of the series because the partial sums of the series form an increasing sequence (the terms of the series are all positive).

3. $\sum_{k=2}^{n} a_k < \int_1^n f(x)\,dx < \sum_{k=1}^{n-1} a_k$ [HINT: Draw pictures like those on pp. 241–243]

4. $\int_{n+1}^{\infty} a(x)\,dx < \sum_{k=n+1}^{\infty} a_k < \int_n^{\infty} a(x)\,dx$

5. Draw a picture illustrating the left sum approximation L_n to the integral $\int_1^{n+1} a(x)\,dx$. Since the integrand is a decreasing function, R_n overestimates the value of the integral.

6. The right Riemann sum approximation R_{n-1} of the integral $\int_1^n a(x)\,dx$ is $\sum_{k=2}^{n} a_k$. The approximation underestimates the value of the integral since the integrand is a decreasing function.

7. Draw a picture that illustrates a right Riemann sum approximation to the integral $\int_n^{\infty} a(x)\,dx$. The right sum is $\sum_{k=n+1}^{\infty} a_k$; it underestimates $\int_n^{\infty} a(x)\,dx$ since the integrand is a decreasing function. For the same reason, $a_{n+1} + \int_{n+1}^{\infty} a(x)\,dx \leq \int_n^{\infty} a(x)\,dx$. Finally, the first inequality established above implies that $\sum_{k=n+2}^{\infty} a_k \leq \int_{n+1}^{\infty} a(x)\,dx$. Adding a_{n+1} to both sides of this inequality establishes that
$$\sum_{k+n+1}^{\infty} a_k \leq a_{n+1} + \int_{n+1}^{\infty} a(x)\,dx.$$

8. $\int_0^{\infty} \frac{dx}{x^2+1} = \arctan x \Big|_0^{\infty} = \pi/2 \implies \sum_{k=0}^{\infty} \frac{1}{k^2+1}$ converges and $1 + \frac{\pi}{4} \leq 1 + \sum_{k=1}^{\infty} \frac{1}{k^2+1} \leq 1 + \frac{1}{2} + \frac{\pi}{4}$.

SECTION 11.3   TESTING FOR CONVERGENCE; ESTIMATING LIMITS   117

9. $\int_1^\infty \frac{dx}{x^{3/2}} = -\frac{2}{\sqrt{x}}\Big|_1^\infty = 2 \implies \sum_{k=1}^\infty \frac{1}{k\sqrt{k}}$ converges and $2 \le \sum_{k=1}^\infty \frac{1}{k\sqrt{k}} \le 3$.

10. $\int_1^\infty xe^{-x}\,dx = -e^{-x}(x+1)\Big|_1^\infty = 2e^{-1} \implies \sum_{j=1}^\infty je^{-j}$ converges and $2e^{-1} \le \sum_{j=1}^\infty je^{-j} \le 3e^{-1}$.

11. Since $\int_2^\infty \frac{dx}{(\ln x)^2} \ge \int_2^\infty \frac{dx}{4x} = \infty$, the series diverges by the integral test.

12. $\sum_{k=3}^\infty \frac{1}{(\ln k)^k} < \sum_{k=3}^\infty \frac{1}{(\ln 3)^k} = \frac{1}{(\ln 3)^3 (1 - 1/\ln 3)} \approx 8.40195$. Thus, the series converges by the comparison test.

13. (a) The integral test can't be used to prove that the series converges because the function $a(x) = \frac{2 + \sin x}{x^2}$ is not decreasing on the interval $[1, \infty)$.

   (b) Since $\frac{2 + \sin k}{k^2} \le \frac{3}{k^2}$ for all $k \ge 1$, and $\sum_{k=1}^\infty \frac{3}{k^2}$ converges, the comparison test implies that the series $\sum_{k=1}^\infty \frac{2 + \sin k}{k^2}$ converges.

14. Let $a_j = \frac{j^2}{j!}$. Since $\lim_{j \to \infty} \frac{a_{j+1}}{a_j} = \lim_{j \to \infty} \frac{\frac{(j+1)^2}{(j+1)!}}{\frac{j^2}{j!}} = \lim_{j \to \infty} \frac{(j+1)^2}{j} \cdot \frac{j!}{(j+1)!} = \lim_{j \to \infty} \frac{j+1}{j^2} = 0 < 1$, $\sum_{j=0}^\infty a_j$ converges.

15. $\lim_{k \to \infty} \frac{\frac{2^{k+1}}{(k+1)!}}{\frac{2^k}{k!}} = \lim_{k \to \infty} \frac{2}{k+1} = 0 \implies \sum_{k=1}^\infty \frac{2^k}{k!}$ converges.

16. $\lim_{n \to \infty} \frac{\frac{(n+1)^2}{2^{n+1}}}{\frac{n^2}{2^n}} = \lim_{n \to \infty} \left(\frac{n+1}{n}\right)^2 \cdot \frac{1}{2} = \frac{1}{2}$ so $\sum_{n=1}^\infty \frac{n^2}{2^n}$ converges.

17. $\lim_{m \to \infty} \frac{\frac{(m+1)!}{(2m+2)!}}{\frac{m!}{(2m)!}} = \lim_{m \to \infty} \frac{m+1}{(2m+2)(2m+1)} = 0$ so $\sum_{m=1}^\infty \frac{m!}{(2m)!}$ converges.

18. (a) Let $M = L + 1$ where $L = \sum_{k=1}^\infty b_k = \lim_{n \to \infty} T_n$. Since $0 \le a_k \le b_k$ for all $k \ge 1$, $S_n \le T_n$ for all $n \ge 1$. Furthermore, since $0 \le b_k$ for all $k \ge 1$, $\{T_n\}$ is an increasing sequence whose limit is $L < M$. It follows that $S_n \le T_n < M$ for all $n \ge 1$.

   (b) S_n is an increasing sequence because $0 \le a_k$ for all $k \ge 1$: $S_{n+1} = S_n + a_{n+1} > S_n$.

   (c) Together, parts (a) and (b) imply that $\{S_n\}$ is an increasing sequence that is bounded above. Since $\{S_n\}$ is a sequence of partial sums of the infinite series $\sum_{k=1}^\infty a_k$, convergence of the sequence implies convergence of the series.

   (d) Divergence of the series implies that the sequence of its partial sums diverges. Since $0 \le a_k$ for all $k \ge$, $S_n \ge 0$ for all $n \ge 1$ and $\{S_n\}$ is an increasing sequence. Such a sequence diverges only if $\lim_{n \to \infty} S_n = 0$.

   (e) Since $S_n \le T_n$ for all $n \ge 1$, $\lim_{n \to \infty} S_n = \infty \implies \lim_{n \to \infty} T_n = \infty$.

19. (a) $S_{n+1} = \sum_{k=1}^{n+1} a_k = S_n + a_k \ge S_n$ since $a_k = a(k) \ge 0$ for all integers $k \ge 1$.

(b) $\int_1^n a(x)\,dx \le \int_1^\infty a(x)\,dx$ because $a(x) \ge 0$ for all $x \ge 1$.

(c) Part (a) implies that $\{S_n\}$ is an increasing sequence. Since $S_n \le \int_1^n a(x)\,dx$, part (b) implies that the sequence of partial sums is bounded above. Thus, the sequence of partial sums converges to a limit.

20. Since $a_n \ge 0$ for all $n \ge 1$, $\sin(a_n) \ge 0$ and $0 \le \sum_{n=1}^\infty \sin(a_n) \le \sum_{n=1}^\infty a_n$. Therefore, $\sum_{n=1}^\infty \sin(a_n)$ converges (by the comparison test).

21. Yes. Let the series in question be $\sum_{k=0}^\infty a_k$ (i.e., $a_0 = 1$, $a_1 = 1/3$, $a_2 = 1/15$, $a_3 = 1/85$, etc.). Then $a_k \le 1/(2k-1)(2k+1) = 1/(4k^2-1) \le 1/3k^2$ for all $k \ge 1$. Since $\sum_{k=1}^\infty \dfrac{1}{3k^2}$ converges, $\sum_{k=0}^\infty a_k$ also converges (by the comparison test).

NOTE: The given series is $\sum_{k=0}^\infty a_k$ with $a_k = \dfrac{2^k k!}{(2k+1)!}$. Therefore, the ratio test can also be used to show that the series converges.

22. $0 < \sum_{k=3}^\infty \dfrac{1}{(\ln k)^{\ln k}} = \sum_{k=3}^{1619} \dfrac{1}{(\ln k)^{\ln k}} + \sum_{k=1620}^\infty \dfrac{1}{(\ln k)^{\ln k}} < \sum_{k=3}^{1619} \dfrac{1}{(\ln k)^{\ln k}} + \sum_{k=1620}^\infty \dfrac{1}{e^{2\ln k}} < \sum_{k=3}^{1619} \dfrac{1}{(\ln k)^{\ln k}} + \sum_{k=1620}^\infty \dfrac{1}{k^2}$.
Since the last series on the right converges (it is a p-series), the desired result follows by the comparison test.

23. (a) H_n is the nth partial sum of the harmonic series. These partial sums form a monotonically increasing, divergent sequence.

(b) $S_n = \sum_{k=0}^n \dfrac{1}{2k+1} \ge \sum_{k=0}^n \dfrac{1}{2k+2} = \dfrac{1}{2}\sum_{k=0}^n \dfrac{1}{k+1} = \dfrac{1}{2}\sum_{k=1}^{n-1} \dfrac{1}{k} = \dfrac{1}{2}H_{n+1} > \dfrac{1}{2}H_n$

(c) $\sum_{k=0}^\infty \dfrac{1}{2k+1}$ diverges.

24. (a) $\ln(n!) = \ln n + \ln(n-1) + \ln(n-2) + \cdots + \ln 2 + \ln 1$. Thus, (since $\ln 1 = 0$) $\ln(n!)$ is a right sum approximation so $\int_1^n \ln x\,dx$. Since $\ln x$ is an increasing function on the interval $[1, n]$, $\ln(n!) > \int_1^n \ln x\,dx = n\ln n - n + 1$. Therefore, $n! > \exp(n\ln n - n + 1) = n^n e^{1-n}$.

(b) Part (a) implies that $\dfrac{a^N}{N!} < \left(\dfrac{ae}{N}\right)^N \dfrac{1}{e}$. Thus, $\dfrac{a^N}{N!} < \dfrac{1}{2}$ when $\left(\dfrac{ae}{N}\right)^N < \dfrac{e}{2}$.

25. The improper integral $\int_2^\infty \dfrac{dx}{x(\ln n)^p}$ converges when $p > 1$ and diverges when $p \le 1$. Thus, by the integral test, the series converges only for $p > 1$.

26. (a) $\ln x/x$ is decreasing on $[3, \infty)$ and $\int_3^\infty \dfrac{\ln x}{x}\,dx$ diverges, so the integral test implies that $\sum_{k=3}^\infty \dfrac{\ln k}{k}$ diverges. Therefore, since $\sum_{k=1}^\infty \dfrac{\ln k}{k} = \dfrac{1}{2}\ln 2 + \sum_{k=3}^\infty \dfrac{\ln k}{k}$, $\sum_{k=1}^\infty \dfrac{\ln k}{k}$ diverges.

(b) Since $1 - 1/k \le \ln k$ for all $k \ge 1$, $1/k - 1/k^2 \le (\ln k)/k$ for all $k \ge 1$. Therefore, since $\sum_{k=1}^\infty \left(\dfrac{1}{k} - \dfrac{1}{k^2}\right)$ diverges, $\sum_{k=1}^\infty \dfrac{\ln k}{k}$ diverges.

Section 11.3 Testing for Convergence; Estimating Limits

(c) No, because $\lim\limits_{k\to\infty} \dfrac{a_{k+1}}{a_k} = 1$.

27. (a) $\lim\limits_{n\to\infty} \dfrac{a_{n+1}}{a_n}$ does not exist. For $m = 1, 2, 3, \ldots$, $\dfrac{a_{2m}}{a_{2m-1}} = \left(\dfrac{2}{3}\right)^m$ and $\dfrac{a_{2m+1}}{a_{2m}} = \dfrac{1}{2}\left(\dfrac{3}{2}\right)^m$.

    (b) Because the limit in part (a) does not exist, the ratio test says nothing about the convergence of the series $\sum\limits_{k=1}^{\infty} a_k$.

    (c) $\sum\limits_{k=1}^{\infty} a_k = \sum\limits_{k=1}^{\infty} 2^{-k} + \sum\limits_{k=1}^{\infty} 3^{-k} = 1 + \dfrac{1}{2} = \dfrac{3}{2}$

28. (a) The ratio $\dfrac{a_{k+1}}{a_k}$ is 1 when k is odd and $1/2$ when k is even. Thus, $\lim\limits_{k\to\infty} \dfrac{a_{k+1}}{a_k}$ doesn't exist.

    (b) Yes, it converges. Let S_n denote the partial sum of the first n terms of the series. When $n = 2m$, $S_n = 2 - \left(\dfrac{1}{2}\right)^{m-1}$; when $n = 2m+1$, $S_n = 2 - 3\left(\dfrac{1}{2}\right)^{m+1}$. It follows that $\lim\limits_{n\to\infty} S_n = 2$.

29. The harmonic series $\sum\limits_{k=1}^{\infty} \dfrac{1}{k}$ has the specified properties.

30. Let $a_n = n^{-n}$. Then $\dfrac{a_{n+1}}{a_n} = \dfrac{(n+1)^{-(n+1)}}{n^{-n}} = \dfrac{n^n}{(n+1)^{(n+1)}} = \left(\dfrac{n}{n+1}\right)^n \dfrac{1}{n+1}$. Since $\lim\limits_{n\to\infty} \left(\dfrac{n}{n+1}\right)^n = \dfrac{1}{e}$ and $\lim\limits_{n\to\infty} \dfrac{1}{n+1} = 0$, $\lim\limits_{n\to\infty} \dfrac{a_{n+1}}{a_n} = 0$. Thus, the ratio test implies that the series $\sum\limits_{n=1}^{\infty} n^{-n}$ converges.

31. Let $a_n = \dfrac{n^n}{n!}$. Then $\dfrac{a_{n+1}}{a_n} = \dfrac{(n+1)^{n+1}}{n^n} \dfrac{n!}{(n+1)!} = \left(\dfrac{n+1}{n}\right)^n$ and $\lim\limits_{n\to\infty} \dfrac{a_{n+1}}{a_n} = e > 1$. Therefore, the series $\sum\limits_{n=1}^{\infty} \dfrac{n^n}{n!}$ diverges.

32. $\sum\limits_{n=1}^{\infty} \dfrac{1}{n^2 + \sqrt{n}} < \sum\limits_{k=1}^{\infty} \dfrac{1}{n^2} \le 2.$

33. $\sum\limits_{j=0}^{\infty} \dfrac{1}{j + e^j} < \sum\limits_{j=0}^{\infty} \dfrac{1}{e^j} = \dfrac{1}{1 - e^{-1}}$. (The series on the right is a convergent geometric series.)

34. $\sum\limits_{m=1}^{\infty} \dfrac{1}{m\sqrt{1 + m^2}} < \sum\limits_{m=1}^{\infty} \dfrac{1}{m^2} \le 2.$

35. $\sum\limits_{k=1}^{\infty} \dfrac{k}{(k^2 + 1)^2} < \sum\limits_{k=1}^{\infty} \dfrac{1}{k^3} \le 1 + \int_1^{\infty} x^{-3}\, dx = \dfrac{3}{2}.$

36. Diverges. Each term of this series is a constant multiple $(1/100)$ of the corresponding term of the harmonic series.

37. Converges—comparison test: $\sum\limits_{n=1}^{\infty} \dfrac{\arctan n}{1 + n^2} < \dfrac{\pi}{2} \sum\limits_{n=1}^{\infty} \dfrac{1}{1 + n^2} < \dfrac{\pi}{2} \sum\limits_{n=1}^{\infty} \dfrac{1}{n^2} = \dfrac{\pi^3}{12}.$

    [NOTE: This series can also be shown to converge via the integral test: $\int_1^{\infty} \dfrac{\arctan x}{1 + x^2}\, dx = \int_{\pi/4}^{\pi/2} u\, du = \dfrac{3\pi^2}{32}$. This leads to the bound $\sum\limits_{n=1}^{\infty} \dfrac{\arctan n}{1 + n^2} \le \dfrac{\pi}{8} + \dfrac{3\pi^2}{32}.$]

38. Converges—comparison test: $\sum_{m=1}^{\infty} \frac{m^3}{m^5+3} < \sum_{m=1}^{\infty} \frac{m^3}{m^5} = \sum_{m=1}^{\infty} \frac{1}{m^2} \le 2$.

39. Diverges—integral test: $\int_{1}^{\infty} \frac{dx}{100+5x} = \frac{1}{5}\int_{105}^{\infty} \frac{du}{u} = \infty$.

40. Diverges—integral test: $\int_{2}^{\infty} \frac{dx}{x \ln x} = \int_{\ln 2}^{\infty} \frac{du}{u} = \infty$.

41. Converges—comparison test: $\sum_{n=1}^{\infty} \frac{1}{n \, 3^n} < \sum_{n=1}^{\infty} \frac{1}{3^n} = \frac{1}{2}$.

42. Diverges. Each term of the series is a constant multiple $(1/\ln 10)$ of the corresponding term of the harmonic series since $\ln(10^k) = k \ln 10$.

43. Converges—integral test: $\sum_{j=1}^{\infty} j 5^{-j} \le \frac{1}{5} + \int_{1}^{\infty} x 5^{-x} \, dx = \frac{1}{5} + \frac{1+\ln 5}{5(\ln 5)^2}$.

NOTE: This series can also be shown to converge via the ratio test:

$$\lim_{j \to \infty} \frac{\frac{j+1}{5^{j+1}}}{\frac{j}{5^j}} = \lim_{j \to \infty} \frac{j+1}{j} \cdot \frac{5^j}{5^{j+1}} = \lim_{j \to \infty} \frac{j+1}{5j} = \frac{1}{5} < 1.$$

Furthermore, since $a_{j+1}/a_j = (j+1)/5j \le 2/5$ when $j \ge 1$, $\sum_{j=1}^{\infty} j 5^{-j} \le \sum_{j=1}^{\infty} \frac{2^{j-1}}{5^j} = \frac{1}{3}$.

44. Diverges—nth term test: $\lim_{k \to \infty} \frac{k^2}{5k^2+3} = \frac{1}{5} \ne 0$.

45. Diverges. The given series is 2 minus the harmonic series: $1 - \frac{1}{2} - \frac{1}{3} - \frac{1}{4} - \frac{1}{5} - \cdots = 2 - \sum_{k=1}^{\infty} \frac{1}{k}$. Since the harmonic series diverges, so does this series.

46. Converges—comparison test: $\sum_{j=1}^{\infty} \frac{j}{j^4+j^2-1} < \sum_{j=1}^{\infty} \frac{j}{j^4} = \sum_{j=1}^{\infty} \frac{1}{j^3}$.

47. Diverges—comparison test: $\sum_{n=2}^{\infty} \frac{1}{\sqrt[3]{n^2-1}} > \sum_{n=2}^{\infty} \frac{1}{n^{2/3}}$.

48. Converges—comparison test: $\sum_{k=1}^{\infty} \frac{\sqrt{k}}{k^2+k+1} < \sum_{k=1}^{\infty} \frac{1}{k^{3/2}} \le 1 + \int_{1}^{\infty} x^{-3/2} \, dx = 3$.

49. Converges—integral test: $\int_{0}^{\infty} e^{-x^2} \, dx = \sqrt{\pi}/2$. $\sum_{m=0}^{\infty} e^{-m^2} \le 1 + \int_{0}^{\infty} e^{-x^2} \, dx = 1 + \sqrt{\pi}/2$.

50. Converges—comparison test: $\sum_{j=0}^{\infty} \frac{j!}{(j+2)!} = \sum_{j=0}^{\infty} \frac{1}{(j+2)(j+1)} = \frac{1}{2} + \sum_{j=1}^{\infty} \frac{1}{(j+2)(j+1)} < \frac{1}{2} + \sum_{j=1}^{\infty} \frac{1}{j^2}$.

51. Converges—ratio test: $\lim_{n \to \infty} \frac{\frac{(n+1)!}{(2n+2)!}}{\frac{n!}{(2n)!}} = \lim_{n \to \infty} \frac{(n+1)!}{n!} \cdot \frac{(2n)!}{(2n+2)!} = \lim_{n \to \infty} \frac{1}{2(2n+1)} = 0$.

Since $\frac{a_{n+1}}{a_n} = \frac{1}{2(2n+1)} \le \frac{1}{2}$ for all $n \ge 0$, $\sum_{n=0}^{\infty} a_n \le a_0 \sum_{n=0}^{\infty} \frac{1}{2^n} = 2a_0 = 2$.

SECTION 11.3   TESTING FOR CONVERGENCE; ESTIMATING LIMITS

NOTE: Since $\dfrac{n!}{(2n)!} \leq \dfrac{1}{(n+1)!}$ when $n \geq 0$, the series can also be shown to converge using the comparison test: $\displaystyle\sum_{n=0}^{\infty} \dfrac{n!}{(2n)!} < \sum_{n=0}^{\infty} \dfrac{1}{(n+1)!} = e - 1.$

52. Diverges—comparison test: $\displaystyle\sum_{m=1}^{\infty} \dfrac{m^3}{m^4 - 7} = -\dfrac{1}{6} + \sum_{m=2}^{\infty} \dfrac{m^3}{m^4 - 7} > -\dfrac{1}{6} + \sum_{m=2}^{\infty} \dfrac{m^3}{m^4} = -\dfrac{1}{6} + \sum_{m=2}^{\infty} \dfrac{1}{m}.$

53. Diverges—comparison test: $\displaystyle\sum_{k=1}^{\infty} \dfrac{k!}{(k+1)! - 1} > \sum_{k=1}^{\infty} \dfrac{k!}{(k+1)!} = \sum_{k=1}^{\infty} \dfrac{1}{k+1} = \dfrac{1}{2} + \dfrac{1}{3} + \dfrac{1}{4} + \cdots$

54. Converges—integral test: $\displaystyle\int_2^{\infty} \dfrac{\ln x}{x^2}\, dx = \int_{\ln 2}^{\infty} u e^{-u}\, du = \dfrac{1 + \ln 2}{2}.$ Thus,
$\displaystyle\sum_{j=2}^{\infty} \dfrac{\ln j}{j^2} \leq \dfrac{\ln 2}{4} + \dfrac{1 + \ln 2}{2} = \dfrac{2 + 3\ln 2}{4}.$

55. Diverges by the nth term test: $\displaystyle\lim_{n \to \infty} \sum_{k=1}^{n} k^{-1} = \infty$

56. Diverges. The partial sum of the first $2n$ terms of the series is one-half the partial sum of the first n terms of the harmonic series.

57. Converges (by the comparison test): $\displaystyle\sum_{k=1}^{\infty} \dfrac{1}{k^2 + 3} \leq \sum_{k=1}^{\infty} \dfrac{1}{k^2}.$ (The series on the right side of the inequality is a convergent p-series.)

Since $R_N = \displaystyle\sum_{k=N+1}^{\infty} \dfrac{1}{k^2 + 3} \leq \sum_{k=N+1}^{\infty} \dfrac{1}{k^2} \leq \int_N^{\infty} \dfrac{dx}{x^2} = \dfrac{1}{N} \leq 0.001$ when $N \geq 1000$, $n \geq 1000$ implies that S_n approximates the sum of the series within 0.001.

58. Diverges (by the comparison test): $\displaystyle\sum_{m=1}^{\infty} \dfrac{\arctan m}{m} \geq \dfrac{\pi}{4} \sum_{m=1}^{\infty} \dfrac{1}{m}.$ (The series on the right side of the inequality is the harmonic series—a series known to diverge.)

$R_n = \displaystyle\sum_{m=1}^{n} \dfrac{\arctan m}{m} \geq \dfrac{\pi}{4} \sum_{m=1}^{n} \dfrac{1}{m} \geq \dfrac{\pi}{4} \int_1^{n+1} \dfrac{dx}{x} = \dfrac{\pi}{4} \ln(n+1) \geq 1000$ when $n \geq e^{4000/\pi} - 1 \approx 9.2 \times 10^{552}.$

59. Converges: $\displaystyle\sum_{j=2}^{\infty} \dfrac{3^j}{4^{j+1}} = \dfrac{1}{4}\sum_{j=2}^{\infty} \left(\dfrac{3}{4}\right)^j = \dfrac{9}{64}\sum_{j=0}^{\infty} \left(\dfrac{3}{4}\right)^j = \dfrac{9}{64}\dfrac{1}{1 - 3/4} = \dfrac{9}{16}.$

$R_N = \displaystyle\sum_{j=N+1}^{\infty} \dfrac{3^j}{4^{j+1}} = \dfrac{3^{N+1}}{4^{N+2}} \sum_{j=0}^{\infty} \left(\dfrac{3}{4}\right)^j = \left(\dfrac{3}{4}\right)^{N+1} \leq 0.001$ when $N \geq -\dfrac{\ln 1000}{\ln(3/4)} - 1 \approx 23.012.$ Thus, $n \geq 24$ implies that S_n approximates the sum of the series within 0.001.

60. Diverges. ($\displaystyle\lim_{k \to \infty} \dfrac{1}{2 + \cos k}$ doesn't exist.)

$S_n = \displaystyle\sum_{k=0}^{n} \dfrac{1}{2 + \cos k} \geq \sum_{k=0}^{n} \dfrac{1}{3} = \dfrac{n+1}{3}$ so $S_n \geq 1000$ when $n \geq 2999.$

61. Converges.

$$0 < R_n = \sum_{m=n+1}^{\infty} \frac{\ln m}{m^3} \leq \int_n^{\infty} \frac{\ln x}{x^3}\,dx \leq \int_n^{\infty} \frac{x}{x^3}\,dx = \frac{1}{n} \leq 0.001 \text{ when } n \geq 1000.$$

$$0 < R_n = \sum_{m=n+1}^{\infty} \frac{\ln m}{m^3} \leq \int_n^{\infty} \frac{\ln x}{x^3}\,dx \leq \int_n^{\infty} \frac{\sqrt{x}}{x^3}\,dx = \frac{2}{3n^{3/2}} \leq 0.001 \text{ when } n \geq 77.$$

$$0 < R_n = \sum_{m=n+1}^{\infty} \frac{\ln m}{m^3} \leq \int_n^{\infty} \frac{\ln x}{x^3}\,dx = \frac{1 + 2\ln n}{4n^2} \leq 0.001 \text{ when } n \geq 47.$$

62. The comparison $\sum_{k=1}^{\infty} \frac{k}{k^6+17} < \sum_{k=1}^{\infty} \frac{k}{k^6} = \sum_{k=1}^{\infty} \frac{1}{k^5}$ shows that the original series converges.

Since $R_n \leq \int_n^{\infty} \frac{x}{x^6+17}\,dx < \int_n^{\infty} \frac{dx}{x^5} = \frac{1}{4n^4} < 0.005$ when $n \geq 3$, the estimate

$$S \approx \sum_{k=1}^{3} \frac{k}{k^6+17} = \frac{2546}{30213} \approx 0.084268 \text{ is guaranteed to be in error by no more than } 0.005.$$

63. Since the improper integral $\int_2^{\infty} \frac{1}{x(\ln x)^5}\,dx = \int_{\ln 2}^{\infty} \frac{1}{u^5} = \frac{1}{4(\ln 2)^4}$ converges, the corresponding series converges. Furthermore, since $\int_n^{\infty} \frac{dx}{x(\ln x)^5}\,dx < 0.005$ when $n \geq 15$, the approximation $L \approx \sum_{k=2}^{15} \frac{1}{k(\ln k)^5} \approx 3.4254$ has the desired accuracy.

64. (a) The assumption that $a(x)$ is decreasing is necessary to ensure that the desired geometric relationships hold (e.g., that $\int_1^{\infty} a(x)\,dx \leq \sum_{k=1}^{\infty} a_k$).

(b) Under the new assumption the inequality in the second bullet must be replaced by

$$\int_{10}^{\infty} a(x)\,dx \leq \sum_{k=10}^{\infty} a_k \leq a_{10} + \int_{10}^{\infty} a(x)\,dx$$

and the condition $n \geq 10$ must be placed on the inequality in the last bullet. No other changes in the conclusion of the theorem are necessary.

65. (a) $k! = \overbrace{k \cdot (k-1) \cdots \cdot 2}^{(k-1)\text{-terms}} \geq \overbrace{2 \cdot 2 \cdots \cdot 2}^{(k-1)\text{-terms}} = 2^{k-1} \implies \frac{1}{k!} \leq \frac{1}{2^{k-1}}$

(b) $k! = \overbrace{k \cdot (k-1) \cdots (k-10)}^{(k-10)\text{-terms}} \cdot 10 \cdots \cdot 2 \geq \overbrace{10 \cdot 10 \cdots \cdot 10}^{(k-10)\text{-terms}} \cdot 10! = 10^{k-1} \cdot 10! \implies \frac{1}{k!} \leq \frac{1}{10! \, 10^{k-10}}$

(c) Let $S_n = \sum_{k=0}^{n} \frac{1}{k!}$. S_{10} underestimates $\sum_{k=0}^{\infty} \frac{1}{k!}$ since S_n is a monotonically increasing sequence.

$$R_{10} = \sum_{k=0}^{\infty} \frac{1}{k!} - S_{10} = \sum_{k=11}^{\infty} \frac{1}{k!} \leq \sum_{k=11}^{\infty} \frac{1}{10! \, 10^{k-10}} = \frac{1}{10!} \sum_{k=1}^{\infty} \frac{1}{10^k} = \frac{1}{10!} \cdot \frac{1}{9} = \frac{1}{32659200} \approx 3.0619 \times 10^{-8}$$

66. (a) The key idea is that $(n+k)! = (n+k) \cdot (n+k-1) \cdot (n+k-2) \cdots (n+1) \cdot n! > n^k \cdot n!$ for all $k \geq 1$.

Therefore, $\sum_{k=0}^{\infty} \frac{1}{(n+k)!} < \sum_{k=0}^{\infty} \frac{1}{n^k \, n!}$.

(b) By part (a), $\sum_{k=n}^{\infty} \frac{1}{k!} = \sum_{k=0}^{\infty} \frac{1}{(k+n)!} < \sum_{k=0}^{\infty} \frac{1}{n^k \, n!} = \frac{1}{n!} \sum_{k=0}^{\infty} \frac{1}{n^k} = \frac{1}{n!} \cdot \frac{1}{1 - \frac{1}{n}}$.

(c) Let $S_N = \sum_{k=1}^{N} \frac{1}{k!}$ denote the partial sum of N terms of the $\sum_{k=1}^{\infty} \frac{1}{k!}$. Since all the terms of the series are positive, the sequence of partial sums $\{S_N\}$ is monotonically increasing. Furthermore,

$$S_N < \sum_{k=1}^{\infty} \frac{1}{k!} = 1 + \frac{1}{2} + \sum_{k=2}^{\infty} \frac{1}{k!} < \frac{3}{2} + \sum_{k=0}^{\infty} \frac{1}{(k+2)!} < \frac{3}{2} + \frac{1}{2} \frac{1}{1-\frac{1}{2}} = \frac{5}{2}$$

so the sequence of partial sums is bounded above. It follows that sequence of partial sums converges (i.e., the series converges).

(d) Let $R_N = \sum_{k=1}^{\infty} \frac{1}{k!} - S_N = \sum_{k=N+1}^{\infty} \frac{1}{k!}$. From part (b), $R_N < \frac{1}{(N+1)!} \frac{1}{1-\frac{1}{N+1}}$. Therefore, $R_N < 0.00001$ when $N \geq 8$.

67. (a) Let $a_k = (1/k!)^2$. Since $a_{k+1}/a_k = \left(k!/(k+1)!\right)^2 = 1/(k+1)^2$, $\lim_{k\to\infty} \frac{a_{k+1}}{a_k} = 0$. Thus, the ratio test implies that $\sum_{k=1}^{\infty} a_k$ converges.

(b) Let $R_N = \sum_{k=N+1}^{\infty} \left(\frac{1}{k!}\right)^2$. Now,

$$R_N = \sum_{k=0}^{\infty} \left(\frac{1}{(N+1+k)!}\right)^2$$
$$< \sum_{k=0}^{\infty} \left(\frac{1}{(N+1)^k (N+1)!}\right)^2 = \frac{1}{((N+1)!)^2} \sum_{k=0}^{\infty} \left(\frac{1}{(N+1)^2}\right)^k$$
$$= \frac{1}{((N+1)!)^2} \frac{1}{1 - \left(\frac{1}{N+1}\right)^2}.$$

Therefore, since $R_N < 5 \times 10^{-6}$ when $N \geq 5$, S_N approximates the sum of this series within 5×10^{-6} when $N \geq 5$.

68. Let $a_k = e^k/k!$. Since $a_k \leq \frac{e^k}{2 \cdot 3^{k-2}}$ when $k \geq 2$,

$$R_N = \sum_{k=N+1}^{\infty} a_k \leq \sum_{k=N+1}^{\infty} \frac{e^k}{2 \cdot 3^{k-2}} = \frac{9}{2} \sum_{k=N+1}^{\infty} \left(\frac{e}{3}\right)^k = \frac{9}{2} \left(\frac{e}{3}\right)^{N+1} \frac{1}{1-\frac{e}{3}}.$$

Therefore, $R_N < 5 \times 10^{-6}$ when $N \geq 163$.

69. (a) If $k \geq N$, then $k! = \overbrace{k \cdot (k-1) \cdot (k-2) \cdots \cdots (N+1)}^{(k-N)\text{-terms}} \cdot N! \geq (N+1)^{k-N} \cdot N!$.

Thus, $\frac{x^k}{k!} = \frac{x^N \cdot x^{k-N}}{k!} \leq \frac{x^N \cdot x^{k-N}}{N! \cdot (N+1)^{k-N}} = \frac{x^N}{N!} \left(\frac{x}{N+1}\right)^{k-N}$

(b) $\sum_{k=N}^{\infty} \frac{x^k}{k!} \leq \sum_{k=N}^{\infty} \frac{x^N}{N!} \left(\frac{x}{N+1}\right)^{k-N} = \frac{x^N}{n!} \sum_{k=N}^{\infty} \left(\frac{x}{N+1}\right)^{k-N} = \frac{x^N}{n!} \sum_{k=0}^{\infty} \left(\frac{x}{N+1}\right)^k = \frac{x^N}{N!} \cdot \frac{1}{1-\frac{x}{N+1}}$

since $x/(N+1) < 1$.

70. (a) $\frac{a_{k+1}}{a_k} \leq r < 1$ for all $k \geq 1$ implies that $a_{k+1} \leq a_k r < a_k$. From this it follows that $a_2 \leq a_1 r$ and $a_3 \leq a_2 r$. Multiplying both sides of the first inequality by r produces $a_2 r \leq a_1 r^2$. Therefore, $a_3 \leq a_2 r \leq a_1 r^2$. [NOTE: Multiplying both sides of the inequality $a_3 \leq a_1 r^2$ by r leads to $a_3 r \leq a_1 r^3$. Since $a_4 \leq a_3 r$, we may conclude that $a_4 \leq a_1 r^3$. Proceeding in a similar fashion, one finds that $a_{k+1} \leq a_1 r^k$.]

(b) $\sum_{k=1}^{\infty} a_k \leq \sum_{k=0}^{\infty} a_1 r^k = \dfrac{a_1}{1-r}$. This implies that the given series converges.

(c) $a_{n+k} \leq a_{n+1} r^{k-1}$ for all $k \geq 1$ so $R_n = \sum_{k=n+1}^{\infty} a_k \leq a_{n+1} \sum_{k=0}^{\infty} r^k = \dfrac{a_{n+1}}{1-r}$.

71. There does not exist a number $r < 1$ such that $\dfrac{a_{k+1}}{a_k} \leq r$ for all $k \geq 1$.

72. Let $a_n = n^2/2^n$. Since $a_{n+1}/a_n = \dfrac{(n+1)^2}{2n^2} \leq 8/9$ when $n \geq 3$,

$$R_N = \sum_{n=N+1}^{\infty} a_n \leq a_{N+1} + \tfrac{8}{9} a_{N+1} + \left(\tfrac{8}{9}\right)^2 a_{N+1} + \cdots = a_{N+1} \sum_{n=0}^{\infty} \left(\tfrac{8}{9}\right)^n = 9 a_{N+1}.$$

Thus, $R_N < 0.0005$ when $N \geq 23$.

73. Let $a_n = (n!)^2/(2n)!$. Since $a_{n+1}/a_n = \dfrac{n+1}{2(2n+1)} \leq \tfrac{1}{3}$ when $n \geq 1$, $R_N = \sum_{n=N+1}^{\infty} a_n \leq a_{N+1} \sum_{n=0}^{\infty} \left(\tfrac{1}{3}\right)^n = \tfrac{3}{2} a_{N+1}$.

Thus, $R_N < 0.0005$ when $N \geq 6$.

74. (a) The following inequalities are apparent from the figures on pp. 241–243:

$$\int_1^{n+1} a(x)\,dx < \sum_{k=1}^{n} a_k < a_1 + \int_1^{n} a(x)\,dx.$$

Taking $a(x) = 1/x$, these inequalities imply that $\ln(n+1) < H_n < 1 + \ln n$.

(b) Since $H_N < 1 + \ln N$, $H_N > 10 \implies 1 + \ln N > 10$. From this it follows that $N \geq 8104$.

(c) Using the inequalities derived in part (a), $\dfrac{\ln(n+1)}{\ln n} < \dfrac{H_n}{\ln n} < \dfrac{1 + \ln n}{\ln n}$. Since $\lim_{n \to \infty} \dfrac{\ln(n+1)}{\ln n} = 1$ and $\lim_{n \to \infty} \dfrac{1 + \ln n}{\ln n} = 1$, $\lim_{n \to \infty} \dfrac{H_n}{\ln n} = 1$ (Theorem 2).

(d) First, observe that $a_n - a_{n+1} = (H_n - \ln n) - (H_{n+1} - \ln(n+1)) = \ln(n+1) - \ln n - \dfrac{1}{n+1}$. Then, note that

$$\int_n^{n+1} x^{-1}\,dx = \ln(n+1) - \ln n > \dfrac{1}{n+1}$$ since x^{-1} is a decreasing function on the interval $[n, n+1]$.

Thus, $a_n - a_{n+1} > 0$.

(e) The sequence a_n is decreasing and bounded below by 0 (since $H_n - \ln n > \ln(n+1) - \ln n > 0$). Thus, it converges (Theorem 3).

75. (a) $\int_x^{\infty} f'(t)\,dt = \lim_{a \to \infty} \int_x^a f'(t)\,dt = \lim_{a \to \infty} (f(a) - f(x)) = -f(x)$.

[NOTE: $f(a) = \ln\left(\dfrac{a+1}{a}\right) - \dfrac{1}{a+1} = \ln\left(1 + \dfrac{1}{a}\right) - \dfrac{1}{a+1}$.]

(b) When $x > 0$, $f(x) = -\int_x^{\infty} f'(t)\,dt > \int_x^{\infty} \dfrac{dt}{(t+1)^3} = \dfrac{1}{2(x+1)^2}$.

(c) Let $S_N = \sum_{k=n}^{N} (a_k - a_{k+1}) = (a_n - a_{n+1}) + (a_{n+1} - a_{n+2}) + \cdots + (a_N - a_{N+1}) = a_n - a_{N+1}$.

Since $\gamma = \lim_{n \to \infty} a_n$, $\sum_{k=n}^{\infty} (a_k - a_{k+1}) = \lim_{N \to \infty} S_N = a_n - \lim_{N \to \infty} a_{N+1} = a_n - \gamma$.

(d) Since f is a decreasing function, the integral test implies that $\int_n^\infty f(x)\,dx \le \sum_{k=n}^\infty f(k)$. Therefore, part (b) implies that $\int_n^\infty f(x)\,dx > \dfrac{1}{2(n+1)}$.

To get the upper bound on $a_n - \gamma$, note that $f(k) < \dfrac{1}{2}\left(\dfrac{1}{k} - \dfrac{1}{k+1}\right)$. (Apply the trapezoid rule to $\int_k^{k+1} dx/x$.) This inequality implies that $\sum_{k=n}^\infty f(k) < \sum_{k=n}^\infty \dfrac{1}{2}\left(\dfrac{1}{k} - \dfrac{1}{k+1}\right) = \dfrac{1}{2n}$.

NOTE: An argument similar to that used in part (b) shows that $f(x) < 1/2x^2$. Therefore, the integral test implies that $\sum_{k=n}^\infty f(k) \le \int_{n-1}^\infty f(x)\,dx < \int_{n-1}^\infty \dfrac{1}{2x^2}\,dx = \dfrac{1}{2(n-1)}$ which is not quite the desired result.

The lower bound on $a_n - \gamma$ can also be derived in the following way: Applying the midpoint rule to $\int_k^{k+1} dx/x$ yields $f(n) > \dfrac{1}{n+\frac{1}{2}} - \dfrac{1}{n+1} = \dfrac{1}{2n^2+3n+1} > \dfrac{1}{2(n^2+3n+2)} = \dfrac{1}{2}\left(\dfrac{1}{n+1} - \dfrac{1}{n+2}\right)$.

Therefore, $\sum_{k=n}^\infty f(k) > \sum_{k=n}^\infty \dfrac{1}{2}\left(\dfrac{1}{k+1} - \dfrac{1}{k+2}\right) = \dfrac{1}{2(n+1)}$.

76. (a) The sum of k numbers is less than or equal to k times the smallest summand; similarly, it is greater than or eqaul to k times the largest summand. Thus, since the sequence $\{a_n\}$ is decreasing,

$$2^{m-1}a_{2^m} \le a_{2^{m-1}+1} + a_{2^{m-1}+2} + \cdots + a_{2^m} \le 2^{m-1}a_{2^{m-1}+1} \le 2^{m-1}a_{2^{m-1}}.$$

(b) $\displaystyle\sum_{k=2}^{2^m} a_k = \sum_{k=1}^{m}\left(\sum_{j=2^{k-1}+1}^{2^k} a_j\right) \le \sum_{k=1}^{m} 2^{k-1}a_{2^k} = \dfrac{1}{2}\sum_{k=1}^{m} 2^k a_{2^k}$

(c) Since the terms of the series are positive, the sequence of partial sums is increasing. Therefore, if $\sum_{k=1}^\infty 2^k a_{2^k}$ diverges, the sequence of its partial sums is unbounded. Since the partial sums of this series are lower bounds for the partial sums of the series $\sum_{k=1}^\infty a_k$, the latter series must also diverge.

(d) Since the terms of the series are positive,

$$\sum_{k=2}^{n} a_k \le \sum_{k=2}^{2^m} a_k = \sum_{k=1}^{m}\left(\sum_{j=2^{k-1}+1}^{2^k} a_j\right) \le \sum_{k=1}^{m} 2^{k-1} a_{2^{k-1}}.$$

(e) Suppose that $\sum_{k=0}^\infty 2^k a_{2^k}$ converges. Part (d) implies that the partial sums of the series $\sum_{k=1}^\infty a_k$ are bounded above by $\sum_{k=0}^\infty 2^k a_{2^k}$. Furthermore, since $a_k > 0$ for all $k \ge 1$, the partial sums of the series $\sum_{k=1}^\infty a_k$ form an increasing sequence. Together, these results imply that the series $\sum_{k=1}^\infty a_k$ converges (the sequence of its partial sums is increasing and bounded above).

77. Let $a_k = 1/k$. Then $\sum_{k=1}^\infty 2^k a_{2^k} = \sum_{k=1}^\infty 1$ which is obviously a divergent series. Therefore, part (c) of the previous exercise implies that $\sum_{k=1}^\infty a_k$ diverges.

78. (a) Let $a_k = 1/k^p$. According to Exercise 76, the series $\sum_{k=1}^{\infty} a_k$ converges if and only if the series

$$\sum_{k=1}^{\infty} 2^k a_{2^k} = \sum_{k=0}^{\infty} \frac{2^k}{2^{kp}} = \sum_{k=0}^{\infty} \left(\frac{1}{2^{p-1}}\right)^k = \sum_{k=0}^{\infty} \left(2^{1-p}\right)^k$$

converges. Since the series on the right is a geometric series, it converges only if $2^{1-p} < 1$.

(b) Let $a_k = \frac{1}{k(\ln k)^p}$. The series $\sum_{k=2}^{\infty} a_k$ converges if and only if the series

$$\sum_{k=1}^{\infty} 2^k a_{2^k} = \sum_{k=1}^{\infty} \frac{2^k}{2^k \left(\ln 2^k\right)^p} = \sum_{k=1}^{\infty} \frac{1}{k^p (\ln 2)^p} = \frac{1}{(\ln 2)^p} \sum_{k=1}^{\infty} \frac{1}{k^p}$$

converges. By part (a), this series converges only when $p > 1$.

11.4 Absolute Convergence; Alternating Series

1. (a) The series converges conditionally. (After the fifth term, the series has the same terms as the alternating harmonic series.)

   (b) $S_{15} = 1 + 2 + 3 + 4 + 5 + \sum_{k=6}^{15} \frac{(-1)^{k+1}}{k} = 15 - \frac{20887}{360360} \approx 14.942$. S_{15} *overestimates* S because the last term included in the alternating series was positive.

   (c) $14.902 < S < 14.902 + \frac{1}{61} \approx 14.918$

   (d) $S = 15 + \left(\ln 2 - \sum_{k=1}^{5} \frac{(-1)^{k+1}}{k} \right) = 15 + \ln 2 - \frac{47}{60} \approx 14.910$

2. (a) $S_{50} \approx 0.23794$

   (b) $|R_{50}| = \left| \sum_{n=51}^{\infty} \frac{\sin n}{n^3 + n^2 + n + 1 + \cos n} \right| \leq \sum_{n=51}^{\infty} \left| \frac{\sin n}{n^3 + n^2 + n + 1 + \cos n} \right| \leq \sum_{n=51}^{\infty} \frac{1}{n^3 + n^2 + n + 1 + \cos n}$
   $\leq \sum_{n=51}^{\infty} \frac{1}{n^3} \leq \int_{50}^{\infty} \frac{dx}{x^3}$

   (The last step follows from the integral test.)

   (c) By part (b), $|R_{50}| \leq \frac{1}{5000} = 0.0002$. Therefore, using the result in part (a), $S_{50} - 0.0002 \approx 0.23774 < S < S_{50} + 0.0002 \approx 0.23814$.

3. (a) $0 < \sum_{k=1}^{\infty} \frac{|a_k|}{k} < \sum_{k=1}^{\infty} |a_k|$ so the series $\sum_{k=1}^{\infty} \frac{a_k}{k}$ converges absolutely. This implies (by Theorem 9) that this series converges.

   (b) No — An example is $a_k = 1/k$. Then, $\sum_{k=1}^{\infty} a_k/k = \sum_{k=1}^{\infty} 1/k^2$ (a convergent series), but $\sum_{k=1}^{\infty} a_k = \sum_{k=1}^{\infty} 1/k$ (the harmonic series).

4. Let $a_k = k/(k^2 - 1)$. Since $\lim_{k \to \infty} a_k = 0$ and $a_{k+1} < a_k$ for all $k \geq 2$, Theorem 10 (p. 255) implies that $\sum_{k=2}^{\infty} (-1)^k a_k$ converges. However, $\sum_{k=2}^{\infty} a_k$ diverges ($a_k \geq 1/k$), so the alternating series converges conditionally.

5. The series converges absolutely by the alternating series test. Since $c_{n+1} = (n+1)^{-4} < 0.005$ when $n = 3$, $\left| S - \sum_{k=1}^{N} (-1)^k/k^4 \right| < 0.005$ when $N \geq 3$. Using $N = 3$, $S \approx -\frac{1231}{1296} \approx -0.94985$.

6. The series converges absolutely by the comparison test using $b_k = 2^{-k}$. $\left| S - \sum_{k=1}^{N} a_k \right| \leq 0.005$ when $N \geq 7$ since $1/(8^2 + 2^8) < 0.005$. Using $N = 7$, $S \approx -\frac{11393057}{45736800} \approx -0.24910$.

7. The series converges absolutely by the comparison test using $b_k = (2/7)^k$. $\left| S - \sum_{k=0}^{N} a_k \right| \leq 0.005$ when $N \geq 4$ since $2^5/(7^5 + 5) < 0.005$. Using $N = 4$, $S \approx \frac{68917177}{84877260} \approx 0.81196$.

8. The series converges absolutely by the ratio test. $\left| S - \sum_{k=0}^{N} a_k \right| \leq 0.005$ when $N \geq 2$ since $3^3/9! < 0.005$. Using $N = 2$, $S \approx -\frac{13}{8} = -1.625$.

9. The series converges absolutely by the ratio test. $\left|S - \sum_{k=5}^{N} a_k\right| \leq 0.005$ when $N \geq 13$ since $14^{10}/10^{14} < 0.005$. Using $N = 13$, $S \approx -\dfrac{573982077919709}{10000000000000} \approx -57.398$.

10. The series converges absolutely by the comparison test: $\sum_{k=0}^{\infty} \dfrac{1}{(k+1)2^k} < \sum_{k=0}^{\infty} \dfrac{1}{2^k} = 2$. $\left|S - \sum_{k=0}^{N} a_k\right| \leq 0.005$ when $N \geq 5$ since $1/(7 \cdot 2^6) = 1/448 < 0.005$. Using $N = 5$, $S \approx \dfrac{259}{320} = 0.809375$.

11. No. Because $\lim\limits_{n \to \infty} (-1)^n \dfrac{n}{2n-1}$ does not exist, the series cannot converge (by the n-th term test).

12. (a) $0 < a_k < 1/k \Longrightarrow \lim\limits_{k \to \infty} a_k = 0$

   (b) No, the series diverges. Since $a_k = \displaystyle\int_k^{\infty} \dfrac{dx}{2x^2 - 1} \geq \int_k^{\infty} \dfrac{dx}{2x^2} = \dfrac{1}{2k}$, $\displaystyle\sum_{k=1}^{\infty} a_k \geq \dfrac{1}{2} \sum_{k=1}^{\infty} \dfrac{1}{k} = \infty$.

   (c) $\displaystyle\sum_{k=1}^{\infty} (-1)^{k+1} a_k$ converges by the alternating series test — the terms of the series alternate in sign and are decreasing in magnitude (i.e., $a_{k+1} < a_k$).

13. (a) When $p \leq 0$, $\lim\limits_{k \to \infty} \dfrac{\ln k}{k^p} = \infty$, so the series diverges by the nth term test (Theorem 5, p. 230).

   When $p > 0$, the function $f(x) = \dfrac{\ln x}{x^p}$ is continuous, positive, and decreasing on $(e^{1/p}, \infty)$ Therefore, the integral test (Theorem 7, p. 243) implies that the series converges if and only if the improper integral $\displaystyle\int_{e^{1/p}}^{\infty} f(x)\, dx$ converges.

   When $p \neq 1$, $\displaystyle\int \dfrac{\ln x}{x^p}\, dx = \dfrac{(1-p) \ln x - 1}{(1-p)^2 x^{p-1}}$ and, $\displaystyle\int \dfrac{\ln x}{x}\, dx = \tfrac{1}{2}(\ln x)^2$. Therefore, $\displaystyle\int_{e^{1/p}}^{\infty} \dfrac{\ln x}{x^p}\, dx$ diverges when $p \leq 1$ and converges when $p > 1$. It follows that the series converges only when $p > 1$.

   (b) The alternating series test (Theorem 10, p. 255) implies that the series converges for every $p > 0$. (When $p > 0$, the terms of the series decrease in magnitude for $k > e^{1/p}$, approach zero, and alternate in sign.)

   (c) The series converges absolutely when $p > 1$ since $\displaystyle\sum_{k=2}^{\infty} \dfrac{\ln k}{k^p}$ converges only when $p > 1$.

   (d) When $0 < p \leq 1$, $\displaystyle\sum_{k=2}^{\infty} \dfrac{\ln k}{k^p}$ diverges but $\displaystyle\sum_{k=2}^{\infty} (-1)^k \dfrac{\ln k}{k^p}$ converges. Therefore, the series converges conditionally when $0 < p \leq 1$.

14. converges conditionally — $\displaystyle\sum_{k=1}^{\infty} \dfrac{1}{\sqrt{k}}$ is a divergent p-series ($p = 1/2$) but the terms of the series form a decreasing sequence and $\lim\limits_{k \to \infty} \dfrac{1}{\sqrt{k}} = 0$. $-1 < \displaystyle\sum_{k=1}^{\infty} \dfrac{(-1)^k}{\sqrt{k}} < -1 + \dfrac{1}{\sqrt{2}}$

15. converges absolutely — $\displaystyle\sum_{j=1}^{\infty} \dfrac{1}{j^2}$ is a convergent p-series ($p = 2$). $\dfrac{3}{4} < \displaystyle\sum_{j=1}^{\infty} \dfrac{(-1)^{j+1}}{j^2} < 1$

16. diverges — nth term test: $\lim\limits_{n \to \infty} \dfrac{(-3)^n}{n^3}$ does not exist

SECTION 11.4 ABSOLUTE CONVERGENCE; ALTERNATING SERIES

17. converges conditionally—$\sum_{k=4}^{\infty} \frac{\ln k}{k}$ diverges by the integral test but the terms of the series form a decreasing sequence and $\lim_{k \to \infty} \frac{\ln k}{k} = 0$. $\frac{\ln 4}{4} - \frac{\ln 5}{5} < \sum_{k=4}^{\infty} (-1)^k \frac{\ln k}{k} < \frac{\ln 4}{4}$

18. converges absolutely—$\sum_{m=8}^{\infty} \left| \frac{\sin m}{m^3} \right| < \sum_{m=8}^{\infty} \frac{1}{m^3}$. Since $|\sin x| \le 1$ for all x and $\sum_{m=8}^{\infty} \frac{1}{m^3} \le \frac{1}{8^3} + \int_8^{\infty} \frac{dx}{x^3} = \frac{5}{512}$, $-\frac{5}{512} \le \sum_{m=8}^{\infty} \frac{\sin m}{m^3} \le \frac{5}{512}$.

19. converges conditionally—$\sum_{n=1}^{\infty} \frac{\cos(n\pi)}{n} = \sum_{n=1}^{\infty} \frac{(-1)^n}{n}$ which is (almost) the alternating harmonic series. $-1 < \sum_{n=1}^{\infty} \frac{\cos(n\pi)}{n} = -\ln 2 < -1/2$

20. diverges—nth term test: $\lim_{k \to \infty} \frac{k}{2k+1} = \frac{1}{2} \ne 0$

21. converges absolutely—Let $a_m = 4m^3/2^m$. Then $\lim_{m \to \infty} \frac{a_{m+1}}{a_m} = \lim_{m \to \infty} \frac{(m+1)^3}{2m^3} = \frac{1}{2} < 1$ so $\sum_{m=0}^{\infty} \frac{4m^3}{2^m}$ converges by the ratio test. The terms of the series are decreasing in absolute value for all $m \ge 4$. Thus, $\sum_{m=0}^{5} (-1)^m a_m = -\frac{57}{8} < \sum_{m=0}^{\infty} (-1)^m \frac{4m^3}{2^m} < \sum_{m=0}^{4} (-1)^m a_m = \frac{17}{2}$

22. converges absolutely—$\sum_{k=0}^{\infty} \frac{2^k}{3^k + k} \le \sum_{k=0}^{\infty} \left(\frac{2}{3} \right)^k = 3$. $\frac{1}{2} < \sum_{k=0}^{\infty} \frac{(-2)^k}{3^k + k} < 1$

23. converges absolutely—$\lim_{j \to \infty} \frac{a_{j+1}}{a_j} = \lim_{j \to \infty} \frac{j+1}{(j^2 + 2j + 1) \cdots (j^2 + 1)} = 0$. The terms of the series are decreasing in absolute value for all $j \ge 1$. Thus, $\sum_{j=0}^{1} (-1)^j a_j = 0 < \sum_{j=0}^{\infty} (-1)^j a_j < \sum_{j=0}^{2} (-1)^j a_j = \frac{1}{12}$.

24. converges conditionally—$\sum_{n=1}^{\infty} \frac{\arctan n}{n} > \frac{\pi}{4} \sum_{n=1}^{\infty} \frac{1}{n}$. $\frac{\pi}{4} - \frac{\arctan 2}{2} < \sum_{n=1}^{\infty} (-1)^{n+1} \frac{\arctan n}{n} < \frac{\pi}{4}$

25. Let $b_k = a_{k+10^9}$. The alternating series test can be used to show that $\sum_{k=1}^{\infty} (-1)^{k+1} b_k$ converges. Since
$\sum_{k=1}^{\infty} (-1)^{k+1} a_k = \sum_{k=1}^{10^9} (-1)^{k+1} a_k + \sum_{k=10^9+1}^{\infty} (-1)^{k+1} a_k = \sum_{k=1}^{10^9} (-1)^{k+1} a_k + \sum_{k=1}^{\infty} (-1)^{k+1} b_k$, the series $\sum_{k=1}^{\infty} (-1)^{k+1} a_k$ converges.

26. The series converges absolutely because $1 + \frac{1}{2^3} + \frac{1}{3^2} + \frac{1}{4^3} + \frac{1}{5^2} + \frac{1}{6^3} + \frac{1}{7^2} + \frac{1}{8^3} + \cdots < \sum_{k=1}^{\infty} \frac{1}{k^2}$.

    NOTE: Theorem 10 can't be used to show that the given series converges because the terms are *not* decreasing in magnitude.

27. (a) Theorem 9 (p. 253) implies that $\sum_{j=1}^{\infty} (-1)^{j+1} b_j$ converges absolutely.

(b) Since the terms of the series defining S are all positive, $S - \sum_{j=1}^{100} b_j \leq 0.005$ implies that $b_j \leq 0.005$ for all $j \geq 101$. Therefore, since $b_{j+1} \leq b_j$, Theorem 10 implies the desired result.

Alternatively,
$$\left| \sum_{j=1}^{\infty} (-1)^{j+1} b_j - \sum_{j=1}^{100} (-1)^{j+1} b_j \right| = \left| \sum_{j=101}^{\infty} (-1)^{j+1} b_j \right| \leq \sum_{j=101}^{\infty} b_j \leq 0.005.$$

28. $a_k = (-1)^k / \sqrt{k}$

29. No. Since $a_k \geq 0$ for all $k \geq 1$, $|a_k| = a_k$ for all $k \geq 1$.

30. No. This would contradict Theorem 9 since $|a_k| = a_k$.

31. (a) Let $a_k = (-1)^{k+1}/k^2$ and $b_k = 1/k^2$. Then $a_{2m-1} = b_{2m-1}$ and $a_{2m} < b_{2m}$ for $m = 1, 2, 3, \ldots$. Thus,
$$\sum_{k=n+1}^{\infty} a_k \leq \sum_{k=n+1}^{\infty} b_k.$$

(b) According to part (a), for each n the tail of the alternating series is smaller than the tail of the series of positive terms. Thus, the error made by approximating the alternating series by its nth partial sum is less than the error made by approximating the series with positive terms by its nth partial sum.

32. (a) For $n = 1, 2, 3, \ldots$, $S_{2n+2} = S_{2n} + c_{2n+1} - c_{2n+2} \geq S_{2n}$ since $c_{2n+1} - c_{2n} \geq 0$.

(b) For $n = 1, 2, 3, \ldots$, $S_{2n+1} = S_{2n-1} - c_{2n} + c_{2n+1} \leq S_{2n-1}$ since $c_{2n+1} - c_{2n} \leq 0$.

(c) $S_{2m} = S_{2m-1} - c_{2m} \implies S_{2m} \leq S_{2m+1}$ since $c_{2m+1} \geq 0$.

(d) It follows from parts (a)–(c) that $S_2 \leq S_{2m} \leq S_{2m-1} \leq S_1$ for $m = 1, 2, 3, \ldots$. Thus, because the sequence of even partial sums is monotone increasing and bounded above by S_1, it converges. Similarly, the sequence of odd partial sums converges because it is monotone decreasing and bounded below by S_2.

(e) $\lim_{m \to \infty} (S_{2m+1} - S_{2m}) = \lim_{m \to \infty} c_{2m+1} = 0$. This implies that the limit of the sequence of even partial sums is the same as the limit of the sequence of odd partial sums.

(f) Since the sequence of even partial sums is monotone increasing, $0 < S - S_{2m}$ is true. Also, since the sequence of odd partial sums is monotone decreasing $S < S_{2m+1} = S_{2m} + c_{2m+1} \implies S - S_{2m} < c_{2m+1}$. Thus, $0 < S - S_{2m} < c_{2m+1}$.

Similarly, since the sequence of odd partial sums is monotone decreasing, $0 < S_{2m+1} - S$. Also, since the sequence of even partial sums is monotone increasing
$S > S_{2m+2} = S_{2m+1} - c_{2m+2} \implies S - S_{2m+1} > -c_{2m+2}$. Thus, $0 < S_{2m+1} - S < c_{2m+2}$.

33. (b) Because of the $\cot k$ term, there is not a value of N such that the ratio is less than 1 for all $k > N$.

(c) The series converges absolutely since $|a_k| \leq 1/k^2$ and $\sum_{k=1}^{\infty} \frac{1}{k^2}$ converges.

11.5 Power Series

1. $P_1(x) = x$, $P_2(x) = P_1(x) + x^2/2$, $P_4(x) = P_2(x) + x^3/3 + x^4/4$, $P_6(x) = P_4(x) + x^5/5 + x^6/6$, $P_8(x) = P_6(x) + x^7/7 + x^8/8$, $P_{10}(x) = P_8(x) + x^9/9 + x^{10}/10$.

2. $\lim_{j \to \infty} \left| \frac{(x/2)^{j+1}}{(x/2)^j} \right| = \lim_{j \to \infty} |x|/2 < 1$ when $|x| < 2$. Thus, the radius of convergence is $R = 2$.

3. $\left| \frac{a_{k+1}}{a_k} \right| = \frac{k}{2(k+1)} \cdot |x| \implies R = 2$

4. $\lim_{k \to \infty} \left| \frac{\frac{x^{k+1}}{\sqrt{k+1}}}{\frac{x^k}{\sqrt{k}}} \right| = \lim_{k \to \infty} \frac{\sqrt{k}}{\sqrt{k+1}} \cdot |x| < 1$ when $|x| < 1$. Thus, the radius of convergence is $R = 1$.

5. $\left| \frac{a_{m+1}}{a_m} \right| = \frac{m^2+1}{(m+1)^2+1} \cdot |x| \implies R = 1$

6. $\left| \frac{a_{n+1}}{a_n} \right| = \left(\frac{n+1}{n} \right)^n \cdot (n+1) \cdot |x| \implies R = 0$

7. $\left| \frac{x^n}{n!+n} \right| \leq \frac{|x|^n}{n!} \implies R = \infty$

8. $R = 1/3$; interval of convergence is $(-1/3, 1/3)$

9. $R = \infty$; interval of convergence is $(-\infty, \infty)$

10. $R = 1/3$; interval of convergence is $[-1/3, 1/3)$

11. $R = 1/3$; interval of convergence is $[-1/3, 1/3]$

12. $\left| \frac{a_{n+1}}{a_n} \right| = |x - 2| \implies R = 1$; interval of convergence is $(1, 3)$

13. $R = 1$; interval of convergence is $[2, 4]$

14. $R = 1$; interval of convergence is $[4, 6)$

15. $R = 1$; interval of convergence is $[-2, 0)$

16. For each series in this problem, the ratio test can be used to prove that the series converges when $-R < x < R$.

    (a) The nth term test can be used to prove that the series diverges when $|x| > R$.

    (b) The alternating series test can be used to show that the series converges when $x = -R$. When $x = R$ the series becomes the harmonic series (i.e., it diverges).

    (c) When $x = R$ the series becomes $\sum_{k=1}^{\infty} \frac{1}{k^2}$ which converges absolutely.

    (d) $\sum_{k=1}^{\infty} \frac{(-x)^k}{kR^k} = \sum_{k=1}^{\infty} (-1)^k \frac{x^k}{kR^k}$

17. (a) $\sum_{k=1}^{\infty} \frac{x^k}{k4^k}$

    (b) $\sum_{k=1}^{\infty} \frac{(x-2)^k}{k^2 3^k}$

(c) $\sum_{k=1}^{\infty} \frac{(x+2)^k}{2^k}$

(d) $\sum_{k=1}^{\infty} \frac{(12-x)^k}{k4^k}$

(e) $\sum_{k=1}^{\infty} \frac{(x+7)^k}{k4^k}$

19. (a) By definition, the radius of convergence of a power series is the largest value of R such that the series converges for all x such that $|x| < R$.

    (b) This power series converges when $|x - 1| < 2$ and diverges when $|x - 1| > 2$. Thus, its radius of convergence $R = 2$.

    (c) Let $z = x - 3$. Since $\sum_{k=0}^{\infty} z^k$ converges only when $-2 < z \le 2$, $\sum_{k=0}^{\infty} a_k(x-3)^k$ converges only when $1 < x \le 5$.

    (d) The power series $\sum_{k=0}^{\infty} a_k(x+1)^k$ converges only when $-2 < x + 1 \le 2$. Thus, its interval of convergence is $(-3, 1]$.

20. (a) $R = 14$

    (b) $b = 3$

21. $[1, 5)$

22. $(-\infty, \infty)$

23. $[-3, 5)$

24. $[0, 2)$

25. $[-6, -4]$

26. $[1/2, 3/2)$

27. The information given implies that the power series converges on the interval $[-3, 3)$, diverges when $x \ge 7$, and diverges when $x < -7$. It does not imply anything about convergence or divergence on the intervals $[-7, -3)$ and $[3, 7)$.

    (a) cannot

    (b) may

    (c) may

    (d) cannot

    (e) may

    (f) may

28. The information given implies that the power series converges on the interval $[-7, 3)$, diverges when $x \ge 7$, and diverges when $x < -11$. It does not imply anything about convergence or divergence on the intervals $[-11, -7)$ and $[3, 7)$.

    (a) may

    (b) must

SECTION 11.5 POWER SERIES 133

   (c) may

   (d) may

   (e) may

   (f) may

   (g) cannot

29. (a) Cannot be true. The interval of convergence of a power series is symmetric around and includes its base point ($b = 1$ in this case).

   (b) May be true. (The statement is true when $a_k = 1/k!$ but it is false when $a_k = 1$.)

   (c) Must be true. If the radius of convergence of the power series is 3, then the interval of convergence includes all values of x such that $|x - 1| < 3$.

   (d) Cannot be true. The interval of convergence of this power series must be symmetric about the point $b = 1$.

   (e) Cannot be true. The interval of convergence of the power series is the solution set of the inequality $|x - 1| < 8$. Thus, the radius of convergence of the power series is 8.

30. (a) $S_{10} = \sum_{k=0}^{10} \frac{(-1)^k}{k!} = \frac{16481}{44800}$. $S_{10} - 1/e \approx 2.31 \times 10^{-8}$.

   (b) The error bound associated with the alternating series test asserts that $|S_{10} - 1/e| < \frac{1}{11!} \approx 2.5052 \times 10^{-8}$. A similar computation shows that $|S_n - 1/e| < 10^{-10}$ when $n \geq 13$.

31. (a) The series $\sum_{n=0}^{\infty} \frac{2 \cdot 10^n}{3^n + 5}$ diverges by the ratio test—the limit of the ratio of successive terms of the series is $10/3 > 1$.

   (b) The power series defining f converges when $-3 < x < 3$. Thus, only 0.5 and 1.5 are in the domain of f.

   (c) $f(1) - \sum_{n=0}^{N} \frac{2}{3^n + 5} = \sum_{n=N+1}^{\infty} \frac{2}{3^n + 5} < \sum_{n=N+1}^{\infty} \frac{2}{3^n} = \frac{1}{3^{N+1}} \sum_{n=0}^{\infty} \frac{2}{3^n} = \frac{1}{3^{N+1}} \cdot 3 = \frac{1}{3^N}$. Thus, since $3^{-5} < 0.01$, $\sum_{n=0}^{5} \frac{2}{3^n + 5} = \frac{367273}{447888} \approx 0.82001$ approximates $f(1)$ within 0.01.

32. (a) The series defining h converges for all values of x.

   (b) $h(0) = \sum_{k=0}^{\infty} \frac{(-2)^k}{k! + k^3}$. Since this is an alternating series whose terms decrease in magnitude for all $k \geq 1$ and since $\frac{2^9}{9! + 9^3} < 0.005$, $h(0) \approx \sum_{k=0}^{8} \frac{(-2)^k}{k! + k^3} = \frac{1825808722}{7031839815} \approx 0.25965$.

   (c) $h(3) = \sum_{k=0}^{\infty} \frac{1}{k! + k^3}$. Since $\frac{1}{k! + k^3} < \frac{1}{k!}$ for all $k \geq 1$,

$$\sum_{k=6}^{\infty} \frac{1}{k! + k^3} < \sum_{k=6}^{\infty} \frac{1}{k!} = e - \frac{163}{60} \approx 0.0016152 < 0.005.$$

   Therefore, $h(3) \approx \sum_{k=0}^{5} \frac{1}{k! + k^3} = \frac{9677}{5880} \approx 1.6457$.

33. (a) The domain of g is $[-9, 1]$.

(b) $g(0) - \sum_{n=1}^{N} \dfrac{4^n}{n^3 5^n} = \sum_{n=N+1}^{\infty} \dfrac{4^n}{n^3 5^n} < \int_{N}^{\infty} \dfrac{dx}{x^3} = \dfrac{1}{2N^2} \le 0.005$ when $N \ge 10$. Thus,

$$g(0) \approx \sum_{n=0}^{10} \dfrac{4^n}{n^3 5^n} = \dfrac{277892997449134}{305233154296875} \approx 0.91043.$$

(c) The approximation $g(-5) \approx -\dfrac{1}{5}$ is correct within 0.005 because the series defining $g(-5)$ is an alternating series. Since the magnitude of the second term in the series is 0.005, the error made by approximating the series by its first term is smaller than 0.005.

34. (a) $\lim_{x \to 1^-} \sum_{k=0}^{\infty} (-1)^k x^k = \lim_{x \to 1^-} \dfrac{1}{1+x} = \dfrac{1}{2}$

(b) An infinite series converges only if the sequence defined by its partial sums converges. Since the partial sums $\sum_{k=0}^{N} (-1)^k$ are alternately 1 and 0, the infinite series does not converge.

11.6 Power Series as Functions

1. (a) Since $\left|\frac{a_{k+1}}{a_k}\right| = \frac{|x|}{2}$, the radius of convergence is 2.

   (b) Since $\left|\frac{a_{k+1}}{a_k}\right| = \frac{(k+1)|x|}{2k}$, the radius of convergence is 2.

   (c) Since $\left|\frac{a_{k+1}}{a_k}\right| = \frac{(k+1)|x|}{2(k+2)}$, the radius of convergence is 2.

2. $f(x) = \frac{x^2}{1+x} = x^2 \sum_{k=0}^{\infty} (-1)^k x^k = \sum_{k=0}^{\infty} (-1)^k x^{k+2}$ [Substitute $u = -x$ into the power series representation of $(1-u)^{-1}$.]

3. $f(x) = \left(1-x^2\right)^{-1} = \sum_{k=0}^{\infty} x^{2k}$ [Substitute $u = x^2$ into the power series representation of $(1-u)^{-1}$.]

4. $f(x) = (1+x)^{-2} = -\frac{d}{dx}\left((1+x)^{-1}\right) = -\frac{d}{dx}\left(\sum_{k=0}^{\infty}(-x)^k\right) = \sum_{k=0}^{\infty} k(-x)^{k-1}$

5. $f(x) = \frac{x}{1-x^4} = x \sum_{k=0}^{\infty} x^{4k} = \sum_{k=0}^{\infty} x^{4k+1}$

6. $\arctan(2x) = \sum_{k=0}^{\infty} (-1)^k \frac{(2x)^{2k+1}}{2k+1}$

7. $\cos(x^2) = \sum_{k=0}^{\infty} (-1)^k \frac{x^{4k}}{(2k)!}$

8. $x^2 \sin x = \sum_{k=0}^{\infty} (-1)^k \frac{x^{2k+3}}{(2k+1)!}$

9. $\ln\left(1 + \sqrt[3]{x}\right) = \sum_{k=1}^{\infty} (-1)^{k+1} \frac{x^{k/3}}{k}$

10. The power series representation of $f(x) = \ln(1+x)$ converges on the interval $(-1, 1]$. The power series representation for $f'(x) = 1/(1+x)$ converges on the interval $(-1, 1)$.

11. $1/\sqrt{e} = e^{-1/2} = \sum_{k=0}^{\infty} \frac{(-1/2)^k}{k!}$. Since $1/(2^4 \cdot 4!) < 0.005$, $\sum_{k=0}^{3} \frac{(-1/2)^k}{k!} = \frac{29}{48} \approx 0.60417 \approx 1/\sqrt{e}$ within 0.005.

12. $\int_0^{0.2} xe^{-x^3} dx = \int_0^{0.2} \left(\sum_{k=0}^{\infty} \frac{(-x)^{3k+1}}{k!}\right) dx = \sum_{k=0}^{\infty} \frac{(-0.2)^{3k+2}}{(3k+2)k!}$. Since $(-0.2)^8/(8 \cdot 2!) < 10^{-5}$, $\int_0^{0.2} xe^{-x^3} dx \approx \sum_{k=0}^{1} \frac{(-0.2)^{3k+2}}{(3k+2)k!} = \frac{623}{31250} = 0.019936$ within 10^{-5}.

13. Let $f(x) = 1/(1-x) = \sum_{n=0}^{\infty} x^n$ if $|x| < 1$. Then $f'(x) = 1/(1-x)^2 = \sum_{n=1}^{\infty} nx^{n-1} = \frac{1}{x}\sum_{n=1}^{\infty} nx^n$ if $|x| < 1$ and $x \neq 0$. Therefore, $\sum_{n=1}^{\infty} \frac{n}{2^n} = (1/2)f'(1/2) = 2$.

14. $(\sin x - x)^3 = \left(\sum_{k=1}^{\infty}(-1)^k \frac{x^{2k+1}}{(2k+1)!}\right)^3 = \left(-\frac{x^3}{3!} + \frac{x^5}{5!} \pm \cdots\right)^3 = -\frac{x^9}{(3!)^3} + \frac{x^{11}}{1440} \mp \cdots$

$(1 - \cos x)^4 = \left(\sum_{k=1}^{\infty}(-1)^k \frac{x^{2k}}{(2k)!}\right)^4 = \left(-\frac{x^2}{2!} + \frac{x^4}{4!} \mp \cdots\right)^4 = \frac{x^8}{2^4} - \frac{x^{10}}{48} \pm \cdots.$

Therefore,

$$\lim_{x \to 0} \frac{(\sin x - x)^3}{x(1-\cos x)^4} = \lim_{x \to 0} \frac{-x^9/216 + x^{11}/1440 \mp \cdots}{x \cdot (x^8/16 - x^{10}/48 \pm \cdots)} = \lim_{x \to 0} \frac{-1/216 + x^2/1440 \mp \cdots}{1/16 - x^2/48 \pm \cdots} - \frac{2}{27}.$$

15. $x - \sin x = \sum_{k=1}^{\infty}(-1)^{k+1}\frac{x^{2k+1}}{(2k+1)!} = \frac{x^3}{3!} - \frac{x^5}{5!} \pm \cdots = x^3\left(\frac{1}{3!} - \frac{x^2}{5!} \pm \cdots\right)$

$(x \sin x)^{3/2} = \left(\sum_{k=0}^{\infty}(-1)^k \frac{x^{2k+2}}{(2k+1)!}\right)^{3/2} = \left(x^2 - \frac{x^4}{3!} \pm \cdots\right)^{3/2}$

$= \left(x^6 - \frac{1}{2}x^8 \pm \cdots\right)^{1/2} = x^3\left(1 - \frac{1}{2}x^2 \pm \cdots\right)^{1/2}$

Thus,

$$\lim_{x \to 0^+} \frac{x - \sin x}{(x \sin x)^{3/2}} = \lim_{x \to 0^+} \frac{x^3\left(\frac{1}{3!} - \frac{x^2}{5!} \pm \cdots\right)}{x^3\left(1 - \frac{1}{2}x^2 \pm \cdots\right)^{1/2}} = \frac{1}{6}.$$

16. $\frac{\sin x}{x} = \sum_{k=0}^{\infty}(-1)^k \frac{x^{2k}}{(2k+1)!} = 1 - x^2/3! + x^4/5! - x^6/7! + \cdots \implies \lim_{x \to 0}\frac{\sin x}{x} = 1$

17. $\frac{e^x - 1}{x} = x^{-1}\left(\sum_{k=0}^{\infty}\frac{x^k}{k!} - 1\right) = \sum_{k=1}^{\infty}\frac{x^{k-1}}{k!} \implies \lim_{x \to 0}\frac{e^x - 1}{x} = 1$

18. $\frac{1 - \cos x}{x} = x^{-1}\left(1 - \sum_{k=0}^{\infty}(-1)^k\frac{x^{2k}}{(2k)!}\right) = \sum_{k=1}^{\infty}(-1)^{k+1}\frac{x^{2k-1}}{(2k)!} \implies \lim_{x \to 0}\frac{1 - \cos x}{x} = 0$

19. $\frac{1 - \cos x}{x} = x^{-2}\left(1 - \sum_{k=0}^{\infty}(-1)^k\frac{x^{2k}}{(2k)!}\right) = \sum_{k=0}^{\infty}(-1)^k\frac{x^{2k}}{(2k+2)!} \implies \lim_{x \to 0}\frac{1 - \cos x}{x^2} = \frac{1}{2}$

20. $\frac{\arctan x}{x} = x^{-1}\sum_{k=0}^{\infty}(-1)^k\frac{x^{2k+1}}{2k+1} = \sum_{k=0}^{\infty}(-1)^k\frac{x^{2k}}{2k+1} \implies \lim_{x \to 0}\frac{\arctan x}{x} = 1$

21. $\frac{e^x - e^{-x}}{x} = 2\sum_{k=0}^{\infty}\frac{x^{2k}}{(2k+1)!} \implies \lim_{x \to 0}\frac{e^x - e^{-x}}{x} = 2.$

22. $\frac{\ln(1+x) - x}{x^2} = \sum k = 0^{\infty}(-1)^{k+1}\frac{x^k}{k+2} \implies \lim_{x \to 0}\frac{\ln(1+x) - x}{x^2} = -\frac{1}{2}$

23. $\frac{x - \arctan x}{x^3} = \sum_{k=0}^{\infty}(-1)^k\frac{x^{2k}}{2k+3} \implies \lim_{x \to 0}\frac{x - \arctan x}{x^3} = \frac{1}{3}$

24. $\lim_{x \to 1}\frac{\ln x}{x - 1} = \lim_{w \to 0}\frac{\ln(1+w)}{w} = \lim_{w \to 0}\left(\sum_{k=0}^{\infty}(-1)^k\frac{w^k}{k+1}\right) = 1$

Section 11.6 Power Series as Functions

25. $\dfrac{1-\cos^2 x}{x} = \dfrac{\frac{1}{2} - \frac{1}{2}\cos(2x)}{x} = \sum_{k=1}^{\infty}(-1)^{k+1}\dfrac{(2x)^{2k-1}}{(2k)!} \Longrightarrow \lim_{x\to 0}\dfrac{1-\cos^2 x}{x} = 0$

26. $\dfrac{1}{2+x} = \dfrac{1}{2}\left(\dfrac{1}{1+(x/2)}\right) = \dfrac{1}{2}\sum_{k=0}^{\infty}(-1)^k\left(\dfrac{1}{2}\right)^k$

27. $\sin(\sqrt{x}) = \sum_{k=0}^{\infty}(-1)^k\dfrac{x^{(2k+1)/2}}{(2k+1)!}$

28. $\sin x + \cos x = \sum_{k=0}^{\infty}(-1)^k\left(\dfrac{x^{2k}}{(2k)!} + \dfrac{x^{2k+1}}{(2k+1)!}\right)$

29. $2^x = e^{x\ln 2} = \sum_{k=0}^{\infty}\dfrac{(x\ln 2)^k}{k!}$

30. $\ln(1+x^2) = \sum_{k=0}^{\infty}(-1)^k\dfrac{x^{2k}}{k}$

31. $(x^2-1)\sin x = (x^2-1)\sum_{k=0}^{\infty}(-1)^k\dfrac{x^{2k+1}}{(2k+1)!} = -x + \sum_{k=1}^{\infty}(-1)^{k+1}\dfrac{((2k+1)(2k)+1)x^{2k+1}}{(2k+1)!} = \sum_{k=0}^{\infty}(-1)^{k+1}\dfrac{(4k^2+2k+1)x^{2k+1}}{(2k+1)!}$

32. $\ln\left(\dfrac{1+x}{1-x}\right) = \ln(1+x) - \ln(1-x) = 2\sum_{k=0}^{\infty}\dfrac{x^{2k+1}}{2k+1}$

33. $\cos^2 x = \tfrac{1}{2}(1+\cos(2x)) = \dfrac{1}{2}\left(1 + \sum_{k=0}^{\infty}(-1)^k\dfrac{(2x)^k}{(2k)!}\right) = \dfrac{1}{2} + \sum_{k=0}^{\infty}(-1)^k\dfrac{2^{2k-1}x^{2k}}{(2k)!}$

34. $\dfrac{5+x}{x^2+x-2} = \dfrac{2}{x-1} - \dfrac{1}{x+2} = -\dfrac{2}{1-x} - \dfrac{1}{2}\dfrac{1}{1+(x/2)} = -2\sum_{k=0}^{\infty}x^k - \dfrac{1}{2}\sum_{k=0}^{\infty}(-1)^k\left(\dfrac{x}{2}\right)^k = -\sum_{k=0}^{\infty}\left(\dfrac{2^{k+2}+(-1)^k}{2^{k+1}}\right)x^k$

35. $f(x) = \sin^3(x) = \tfrac{1}{4}(3\sin x - \sin(3x)) = \dfrac{3}{4}\sum_{k=0}^{\infty}(-1)^k\dfrac{x^{2k+1}}{(2k+1)!} - \dfrac{1}{4}\sum_{k=0}^{\infty}(-1)^k\dfrac{(3x)^{2k+1}}{(2k+1)!} = \sum_{k=1}^{\infty}(-1)^{k+1}\dfrac{3^{2k+1}-3}{4\cdot(2k+1)!}x^{2k+1}$

36. $\dfrac{1}{x-1} = \dfrac{x}{x-1} - 1 = \dfrac{1}{1-(1/x)} - 1 = \sum_{k=0}^{\infty}\left(\dfrac{1}{x}\right)^k - 1 = \sum_{k=1}^{\infty}\dfrac{1}{x^k}$ since $|1/x| < 1$ if $|x| > 1$.

37. (a) Integrating term by term, $\dfrac{1}{1-x} = \sum_{k=0}^{\infty} x^k \Longrightarrow -\ln|1-x| = \sum_{k=1}^{\infty}\dfrac{x^k}{k}$.

(b) The series converges on the interval $[-1, 1)$.

(c) When $x = 1/2$, part (a) implies that $-\ln(1/2) = \ln 2 = \sum_{k=1}^{\infty} \frac{1}{k\,2^k}$. Since the terms of this series are all positive, the partial sums $S_N = \sum_{k=1}^{N} \frac{1}{k\,2^k}$ form an increasing sequence that is bounded above by the sum of the series. Thus, $\ln 2 - S_N > 0$.

$$\ln 2 - S_N = \sum_{k=1}^{\infty} \frac{1}{k\,2^k} - \sum_{k=1}^{N} \frac{1}{k\,2^k} = \sum_{k=N+1}^{\infty} \frac{1}{k\,2^k} \leq \frac{1}{N+1} \sum_{k=N+1}^{\infty} \frac{1}{2^k}$$

$$= \frac{1}{(N+1)2^{N+1}} \sum_{k=0}^{\infty} \frac{1}{2^k} = \frac{1}{(N+1)2^N}$$

38. The power series representation of $\ln(1+x)$ is a convergent alternating series if $0 < x < 1$:

$$\ln(1+x) = \sum_{k=1}^{\infty} \frac{(-1)^{k+1} x^k}{k}.$$

Because the partial sums of a convergent alternating series alternately overestimate and underestimate the limit, the first two partial sums bracket the value of $\ln(1+x)$.

39. The power series representations of each of the three functions is an alternating series if $x > 0$, so the following inequalities are valid if $0 < x < 1$: $x - x^2/2 < \ln(1+x) < x - x^2/2 + x^3/3$, $x - x^3/3! < \sin x < x$, and $x^2/2 - x^4/4! < 1 - \cos x < x^2/2$.

Since $(x - x^3/3!) - (x - x^2/2 + x^3/3) = x^2(1-x)/2 > 0$ if $0 < x < 1$, the lower bound on $\sin x$ is greater than the upper bound on $\ln(1+x)$ so $\ln(1+x) < \sin x$ if $0 < x < 1$. Also, since $(x - x^2/2) - x^2/2 = x(1-x) > 0$ if $0 < x < 1$, the lower bound on $\ln(1+x)$ is greater than the upper bound on $1 - \cos x$ so $1 - \cos x < \ln(1+x)$ if $0 < x < 1$.

40. (a) No. Using the series representation of $\sin x$ and the alternating series test, $\sin(1/n) > 1/n - 1/6n^3 = (6n^2 - 1)/6n^3 > 5/6n$ for all $n \geq 1$. Thus, $\sum_{n=1}^{\infty} \sin(1/n) > \frac{5}{6} \sum_{n=1}^{\infty} \frac{1}{n} = \infty$.

(b) Yes. Since $0 < \sin(1/n) < 1/n$ for all $n \geq 1$, $0 < \sum_{n=1}^{\infty} \frac{1}{n} \sin(1/n) < \sum_{n=1}^{\infty} \frac{1}{n^2}$. Thus, the given series converges by the comparison test.

41. (a) No. Since $\lim_{n \to \infty} e^{-1/n} = 1 \neq 0$, the series diverges by the n-th term test.

(b) No. Using the power series representation of e^x and the alternating series theorem,

$$1 - e^{-1/n} > \frac{1}{n} - \frac{1}{2n^2} \cdot \frac{2n-1}{2n^2} \geq \frac{2n-n}{2n^2} = \frac{1}{2n}$$

for all $n \geq 1$. Therefore, $\sum_{n=1}^{\infty} \left(1 - e^{-1/n}\right) \geq \frac{1}{2} \sum_{n=1}^{\infty} \frac{1}{n}$.

42. $\int_0^{\infty} e^{-t} \sin(xt)\, dt = \int_0^{\infty} e^{-t} \left(\sum_{k=0}^{\infty} (-1)^k \frac{(xt)^{2k+1}}{(2k+1)!} \right) dt = \sum_{k=0}^{\infty} (-1)^k \frac{x^{2k+1}}{(2k+1)!} \left(\int_0^{\infty} e^{-t} t^{2k+1}\, dt \right) = \sum_{k=0}^{\infty} (-1)^k x^{2k+1} = \frac{x}{1+x^2}$ when $|x| < 1$.

SECTION 11.6 POWER SERIES AS FUNCTIONS

43. (a) Despite first appearances, I is not a doubly improper integral: $\lim_{x\to 0^+}\dfrac{xe^{-x}}{1-e^{-x}} = 1$. Therefore, since

$$I = \int_0^\infty \frac{xe^{-x}}{1-e^{-x}}\,dx = \int_0^1 \frac{xe^{-x}}{1-e^{-x}}\,dx + \int_1^\infty \frac{xe^{-x}}{1-e^{-x}}\,dx,$$

I converges if and only if the improper integral on right above converges. To show this, note that $1-e^{-x} > 1/2$ for all $x \geq 1$. From this it follows that

$$\int_1^\infty \frac{xe^{-x}}{1-e^{-x}}\,dx < 2\int_1^\infty xe^{-x}\,dx = 4e^{-1}.$$

Therefore, I converges.

(b) If $u = 1 - e^{-x}$, $du = e^{-x}\,dx$, and $x = -\ln(1-u)$. Furthermore, $1 - e^{-0} = 0$ and $\lim_{x\to\infty} 1 - e^{-x} = 1$, so

$$I = -\int_0^1 \frac{\ln(1-u)}{u}\,du.$$

(c) When $|u| < 1$, $\ln(1-u) = -\sum_{k=1}^\infty \dfrac{u^k}{k}$. Therefore,

$$I = -\int_0^1 \frac{\ln(1-u)}{u}\,du = \int_0^1\left(\sum_{k=1}^\infty \frac{u^{k-1}}{k}\right)du = \sum_{k=1}^\infty\left(\int_0^1 \frac{u^{k-1}}{k}\,du\right) = \sum_{k=1}^\infty \frac{1}{k^2} = \frac{\pi^2}{6}.$$

44. (a) $\dfrac{1}{1+x^4} = \sum_{k=0}^\infty (-1)^k x^{4k}$

(b) $(-1, 1)$

(c) $\displaystyle\int f(x)\,dx = \int \frac{dx}{1+x^4} = \int\left(\sum_{k=0}^\infty (-1)^k x^{4k}\right)dx = \sum_{k=0}^\infty (-1)^k \frac{x^{4k+1}}{4k+1}$. Therefore,

$$\int_0^{0.5} f(x)\,dx = \sum_{k=0}^\infty (-1)^k \frac{(0.5)^{4k+1}}{4k+1}.$$

Since $\dfrac{(0.5)^9}{9} < 0.001$, $\displaystyle\int_0^{0.5} f(x)\,dx \approx \sum_{k=0}^1 (-1)^k \frac{(0.5)^{4k+1}}{4k+1} = \frac{79}{160} \approx 0.49375$.

45. (a) $\displaystyle\int e^{-x^2}\,dx = \int\left(\sum_{k=0}^\infty \frac{(-x^2)^k}{k!}\right)dx = \sum_{k=0}^\infty \frac{(-1)^k}{(2k+1)\cdot k!}x^{2k+1}$

(b) The approximation $\displaystyle\int_0^1 e^{-x^2}\,dx \approx \sum_{k=0}^3 \frac{(-1)^k}{(2k+1)\cdot k!} = \frac{26}{35}$ has the desired accuracy because $\dfrac{1}{9\cdot 4!} < 0.005$.

46. $\displaystyle\int_0^1 \cos(x^2)\,dx = \int_0^1\left(\sum_{k=0}^\infty \frac{(-1)^k x^{4k}}{(2k)!}\right)dx = \sum_{k=0}^\infty \frac{(-1)^k x^{4k+1}}{(4k+1)\cdot(2k)!}\bigg|_0^1 = \sum_{k=0}^\infty \frac{(-1)^k}{(4k+1)\cdot(2k)!} \approx$ $ds\sum_{k=0}^1 \frac{(-1)^k}{(4k+1)\cdot(2k)!} = \frac{9}{10}$ within 0.005.

47. $\displaystyle\int_0^1 \sqrt{x}\sin x\,dx = \int_0^1\left(\sum_{k=0}^\infty (-1)^k \frac{x^{(4k+3)/2}}{(2k+1)!}\right)dx = \sum_{k=0}^\infty \frac{(-1)^k 2}{(4k+5)(2k+1)!} \approx$
$\displaystyle\sum_{k=0}^2 \frac{(-1)^k 2}{(4k+5)(2k+1)!} = \frac{2557}{7020} \approx 0.36425$

48. Use the fact that $\int_0^{1000} x^n e^{-10x}\,dx \le \int_0^\infty x^n e^{-10x}\,dx = n!/10^{n+1}$ to show that the terms of the alternating series are decreasing in magnitude. Therefore, we may truncate the partial sum just before the first term that is less than 0.00005. It follows that approximating $sin(x)/x$ by $1 - x^2/6$ is adequate.

49. $\dfrac{e^x}{1-x} = \left(\sum\limits_{k=0}^\infty \dfrac{x^k}{k!}\right)\left(\sum\limits_{k=0}^\infty x^k\right) = 1 + 2x + \dfrac{5}{2}x^2 + \dfrac{8}{3}x^3 + \cdots$

50. $\dfrac{\cos x}{1+x^2} = \left(\sum\limits_{k=0}^\infty (-1)^k \dfrac{x^{2k}}{(2k)!}\right)\left(\sum\limits_{k=0}^\infty (-1)^k x^{2k}\right) = 1 - \dfrac{3}{2}x^2 + \dfrac{37}{24}x^4 - \dfrac{1111}{720}x^6 + \cdots$

51. $e^{2x}\ln(1+x^3) = \left(\sum\limits_{k=0}^\infty \dfrac{(2x)^k}{k!}\right)\left(\sum\limits_{k=0}^\infty (-1)^{k+1} \dfrac{x^{3k}}{k}\right) = x^3 + 2x^4 + 2x^5 + \dfrac{5}{6}x^6 + \cdots$

52. $\arctan x\,\sin(4x) = \left(\sum\limits_{k=0}^\infty (-1)^k \dfrac{x^{2k+1}}{2k+1}\right)\left(\sum\limits_{k=0}^\infty (-1)^k \dfrac{x^{2k+1}}{(2k+1)!}\right) = 4x^2 - 12x^4 + \dfrac{116}{9}x^6 - \dfrac{44}{5}x^8 + \cdots$

53. $e^{\sin x} = \sum\limits_{k=0}^\infty \dfrac{(\sin x)^k}{k!} = 1 + \sin x + \dfrac{\sin^2 x}{2} + \dfrac{\sin^3 x}{3!} + \dfrac{\sin^4 x}{4!} + \cdots$

$= 1 + \left(x - x^3/3! + \cdots\right) + \dfrac{1}{2}\left(x - x^3/3! + \cdots\right)^2 + \dfrac{1}{3!}(x - \cdots)^3 + \dfrac{1}{4!}(x - \cdots)^4$

$= 1 + x + x^2/2 - x^4/8 + \cdots$

54. $\ln(1 + \sin x) = \sum\limits_{k=1}^\infty (-1)^{k+1} \dfrac{(\sin x)^k}{k}$

$= \left(x - x^3/6 + \cdots\right) - \dfrac{1}{2}\left(x^2 - x^4/3 + \cdots\right) + \dfrac{1}{3}\left(x^3 + \cdots\right) - \dfrac{1}{4}\left(x^4 + \cdots\right)$

$= x - x^2/2 + x^3/6 - x^4/12 + \cdots$

55. $-\cos x = \cos(x - \pi) = \sum\limits_{k=0}^\infty (-1)^k \dfrac{(x-\pi)^{2k}}{(2k)!} \implies \cos x = \sum\limits_{k=0}^\infty (-1)^{k+1} \dfrac{(x-\pi)^{2k}}{(2k)!}$

56. $\dfrac{1}{1-x} = -\dfrac{1}{1+(x-2)} = -\sum\limits_{k=0}^\infty (-1)^k (x-2)^k \implies a_k = (-1)^{k+1}$.

57. $\sum\limits_{k=1}^\infty k x^{k-1} = \left(\dfrac{1}{1-x}\right)' = \dfrac{1}{(1-x)^2}$

58. $\sum\limits_{k=0}^\infty \dfrac{x^k}{(k+1)!} = \dfrac{e^x - 1}{x}$

59. $\sum\limits_{k=1}^\infty (-1)^{k+1} x^k = 1 + \sum\limits_{k=0}^\infty (-1)^{k+1} x^k = 1 - \sum\limits_{k=0}^\infty (-1)^k x^k = 1 - \dfrac{1}{1+x} = \dfrac{x}{1+x}$

60. $\sum\limits_{k=1}^\infty \dfrac{(2x)^k}{k} = -\ln(1-2x)$ [NOTE: $-\ln(1-x) = \int (1-x)^{-1} dx = \int \left(\sum\limits_{k=0}^\infty x^k\right) dx = \sum\limits_{k=1}^\infty \dfrac{x^k}{k}$.]

62. $y = e^x = \sum\limits_{k=0}^\infty \dfrac{x^k}{k!} \implies y' = \sum\limits_{k=0}^\infty \dfrac{k x^{k-1}}{k!} = \sum\limits_{k=1}^\infty \dfrac{x^{k-1}}{(k-1)!} = \sum\limits_{k=0}^\infty \dfrac{x^k}{k!} = y$

SECTION 11.6 POWER SERIES AS FUNCTIONS

63. $y = 2e^x = \sum_{k=0}^{\infty} \frac{2x^k}{k!} \implies y' = \sum_{k=0}^{\infty} \frac{2kx^{k-1}}{k!} = \sum_{k=1}^{\infty} \frac{2x^{k-1}}{(k-1)!} = \sum_{k=0}^{\infty} \frac{2x^k}{k!} = y$

64. $y = e^{3x} = \sum_{k=0}^{\infty} \frac{(3x)^k}{k!} \implies y' = \sum_{k=0}^{\infty} \frac{k(3x)^{k-1}}{k!} = \sum_{k=1}^{\infty} \frac{(3x)^{k-1}}{(k-1)!} = \sum_{k=0}^{\infty} \frac{(3x)^k}{k!} = y$

65. $y = \sin x = \sum_{k=0}^{\infty} (-1)^k \frac{x^{2k+1}}{(2k+1)!}$ and $y' = \sum_{k=0}^{\infty} (-1)^k \frac{(2k+1)x^{2k}}{(2k+1)!} = \sum_{k=0}^{\infty} (-1)^k \frac{x^{2k}}{(2k)!}$, so

    $y'' = \sum_{k=0}^{\infty} (-1)^k \frac{(2k)x^{2k-1}}{(2k)!} = \sum_{k=1}^{\infty} (-1)^k \frac{x^{2k-1}}{(2k-1)!} = \sum_{k=0}^{\infty} (-1)^{k+1} \frac{x^{2k+1}}{(2k+1)!} = -y$.

66. $y = (1-x)^{-1} = \sum_{k=0}^{\infty} x^k \implies y' = \sum_{k=0}^{\infty} kx^{k-1} = \sum_{k=0}^{\infty} (k+1)x^k = \left(\sum_{k=0}^{\infty} x^k \right)^2 = y^2$

67. (a) $1 + (f(x))^2 = 1 + \tan^2 x = \sec^2 x = f'(x)$

    (b) Since $f(0) = 0$, we assume that $f(x) = \sum_{k=1}^{\infty} a_k x^k$. Inserting this power series into the identity from part (a): $a_1 + 2a_2 x + 3a_3 x^2 + \cdots = 1 + a_1^2 x^2 + 2a_1 a_2 x^3 + (2a_1 a_3 + a_2^2)x^4 + (2a_1 a_4 + 2a_2 a_3)x^5 + (2a_1 a_5 + 2a_2 a_4 + a_3^2)x^6 + \cdots$. Equating powers of x on both sides and solving, we find that $a_1 = 1$, $a_2 = 0$, $a_3 = 1/3$, $a_4 = 0$, $a_5 = 2/15$, $a_6 = 0$, and $a_7 = 17/315$. Thus, $\tan x = 1 + \frac{x^3}{3} + \frac{2x^5}{15} + \frac{17x^7}{315} + \cdots$.

68. (a) Since $\lim_{n \to \infty} \left| \frac{a_{n+1}}{a_n} \right| = \lim_{n \to \infty} \frac{|r-n| \cdot |x|}{n+1} < 1$ if $|x| < 1$, the series defining f converges if $|x| < 1$.

    (b) $f'(x) = \sum_{n=1}^{\infty} \frac{r(r-1)(r-2)\cdots(r-n+1)}{(n-1)!} x^{(n-1)}$. Now, the coefficient of x^n in the series for $(1+x)f'(x)$ is

    $\frac{r(r-1)(r-2)\cdots(r-n+1)(r-n)}{n!} + \frac{r(r-1)(r-2)\cdots(r-n+1)}{(n-1)!}$

    $= \frac{(r-n) \cdot r(r-1)(r-2)\cdots(r-n+1)}{n!} + \frac{n \cdot r(r-1)(r-2)\cdots(r-n+1)}{n!}$

    $= \frac{r \cdot r(r-1)(r-2)\cdots(r-n+1)}{n!}$.

    Since this is also the coefficient of x^n in the series for $rf(x)$, the result follows.

    (c) $g'(x) = \frac{f'(x)}{(1+x)^r} - \frac{rf(x)}{(1+x)(1+x)^r} = \frac{f'(x)}{(1+x)^r} - \frac{(1+x)f'(x)}{(1+x)(1+x)^r} = 0$

    (d) The result in part (c) implies that g is a constant function. Since $g(0) = 1$, $g(x) = 1$ and so $f(x) = (1+x)^r$.

69. Let $r = 1/2$.

70. The binomial series for $(1+u)^3$ terminates after a finite number of terms since $r = 3$ is an integer: $(1+u)^3 = 1 + 3u + 3u^2 + u^3$. Therefore, $f(x) = (1+x^4)^3 = 1 + 3x^4 + 3x^8 + x^{12}$.

71. $g(x) = \sqrt[3]{1-x^2} \approx 1 - x^2/3 - x^4/9 - 5x^6/81$

72. $g(x) = (1+x^2)^{-3/2} \approx 1 - 3x^2/2 + 15x^4/8 - 35x^6/16$

73. $g(x) = \arcsin x = \int \frac{dx}{\sqrt{1-x^2}} = \int \left(1 + \frac{x^2}{2} + \frac{3x^4}{8} + \frac{5x^6}{16} + \cdots\right) dx = x + \frac{x^3}{6} + \frac{3x^5}{40} + \frac{5x^7}{112} + \cdots.$

74. Since $\sqrt{1+u} = 1 + \frac{u}{2} - \frac{u^2}{8} + \frac{u^3}{16} - \frac{5u^4}{128} \pm \cdots$, $\sqrt{1+x^3} = 1 + \frac{x^3}{2} - \frac{x^6}{8} + \frac{x^9}{16} - \frac{5x^{12}}{128} \pm \cdots$ and so

$$\int_0^{2/5} \sqrt{1+x^3}\, dx = \int_0^{2/5} \left(1 + \frac{x^3}{2} - \frac{x^6}{8} + \frac{x^9}{16} - \frac{5x^{12}}{128} \pm \cdots\right) dx$$

$$= \frac{2}{5} + \frac{1}{8}\left(\frac{2}{5}\right)^4 - \frac{1}{56}\left(\frac{2}{5}\right)^7 + \frac{1}{160}\left(\frac{2}{5}\right)^{10} - \frac{5}{1664}\left(\frac{2}{5}\right)^{13} \pm \cdots.$$

Now, since this is an alternating series (after the first term) and $2^7/56 \cdot 5^7 < 5 \times 10^{-4}$, the value of the integral is approximated to the desired accuracy by $2/5 + 2^4/8 \cdot 5^4 = 252/625 = 0.4032$.

11.7 Maclaurin and Taylor Series

1. (a) The Maclaurin series representation of f is the polynomial expression used to define f: $1 + 2x + 44x^2 - 12x^3 + x^4$.

   (b) $f(x) = 160 + 50(x-3) - 10(x-3)^2 + (x-3)^4$.

2. Since $K_{n+1} = 0$, Theorem 13 implies that $|p(x) - P_n(x)| = 0$.

3. (a) $f'(x) = \sqrt{x}e^{-x}$, $f''(x) = e^{-x}\left(\frac{1}{2\sqrt{x}} - \sqrt{x}\right)$, and $f'''(x) = e^{-x}\left(\sqrt{x} - \frac{1}{\sqrt{x}} - \frac{1}{4x^{3/2}}\right)$. Therefore, $f(3) = 0$, $f'(3) = \sqrt{3}e^{-3}$, $f''(3) = -\frac{5\sqrt{3}}{6}$, and $f'''(3) = \frac{23\sqrt{3}}{36}$. It follows from Taylor's Theorem that

$$f(x) \approx \sqrt{3}e^{-3}(x-3) - \frac{5}{12}\sqrt{3}e^{-3}(x-3)^2 + \frac{23}{216}\sqrt{3}e^{-3}(x-3)^3.$$

   (b) If $3 \leq x \leq 3.5$, $|f^{(4)}(x)| \leq 0.035$, so the approximation error is bounded by
   $$\frac{0.035 \cdot (0.5)^4}{4!} \approx 9.1146 \times 10^{-5}.$$

4. (a) $f(0) = 1$, $f'(0) = 1/2$, and $f''(0) = -1/4$ so $\sqrt{1+x} = 1 + \frac{x}{2} - \frac{x^2}{4 \cdot 2!} + \cdots$.

   (b) The magnitude of the approximation error is bounded by $\frac{K_3}{3!} = \frac{3/8}{3!} = \frac{1}{16} \approx 0.0625$.

5. Since $f'(0) > 0$, the coefficient of x in the Maclaurin series representation of f must be positive; the coefficient of x in the series given is negative.

6. (a) $\sqrt{1+x} \approx 1 + \frac{1}{2}x - \frac{1}{8}x^2$.

   (b) $V = 2\pi\sigma\left(\sqrt{r^2+a^2}-r\right) = 2\pi\sigma r\left(\sqrt{1+(a/r)^2}-1\right) \approx 2\pi\sigma r\left(\frac{a^2}{2r^2} - \frac{a^4}{8r^4}\right) = 2\pi\sigma\left(\frac{a^2}{2r} - \frac{a^4}{8r^3}\right)$.

7. Let $f(x) = x^4 - 4x^3 + 5x$. Then $f(1) = 2$, $f'(1) = -3$, $f''(1) = -12$, $f'''(1) = 0$, $f^{(4)}(1) = 24$, and $f^{(k)}(1) = 0$ for all $k \geq 5$. Therefore, Taylor's Theorem implies that $f(x) = 2 - 3(x-1) - 6(x-1)^2 + (x-1)^4$. It follows that $a_0 = 2$, $a_1 = -3$, $a_2 = -6$, $a_3 = 0$, and $a_4 = 1$.

8. Differentiating, $f''(x) = 10(f(x))^9 f'(x)$ and $f'''(x) = 90(f(x))^8 (f'(x))^2 + 10(f(x))^9 f''(x)$. Therefore, $f'(0) = 2$, $f''(0) = 20$, and $f'''(0) = 560$. From this we can compute the first four terms of the Maclaurin series for f:

$$f(x) \approx 1 + 2x + \frac{20}{2}x^2 + \frac{560}{6}x^3.$$

9. (a) No. The Maclaurin series representation of f is $f(x) = f(0) + f'(0)x + \frac{f''(0)}{2}x^2 + \cdots$. Since f is concave down at $x = 0$, the coefficient of x^2 in the Maclaurin series representation of f is negative.

   (b) Yes. $g''(0) = \frac{3}{4}(f'(0))^2(g(0))^5 - \frac{1}{2}f''(0)(g(0))^3 > 0$.

10. The Maclaurin series representation of f is

$$f(x) = \sum_{k=0}^{\infty} \frac{(2x)^k}{k!} = \sum_{k=0}^{\infty} \frac{f^{(k)}(0)}{k!}x^k.$$

Thus, the coefficient of x^{100} in the Maclaurin series representation of f is $2^{100}/100!$.

11. (a) Using the fact that the Maclaurin series for $(1-u)^{-1}$ is $\sum_{k=0}^{\infty} u^k$, the MacLaurin series for f is $\sum_{k=0}^{\infty} x^{3k+1}$.

   (b) The interval of convergence of the power series for f is $(-1, 1)$.

   (c) Differentiating the power series for f term by term, $f''(x) = \sum_{k=1}^{\infty} (3k+1)(3k)x^{3k-1}$.

   (d) Integrating the power series for f term by term, $\int_0^x f(t)\,dt = \sum_{k=0}^{\infty} \frac{x^{3k+2}}{3k+2}$.

12. (a) $\dfrac{1}{2+x} = \dfrac{1}{2} \cdot \dfrac{1}{1+(x/2)} = \dfrac{1}{2} \cdot \dfrac{1}{1-(-x/2)} = \dfrac{1}{2}\sum_{k=0}^{\infty}\left(\dfrac{-x}{2}\right)^k = \sum_{k=0}^{\infty}\dfrac{(-1)^k x^k}{2^{k+1}}$

   (b) The coefficient of x^{259} in the Maclaurin series representation of $f(x)$ is $f^{(259)}(0)/259!$. Thus, $f^{(259)}(0) = -259!/2^{260}$.

13. The Maclaurin series representation of f is $f(x) = \sum_{k=0}^{\infty}\dfrac{f^{(k)}(0)}{k!}x^k = \sum_{k=0}^{\infty}(-1)^k\dfrac{1}{(2k+2)!}x^{2k}$.
   Thus, $f^{(100)}(0)/100! = 1/102! \implies f^{(100)}(0) = 100!/102! = 1/102 \cdot 101 = 1/10302$.

14. (a) $f(1.5) \approx f(1) + f'(1) \cdot (1.5 - 1) + \dfrac{f''(1)}{2}(1.5-1)^2 = 1 + 2 \cdot 0.5 + \dfrac{1}{4}(0.5)^2 = \dfrac{33}{16} = 2.0625$

   (b) Since $f'''(x) = -\dfrac{3x^2}{(1+x^3)^2}$, $K_3 = 3/4$. Therefore, the approximation error is less than or equal to $\dfrac{K_3}{3!}(1.5-1)^3 = \dfrac{1}{64} = 0.015625$.

15. (a) $f(x) = x^{-1}\sin x = x^{-1}\sum_{k=0}^{\infty}\dfrac{(-1)^k x^{2k+1}}{(2k+1)!} = \sum_{k=0}^{\infty}\dfrac{(-1)^k x^{2k}}{(2k+1)!}$

   (b) The power series in part (a) converges for values of x in the interval $(-\infty, \infty)$.

   (c) $f'''(x) = \sum_{k=2}^{\infty}\dfrac{(-1)^k(2k)(2k-1)(2k-2)x^{2k-3}}{(2k+1)!}$ for any $x \in (-\infty, \infty)$. Since $f'''(1)$ is represented by an alternating series, $\left|f'''(1) - \sum_{k=2}^{3}\dfrac{(-1)^k(2k)(2k-1)(2k-2)}{(2k+1)!}\right| < \dfrac{8 \cdot 7 \cdot 6}{9!} = \dfrac{1}{1080} < 0.005$; $f'''(1) \approx \dfrac{37}{210} \approx 0.17619$.

16. From the information given, $f(x) = 26 + 22x - \dfrac{16}{2!}x^2 + \dfrac{12}{3!}x^3 + \cdots = 26 + 22x - 8x^2 + 2x^3 + \ldots$.
    Since $P_3(1) = 42$, Theorem 13 implies that $|f(x) - 42| < \dfrac{7}{4!} = \dfrac{7}{24}$. From this it follows that

   (a) $42 - \dfrac{7}{24} = \dfrac{1001}{24} \le f(1) \le \dfrac{1015}{24} = 42 + \dfrac{7}{24}$ (b).

17. $K_{n+1} = e^x$ so $\left|e^x - P_n(x)\right| \le \dfrac{e^x \cdot |x|^{n+1}}{(n+1)!} \to 0$ as $n \to \infty$.

18. When $0 < x < 1$, $K_{n+1} = (n+1)!$ so $\left|\dfrac{1}{1+x} - P_n(x)\right| \le \dfrac{(n+1)! \cdot |x|^{n+1}}{(n+1)!} = |x|^{n+1} \to 0$ as $n \to \infty$.

   When $-1/2 < x < 0$, $K_{n+1} = 2^{n+2}(n+1)!$ so $\left|\dfrac{1}{1+x} - P_n(x)\right| \le \dfrac{2^{n+2}(n+1)! \cdot |x|^{n+1}}{(n+1)!} = 2^{n+2}|x|^{n+1} \to 0$ as $n \to \infty$.

SECTION 11.7   MACLAURIN AND TAYLOR SERIES

19. (a) If $\left|f^{(n)}(x)\right| \le n$ for all $n \ge 1$, then $K_{n+1} \le n+1$ and $|f(x) - P_n(x)| \le \dfrac{(n+1)|x|^{n+1}}{(n+1)!} \to 0$ as $n \to \infty$.

   (b) Yes, because $\lim\limits_{n\to\infty} \dfrac{x^n}{n!} = 0$ for all x.

20. (a) Using the definition of the derivative: $f'(0) = \lim\limits_{h\to 0} \dfrac{f(h) - f(0)}{h} = \lim\limits_{h\to 0} \dfrac{e^{-1/h^2}}{h}$. Now, letting $x = 1/h$, $\lim\limits_{h\to 0^+} \dfrac{e^{-1/h^2}}{h} = \lim\limits_{x\to\infty} xe^{-x^2} = 0$ (by l'Hôpital's rule). Similarly, $\lim\limits_{h\to 0^-} \dfrac{e^{-1/h^2}}{h} = \lim\limits_{x\to -\infty} xe^{-x^2} = 0$. Therefore, $f'(0) = 0$.

   (b) The Maclaurin series for f is the constant function 0.

   (c) The radius of convergence of the series in part (b) is $R = \infty$.

   (d) The series in part (b) converges to $f(x)$ only when $x = 0$.

21. Start by writing $\cos x = \cos(x - \pi/3 + \pi/3) = \dfrac{1}{2}\cos(x - \pi/3) - \dfrac{\sqrt{3}}{2}\sin(x - \pi/3)$. Now, using the Maclaurin series for $\cos u$ and $\sin u$ (with $u = x - \pi/3$), the result is

$$\cos x = \frac{1}{2}\sum_{k=0}^{\infty} \frac{(-1)^k}{(2k)!}(x - \pi/3)^{2k} - \frac{\sqrt{3}}{2}\sum_{k=0}^{\infty} \frac{(-1)^k}{(2k+1)!}(x - \pi/3)^{2k+1}.$$

22. $\displaystyle\sum_{k=0}^{\infty} (-1)^k \frac{\sqrt{2}}{2}\left(\frac{(x - \pi/4)^{2k}}{(2k)!} + \frac{(x - \pi/4)^{2k+1}}{(2k+1)!}\right).$

12.1 Differential equations: the basics

1. (a) If $y = Ce^x$ and $x = 0$, then $y = Ce^0 = C$. This means that the graph of $y = Ce^x$ has y-intercept C. The fact that $y'(0) = C$ means that the each graph has slope C at $x = 0$.

   (b) Each of the graphs has the appropriate y-intercept and slope at $x = 0$.

2. (a) Choose any point on any curve (except C2) and estimate the slope there. Consider, e.g., the point on C4 with y-coordinate 120. At this point the slope appears to be about -6.5. Plugging this information into the DE $y' = k(y - 70)$ gives
$$-6.5 \approx k(120 - 70) \implies k \approx -\frac{6.5}{50} = -0.13.$$

   (b) All curves satisfy the same DE, so all that remains is to specify an initial condition for each. One easy way to do so is to read values at $t = 0$. Doing so gives the following initial conditions: C1: $y(0) = 40$; C2: $y(0) = 70$; C3: $y(0) = 120$; C4: $y(0) = 160$; C5: $y(0) = 200$.

3. $y' = 3x^2$ so $xy' = x \cdot 3x^2 = 3x^3 = 3y$

4. $y' = 2e^{-2x}$ so $y' + 2y = 2e^{-2x} + 2(1 - e^{-2x}) = 2$.

5. (a) It sounds more or less reasonable. As the goal approaches, people's ardor to contribute seems likely to cool somewhat. On the other hand, as the goal becomes really near, some people might give money to get it over with.

   (b) As a DE, Neuman's Law of Cooling says $y' = k(y - 65)$.

   (c) "Neuman's Law of Cooling" suggests the analogy to Newton's Law of Cooling.

6. (a) Check directly that $y' = k(y - T)$:
$$\begin{aligned} y'(t) &= Ake^{kt}; \\ k(y - T) &= k(T + Ae^{kt} - T) = Ake^{kt}, \end{aligned}$$
as claimed.

   (b) The situation described in the exercise says, in DE language, that
$$y'(t) = k(y(t) - 65); \qquad y(0) = 10; \qquad y(5) = 30.$$

By (a), $y(t) = 65 + Ae^{kt}$ solves our DE. Now we'll use what we know to find values for A and k:
$$\begin{aligned} y(0) = 10 &\implies 10 = 65 + A \implies A = -55; \\ y(5) = 45 &\implies 45 = 65 - 55e^{5k} \implies k \approx -0.2023. \end{aligned}$$

Thus $y(t) = 65 - 55e^{-0.2023t}$ solves our DE. To answer the questions raised above we solve $y(t) = 60$ and $y = 64.9$ for t. The answers are (approximately) 11.85 weeks and 31.19 weeks, respectively. The following graph shows the initially promising but ultimately dispiriting progress of the campaign:

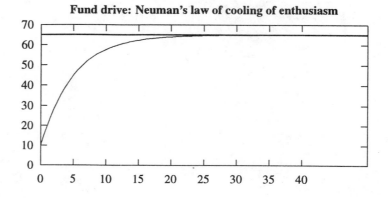

SECTION 12.1 DIFFERENTIAL EQUATIONS: THE BASICS

7. $y' = -1/x^2$ and $y'' = 2/x^3$ so $x^3 y'' + x^2 y' - xy = x$.

8. (a) $y = y' = y'' = e^x$. Thus, $xy'' - (2x+1)y' + (x+1)y = xe^x - (2x+1)e^x + (x+1)e^x = 0$.

   (b) $y' = (2x + x^2)e^x$ and $y'' = (2 + 4x + x^2)e^x$. Thus, $xy'' - (2x+1)y' + (x+1)y = x(2 + 4x + x^2)e^x - (2x+1)(2x + x^2)e^x + (x+1)x^2 e^x = (2x + 4x^2 + x^3)e^x - (2x + 5x^2 + 2x^3)e^x + (x^3 + x^2)e^x = 0$.

9. $y' = -C_1 \sin x + C_2 \cos x$ and $y'' = -C_1 \cos x - C_2 \sin x$. Thus, $y'' + y = (-C_1 \cos x - C_2 \sin x) + 1 + C_1 \cos x + C_2 \sin x = 1$.

10. If $y = e^x$ is a solution of the differential equation, $(a + b + 1)e^x = 0 \implies a + b = -1$ must hold. Similarly, if $y = e^{2x}$ is a solution of the differential equation, $(4a + 2b + 1)e^{2x} = 0 \implies 4a + 2b = -1$ must hold. Together, these two conditions imply that $a = 1/2$ and $b = -3/2$.

11. If $y = x$ and $y = x^2$ are both solutions of the differential equation, the following conditions must hold: $g(x) + x = 0$ and $2f(x) + g(x) \cdot x + x^2 = 0$. Both conditions are satisfied if $f(x) = x^2/2$ and $g(x) = -x$.

12. Observe that the graph of $y = f(x)$ is decreasing when $1 < x < 2$ and increasing when $2 < x < 5$. Thus, y' must be negative on $(1, 2)$ and positive on $(2, 5)$. Also, observe that $y > x$ on $(1, 2)$ and $y < x$ on $(2, 5)$. Furthermore, $y < x^2$ on $(2, 5)$.

The differential equation (a) cannot be the correct answer since the expression $(y - x)/x$ is negative on the interval $(2, 5)$. Similarly, the differential equation (c) cannot be the correct answer because the expression $(x^2 - y)/x$ is not negative over the entire interval $(1, 2)$. On the other hand, the expression $(x - y)/y$ is negative on the interval $(1, 2)$ and positive on the interval $(2, 5)$. Therefore, the correct differential equation is (b).

12.2 Slope fields: solving DE's graphically

1. (a) The solution curves are "parallel" to each other in the sense that they differ from each other only in their *horizontal* position. Thus, e.g., all the curves have the same slope where $y = 2$.

    (b) It *does* appear that each of the five "upper" curves has the same slope when $y = 3$. Carefully draw a tangent line to any one of the curves at the appropriate point; measure its slope. The result should be 3 (or very close to 3).

    The answer *could* have been predicted in advance. The fact that each curve is a solution to the DE $y' = y$ means precisely that when $y = 3$, $y' = 3$, too.

    (c) At the level $y = -4$, each curve has slope -4. Again, this is exactly what the DE predicts.

    (d) All curves appear to be very nearly *horizontal* near $y = 0$. The only solution curve that actually touches the line $y = 0$ is the solution curve $y = 0$ itself. Appropriately, this curve has slope 0 everywhere.

2. (a) We'll check the claim directly, by computation. Since $y = Ce^t - t - 1$,
    $$y' = Ce^t - 1; \quad y + t = Ce^t - t - 1 + t = Ce^t - 1.$$
    Thus $y' = y + t$, as claimed.

    (b) The previous part says that our solution is of the form $y = Ce^t - t - 1$; we need only choose C properly. Here goes:
    $$y(0) = 1 \implies Ce^0 - 0 - 1 = 1 \implies C - 1 = 1 \implies C = 2.$$
    Therefore $y = 2e^t - t - 1$ is our solution, and $y(2) = 2e^2 - 2 - 1 = 2e^2 - 3 \approx 11.778$.

    (c) The solution function just found (for which $C = 2$) is *not* shown in Example 3. To draw this solution curve we would squeeze a new curve in between those labeled $C = 1$ and $C = 10$.

3. (a) The straight line is $y = -t - 1$; it corresponds to $C = 0$.

    (b) The curve $y = 5e^t - t - 1$ passes through $(0, 4)$. One way to tell this is to solve the equation $5 = Ce^0 - 0 - 1$ for C.

    (c) At $(0, 4)$ the curve mentioned above has slope 4. This can be found (approximately) graphically by looking at slope, or symbolically by reading the DE.

    (d) The line $y + t = 0$ (aka $y = -t$) crosses each of the "upper" four solution curves at a *stationary point*, i.e., a point where the slope is zero. This happens because, as the DE demands, $y' = 0$ whenever $y + t = 0$.

    (e) The line $y + t = -3$ (aka $y = -t - 3$) crosses the "lower" four solution curves at four points. At each of these points, the solution curve in question has the same slope: -3. This is as it should be. As the DE requires at such points, $y' = y + t = -3$.

    (f) Any line with slope -1 crosses the solution curves at points of *equal slope*. As in the previous two parts, this is because the DE requires it.

    (g) Moving clockwise from upper left, the curves correspond to the C-values 500, 50, 5, 0.2, 0, -0.2, -5, -50, -500.

4. All solutions are of the form $y(t) = Ce^t - t - 1$. Therefore
    $$\lim_{t \to -\infty} (y(t) - (-t - 1)) = \lim_{t \to -\infty} \left(Ce^t - t - 1 - (-t - 1)\right) = \lim_{t \to -\infty} Ce^t = 0.$$
    This means that, indeed, every solution curve tends toward the line $y = -t - 1$ as $t \to -\infty$.

5. (a) In the slope field for $y' = 2t - 5$, all ticks at the same *horizontal position* are parallel. (That's because all such ticks have the same t-value.)

    (b) In the slope field for $y' = y$, all ticks at the same *vertical position* are parallel. (That's because all such ticks have the same y-value.)

SECTION 12.2 SLOPE FIELDS: SOLVING DE'S GRAPHICALLY

6. (a) For each value of x, the slopes do not depend on y (i.e., all ticks at the same horizontal position are parallel).

   (b) For each value of y, the slopes do not depend on x (i.e., all ticks at the same vertical position are parallel).

7. (b) $y(x) = \sqrt{x^2 + 1}$

   (c) $y(x) = \pm\sqrt{x^2 - 1}$

8. (b) $y(x) = \frac{1}{2}e^{x^2}\sqrt{\pi}\,\text{erf}(x) + \frac{1}{2}e^{x^2}$

   (c) $y(x) = \frac{1}{2}e^{x^2}\sqrt{\pi}\,\text{erf}(x) - e^{x^2}$

9. (b) $y(x) = \frac{3}{2}e^{x^2} - \frac{1}{2}$

   (c) $y(x) = -\frac{1}{2e}e^{x^2} - \frac{1}{2} = -\frac{1}{2}\left(e^{x^2-1} + 1\right)$

12.3 Euler's method: solving DE's numerically

1. (a) A table helps keep track of results:

step	t	y'	y
0	0.00	0	0
1	0.25	0.2474	0
2	0.50	0.4794	0.06185
3	0.75	0.6816	0.1817
4	1.00	0.8415	0.3521

   (b) The left rule with 4 subdivisions, applied to $I = \int_0^1 \sin(t)\,dt$, gives
   $$L_4 = \frac{\sin(1/4)}{4} + \frac{\sin(1/2)}{4} + \frac{\sin(3/4)}{4} \approx 0.3521.$$

   (c) The function $y(t) = 1 - \cos t$ solves the DE exactly. Thus, exactly, $y(1) = 1 - \cos 1 \approx 0.4597$.

   (d) $\int_0^1 \sin(t)\,dt = -\cos t\big]_0^1 = -\cos 1 + \cos 0 = 1 - \cos 1 \approx 0.4597.$

   (e) The error committed by L_4 is $|I - L_4| \approx 0.4597 - 0.3521 = 0.1076$.

   (f) Here $y(t) = 1 - \cos t$, and $Y(t)$ is the function tabulated above. Thus we get:

t	0.00	0.25	0.50	0.75	1.00
$Y(t)$	0	0	0.0618	0.1817	0.3521
$y(t)$	0	0.0311	0.1224	0.2683	0.4597

   (g) To plot $y(t)$ and $Y(t)$ on one pair of axes, we use the formula $1 - \cos t$ for y, and "connect the dots" for Y:

   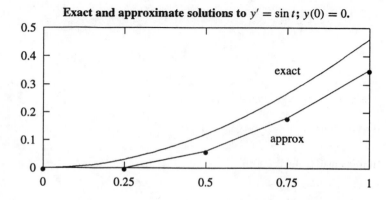

   Notice that the two functions start out together, but spread apart as t increases.

Section 12.3 Euler's Method: Solving DE's Numerically

2. (a) A table helps keep track of results:

step	t	y'	y
0	0.00	1.0000	0.0000
1	0.25	1.2840	0.2500
2	0.50	1.6487	0.5710
3	0.75	2.1170	0.9832
4	1.00	2.7183	1.5124

(b) The left rule with 4 subdivisions, applied to $I = \int_0^1 e^t\, dt$, gives

$$L_4 = \frac{\exp(0)}{4} + \frac{\exp(1/4)}{4} + \frac{\exp(1/2)}{4} + \frac{\exp(3/4)}{4} \approx 1.5124.$$

(c) The function $y(t) = \exp(t) - 1$ solves the DE exactly. Thus, exactly, $y(1) = e - 1 \approx 1.7183$.

(d) $\int_0^1 \exp(t)\, dt = \exp(t)\big]_0^1 = e - 1 \approx 1.7183.$

(e) The error committed by L_4 is $|I - L_4| \approx |1.7183 - 1.5124| = 0.2059$.

(f) Here $y(t) = \exp(t) - 1$, and $Y(t)$ is the function tabulated above. Thus we get:

t	0.00	0.25	0.50	0.75	1.00
$Y(t)$	0	0.2500	0.5170	0.9832	1.5124
$y(t)$	0	0.2840	0.6487	1.1170	1.7183

(g) To plot $y(t)$ and $Y(t)$ on one pair of axes, we use the formula $\exp(t) - 1$ for y, and "connect the dots" for Y:

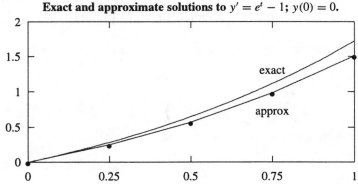

Exact and approximate solutions to $y' = e^t - 1$; $y(0) = 0$.

Notice that the two functions start out together, but spread apart as t increases.

3. (a) It's not hard to guess that the function $y(t) = 3t$ solves the IVP above. Thus $y(1) = 3$, $y(2) = 6$, $y(3) = 9$, $y(4) = 12$, $y(5) = 15$.

(b) In this case, Euler's method gives *exact* values; the Euler estimates commit *no* error.

(c) Euler's method pretends, in effect, that y' remains constant over small intervals. In this case, y' *is* constant, so Euler's method commits no error.

Euler's method will behave this way (i.e., commit *no* error) whenever y' is a constant function.

4. (a) With just 1 subdivision, Euler's method gives

$$e = y(1) \approx y(0) + y'(0) \cdot 1 = 1 + 1 \cdot 1 = 2.$$

This answer *underestimates* e. We can tell this even without knowing the true value for e. The DE $y' = y$ means that $y'(t)$ increases over the interval $0 \le t \le 1$; taking just one Euler step over the interval amounts to pretending that y' is *constant* over this interval.

(b) Carrying out Euler's method with 4 subdivisions, from $t = 0$ to $t = 1$, gives the following table:

step	t	y'	y
0	0.00	1.0000	1.0000
1	0.25	1.2500	1.2500
2	0.50	1.5626	1.5625
3	0.75	1.9351	1.9351
4	1.00	2.4414	2.4414

In particular, $e = y(1) \approx 2.4414$. For the same reason as in the previous part of this exercise, this result *underestimates* the true value of e.

(c) A computer says that with 100 subdivisions, Euler's method gives $e = y(1) \approx 2.7044$. This is *still* an underestimate.

(d) Below are tabulated the results of the first three steps for Euler's method with 10000 subdivisions, applied to $0 \le t \le 1$. Watch the last column:

step	t	y'	y
0	0.0000	1.0000	1.0000
1	0.0001	1.0001	$1 + 0.0001 \cdot 1.0000 = 1.0001$
2	0.0002	1.0001^2	$1.0001 + 0.0001 \cdot 1.0001 = 1.0001^2$
3	0.0003	1.0001^3	$1.0001^2 + 0.0001 \cdot 1.0001^2 = 1.0001^3$

The same pattern persists; in particular, the last line of the table would be

step	t	y'	y
10000	1.0000	1.0001^{10000}	$1.0001^{9999} + 0.0001 \cdot 1.0001^{9999} = 1.0001^{10000}$

Evaluating the last expression with any calculator shows

$$e = y(1) \approx 1.0001^{10000} \approx 2.718145927.$$

This is *still* an underestimate, but not badly so. The "true value" of e is (to 10 digits' accuracy) 2.718281828.

5. One does the "easy calculation" by checking explicitly that both the DE and the initial condition are satisfied. Since $y(t) = 70 + 120e^{-0.1t}$, it's easy to check that

$$y'(t) = -12e^{-0.1t} = -0.1 \cdot 120 \cdot e^{-0.1t}) = -0.1 \cdot (y - 70).$$

Thus, the DE does hold as advertised. Also, $y(0) = 70 + 120 = 190$, so the initial condition holds, too. Finally,

$$y(5) = 70 + 120e^{-0.1 \cdot 5} = 70 + 120e^{-0.5} \approx 142.78.$$

6. (a) Results of Euler's method for 3 one-day steps are shown below:

step	t	y'	y
0	0	5.5044	100
1	1	5.8041	105.50
2	2	6.1199	111.31
3	3	6.4524	117.43

    (b) The original fly population—which started with 1000 members—was about 9600 on day 100. By comparison, using 100 Euler steps of size 1 day to estimate the new population—which starts with 100 members—gives (thanks to a computer) $P(100) \approx 7069$ flies. The point to notice is that the new population is "catching up," despite having started with many fewer members.

7. (a) Work quickly shows that given *initial* population 0, the population *remains* at 0. Nothing happens.
   (b) In biological terms, the situation is simply that without any initial breeding members a population can't grow. No parents; no children.

8. (a) $y(2) = 2$
   (b) $y(2) = 2$
   (c) $y'(t) = 1 = 1 + t - y = 1 + t - t$ and $y(0) = 0$
   (d) Euler's method produces the exact solution.

9. (a) $y(1) \approx -0.59374$
   (b) $y(1) \approx -0.65330$
   (c) $y'(t) = -e^t = (2 - e^t) - 2 = y(t) - 2$ and $y(0) = 1$
   (d) $y(1) = 2 - e \approx -0.71828$

10. (a) $y(1) \approx 1$
    (b) $y(1) \approx 1.25$
    (c) $y(1) \approx 1.4194$
    (d) $y(1) \approx 1.5240$
    (e) $y'(x) = xe^{x^2/2} = xy(x)$ and $y(0) = 1$
    (f) $y(1) = e^{1/2} \approx 1.64872$
    (g) The approximation error is (approximately) halved when the number of steps is doubled. This implies that the error made is proportional to $1/n$.

11. (a) $y(0.8) \approx 2.6764$
    (b) $y'(t) = (1 - t)^{-2} = (y(t))^2$ and $y(0) = 1$
    (c) The derivative is changing rapidly and is becoming large. Thus, very small steps are required to achieve an accurate result.

12. (a) The computations are identical. That is, Euler's method reproduces the left-sum estimate.
    (b) The computations are identical. That is, Euler's method reproduces the left-sum estimate.

12.4 Separating variables: solving DE's symbolically

1. (a) $f(y) = 1/y$, $g(x) = x$
   (b) $F(y) = \ln y$, $G(x) = x^2/2$
   (c) Differentiating both sides of the equation $G(x) = F(y) + C$ with respect to x leads to the equation $G'(x) = F'(y)y'$ or $g(x) = f(y)y'$. Thus, if y is defined by the equation $G(x) = F(y) + C$, y also is a solution of the DE.
   (d) $G(x) = F(y) + K \implies x^2/2 = \ln y + K \implies e^{x^2/2} = e^K y \implies y = Ce^{x^2/2}$

2. (a) The Fundamental Theorem of Calculus implies that $y'(x) = \left(C + \int_a^x g(t)\,dt\right)' = g(x)$. Furthermore, $y(a) = C + \int_a^a g(t)\,dt = C$ since $\int_a^a g(t)\,dt = 0$.
   (b) By the chain rule and the FTC, $(F(y))' = f(y)y'$ and $\left(F(C) + \int_a^x g(t)\,dt\right)' = g(x)$. Furthermore, $F(y(a)) = F(C)$ and $\int_a^a g(t)\,dt = 0$.

Find a solution of each of the following separable differential equations.

3. $\arctan y = x + C$ or $y = \tan(x + C)$
4. $-1/y = x + C$ or $y = -1/(x + C)$
5. $\arcsin y = x + C$ or $y = \sin(x + C)$
6. $e^y = 2x + C$ or $y = \ln(2x + C)$
7. $-\ln|1 - y| = x + C$ or $y = 1 - Ae^{-x}$
8. $\frac{1}{4}\ln\left|\frac{y+2}{y-2}\right| = x + C$ or $y = \frac{2Ae^{4x} + 2}{Ae^{4x} - 1}$
9. $\ln(\ln y) = x + C$ or $y = e^{Ae^x}$
10. $\ln y = x^3/3 + C$ or $y = Ae^{x^3/3}$
11. $y + y^3/3 = x^2/2 + C$
12. $\ln(\sec y + \tan y) = x^3/3 + C$
13. $-1/y = x^2/2 + C$ or $y = 2/(A - x^2)$
14. $y^2/2 = x^2/2 + C$ or $y = \pm\sqrt{x^2 + A}$

15. Separating variables in the DE gives
$$\frac{dP}{dt} = -0.00000556 P(P - 10000) \implies \int \frac{dP}{P(P - 10000)} = -\int 0.00000556\,dt.$$

Integrating both sides gives
$$-\frac{\ln|P|}{10000} + \frac{\ln|P - 10000|}{10000} = -0.00000556t + C,$$

for some constant C. Setting $P = 1000$ and $t = 0$ gives
$$C = -\frac{\ln 1000}{10000} + \frac{\ln 9000}{10000} = \frac{\ln 9}{10000} \approx 0.0002197.$$

Thus, $P(t) = \dfrac{10000}{9e^{-0.0556t} + 1}$. Setting $t = 10$ and solving for P gives $P(10) \approx 1623$—not much different from what we've seen in other sections.

16. (a) That all curves "start" at $(0, 190)$ means that all cups of coffee are at 190 degrees at time 0.

    (b) In each case, T_r describes the (stable) *room* temperature. Thus the nine coffee cups are in rooms of different temperatures.

    (c) One way to find the values of T_r that correspond to C_1, C_4, and C_9 is just to observe that T_r represents the "long-run" temperature—the temperature to which the coffee tends over a long period of time. Looking at the right-hand parts of the graphs lets us read off these numbers for C_1 and C_4. Thus C_1 corresponds to $T_r = 100$; C_4 corresponds to $T_r = 70$. Not enough of C_9 appears for this to work very well. Instead, we can use the fact that for C_9, $y(10) = 70$. From this it follows:

    $$y(10) = (190 - T_r)e^{-1} + T_r = 70 \implies T_r \approx 0.$$

    (d) C_9 is cooling fastest at $t = 10$; of all the curves, C_9 has steepest downward slope at $t = 10$.

    (e) Curve C_4 solves the IVP $y' = -0.1(y - 70)$; $\quad y(0) = 190$. In the same spirit, C_1 solves the IVP $y' = -0.1(y - 100)$; $\quad y(0) = 190$, and C_9 solves the IVP $y' = -0.1(y - 0)$; $\quad y(0) = 190$. (The DE's differ only in the values of T_r.)

    (f) Estimating the *slope* of the curve C_1 at the point $t = 10$ gives $y'(10) \approx -3.5$ degrees per minute.

    (g) The DE for C_1 is (as we found above) $y' = -0.1(y - 100)$. The graph C_1 shows that when $t = 10$, $y \approx 135$, so, according to the DE, $y' = -0.1(135 - 100) = -3.5$ degrees per minute. (This agrees with the previous part, of course.)

    (h) Estimating the *slope* of the curve C_9 at the point $t = 10$ gives $y'(10) \approx -7$ degrees per minute.

    (i) The DE for C_9 is (as we found above) $y' = -0.1y$. The graph C_9 shows that when $t = 10$, $y \approx 70$, so, according to the DE, $y' = -0.1 \cdot 70 = -7$ degrees per minute. (This agrees with the previous part, as it should.)

    (j) Looking carefully at the picture. A curve that represents coffee that starts at 190 degrees and is at 100 degrees after 20 minutes would fall about *halfway* between C_2 and C_3. In particular, C_2 and C_3 correspond to room temperatures of 90 and 80 degrees, respectively. Thus the desired curve would correspond to room temperature of about 85 degrees.

17. (a) We need to check (1) that $y(t) = (T_0 - T_r)e^{-0.1t} + T_r$ solves the DE $y' = -0.1(y - T_r)$; and (2) that $y(0) = T_0$. The latter is easy: $y(0) = (T_0 - T_r)e^0 + T_r = T_0$, as claimed.
    To check (1), we calculate both sides of the DE and see that they agree:

    $$y(t)' = (T_0 - T_r) \cdot e^{-0.1t} \cdot (-0.1);$$
    $$-0.1(y - T_r) = (-0.1) \cdot (T_0 - T_r)e^{-0.1t}.$$

    (b) If $T_0 = 200$ and $y(10) = 100$, then

    $$y(10) = (200 - T_r)e^{-1} + T_r = 100 \implies T_r = -\dfrac{200e^{-1} - 100}{-e^{-1} + 1} \approx 41.802.$$

    Given this value of T_r,

    $$y(20) \approx (200 - 41.802)e^{-2} + 41.802 \approx 63.212.$$

    In coffee language, this means that the coffee temperatures were 200 degrees, 100 degrees, and about 63 degrees at times 0, 10, and 20, respectively.

(c) If $T_r = 80$ and $y(10) = 120$, then

$$y(10) = (T_0 - 80)e^{-1} + 80 = 120 \implies T_0 = -\left(-\frac{80}{e} - 40\right)e \approx 188.731.$$

With this value for T_0, we have

$$y(20) \approx (188.731 - 80)e^{-2} + 80 \approx 94.715.$$

In coffee language, this means that the coffee temperatures were around 189 degrees, 120 degrees, and 95 degrees at times 0, 10, and 20, respectively.

(d) We can use the information $y(10) = 100$ and $y(20) = 80$ to solve for *both* constants T_0 and T_r, as follows:

$$y(10) = 100 \implies 100 = (T_0 - T_r) \cdot e^{-1} + T_r;$$
$$y(20) = 80 \implies 80 = (T_0 - T_r) \cdot e^{-2} + T_r.$$

Solving these two equations gives $T_r \approx 68.36$, $T_0 \approx 154.36$. In other words, the coffee was about 154 degrees at time 0; room temperature is about 68 degrees. When $t = 40$, the temperature will be $y(40) = (154.36 - 68.36)e^{-4} + 68.36 \approx 69.94$ degrees.

(e) Since $\lim_{t \to \infty} e^{-0.1t} = 0$,

$$\lim_{t \to \infty} y(t) = \lim_{t \to \infty} (T_0 - T_r)e^{-0.1t} + T_r = T_r.$$

In coffee language, this means (as experience suggests!) that in the long run coffee cools to room temperature.

18. (a) If $P < C$, then $P' > 0$. This means that the population *increases*—as it should, since carrying capacity hasn't been reached.

(b) If $P > C$, then $P' < 0$. This means that the population *decreases*—as it should, since carrying capacity has been exceeded.

(c) If $P = C$, then $P' = 0$. This means that the population is *stable*—as it should be, when exactly at capacity.

19. The point is that $e^{-x} \to 0$ as $x \to \infty$. Now if k and C are positive, $Ckt \to \infty$ as $t \to \infty$. Therefore $e^{-Ckt} \to 0$ as $t \to \infty$. Thus

$$\lim_{t \to \infty} P(t) = \lim_{t \to \infty} \frac{C}{Ke^{-Ckt} + 1} = \frac{C}{K \cdot 0 + 1} = C.$$

In biological terms, this means that in the long run, the population tends toward C, its carrying capacity.

20. Solving algebraically for k gives $k = -\frac{\ln(\frac{19}{49})}{1000}$; an approximate decimal equivalent is 0.000947, as claimed.

21. (a) Differentiating the right side of $P' = kP(C - P)$ with respect to P gives

$$\frac{d}{dP}(kP(C - P)) = \frac{d}{dP}\left(kCP - kP^2\right) = kC - 2KP = 0 \iff C = \frac{P}{2}.$$

Thus P' has its maximum value where $P = C/2$, as claimed. (In rumor language: The rumor spreads fastest when *half* the people know it.)

(b) Let's solve $P(t) = 500$ for t (check the steps):

$$P(t) = \frac{1000}{49e^{-0.947t} + 1} = 500 \implies 2 = 49e^{-0.947t} + 1 \implies t = \frac{\ln 49}{0.947} \approx 4.11.$$

(In rumor language: After 4.11 days, half the people know the rumor.) (Notice that one could also have read this from the graph.)

SECTION 12.4 SEPARATING VARIABLES: SOLVING DE'S SYMBOLICALLY

22. (a) The exact values of k for curves P_1 to P_4 are, respectively, 2, 1.5, 1, and 0.5. (Thus the values of $-k$ are, respectively, -2, -1.5, -1, and -0.5.)

   The k-values can be found—approximately—by reading the graphs. The graph of P_1, for example, seems to pass through the point $(2, 750)$. This gives an equation we can solve for k, as follows:

   $$P_2(2) = \frac{1000}{19e^{-2k} + 1} \approx 750 \implies k = \frac{\ln 57}{2} \approx 2.02.$$

   Values of k for the other curves are found similarly.

   (b) P_1 corresponds to the "hottest" rumor, P_4 to the "coolest." The graph shapes agree—hotter rumors spread faster.

23. For each part, we know two things:

    $$P'(t) = \frac{K}{1+t} P(1000 - P) \quad \text{and} \quad P(t) = \frac{1000}{d(1+t)^{-1000K} + 1}.$$

    The point is to use given information to find explicit values for K and d.

    (a) If $P(0) = 100$ and $P'(0) = 50$, then

    $$P(0) = \frac{1000}{d+1} = 100 \implies d = 9;$$

    $$P'(0) = K \cdot 100 \cdot 900 = 50 \implies K = \frac{1}{1800}.$$

    With these values for K and d, we have

    $$P(t) = \frac{1000}{9(1+t)^{-1000/1800} + 1} = \frac{1000}{9(1+t)^{-5/9} + 1}.$$

    (b) If $P(0) = 50$ and $P'(0) = 100$, then

    $$P(0) = \frac{1000}{d+1} = 50 \implies d = 19;$$

    $$P'(0) = K \cdot 100 \cdot 900 = 100 \implies K = \frac{1}{900}.$$

    Therefore

    $$P(t) = \frac{1000}{19(1+t)^{-1000/900} + 1} = \frac{1000}{19(1+t)^{-10/9} + 1}.$$

    (c) If $P(0) = 800$ and $P'(0) = 50$, then (as above) $d = 1/4$, $k = 1/3200$, so $P(t) = \dfrac{4000}{(1+t)^{-5/16} + 4}$.

    (d) If $P(0) = 800$ and $P'(0) = 100$, then (as above) $d = 1/4$, $k = 1/1600$, so $P(t) = \dfrac{4000}{(1+t)^{-5/8} + 4}$.

    (e) Here are graphs, machine drawn:

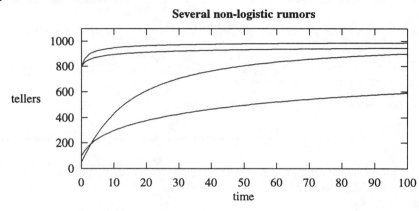

24. From the problem statement, $\int_a^x \sqrt{1 + (f'(t))^2} \, dt = \int_a^x f(t) \, dt$. Differentiating both sides of this equation with respect to x leads to the equation $\sqrt{1 + (f'(x))^2} = f(x)$. Solving this equation for $f'(x)$, we find that $y = f(x)$ must be a solution of the DE $y' = \sqrt{y^2 - 1}$. Since $\int \frac{dx}{\sqrt{x^2 - 1}} = \cosh^{-1} x$ and $\cosh 0 = 1$, $y(x) = \cosh x$ is a function with the desired properties.

25. Amy drinks the hotter coffee. Let $A(t)$ be the temperature of Amy's coffee and $J(t)$ be the temperature of Joan's coffee. Also, let $J(0)$ be the initial temperature of both cups of coffee (before any cream is added) and $A(0)$ be the temperature of Amy's coffee after the cream is added. The effect of adding the cream to a cup of coffee is to reduce the temperature of the coffee by by $J(0) - A(0)$. Now, because of Newton's law of cooling, we know that $J(10) - A(10) < J(0) - A(0)$ so $J(10) - (J(0) - A(0)) < A(10)$.

    NOTE: This argument is only qualitatively correct. A more careful analysis requires taking the relative amounts of the coffee and the cream into account.

26. (a) Using the chain rule and the product rule, $(e^{F(x)} y(x))' = e^{F(x)}(y'(x) + F'(x)y(x))e^{F(x)}(y'(x) + f(x)y(x))$.

    (b) $e^{F(x)} y(x) = C + \int_a^x e^{F(t)} g(t) \, dt$ so $(e^{F(x)} y(x))' = e^{F(x)} g(x) = e^{F(x)}(y'(x) + f(x)y(x))$ — the last equality from part (a). Since $e^{F(x)} \neq 0$ for any x, $y'(x) + f(x)y(x) = g(x)$.

    ALTERNATIVE SOLUTION: $y'(x) = -f(x)e^{-F(x)} \left(C + \int_a^x e^{F(t)} g(t) \, dt \right) + e^{-F(x)} e^{F(x)} g(x) = -f(x)y(x) + g(x)$. Therefore, $y'(x) + f(x)y(x) = g(x)$. Furthermore,
    $$y(a) = e^{-F(a)} \left(C + \int_a^a e^{F(t)} g(t) \, dt \right) = Ce^{-F(a)}.$$

    (c) Let $f(x) = -1$ and $g(x) = x$. Then $F(x) = -x$ and $\int_0^x e^{F(t)} g(t) \, dx = 1 - (x + 1)e^{-x}$. Therefore, since $e^{-F(0)} = 1$, $y(x) = e^x \left(C + 1 - (x + 1)e^{-x} \right) = (C + 1)e^x - x - 1$.

27. Let $f(x) = 2x$ and $g(x) = x$. Then $F(x) = x^2$ and $\int_0^x e^{F(t)} g(t) \, dx = e^{x^2}/2 - 1/2$. Therefore, the solution of the IVP is $y(x) = 1/2 - e^{-x^2}/2$.

Use part (b) of Exercise 26 to find the solution of each of the IVPs in Exercises 28–31.

28. Here $f(x) = -2x$ and $g(x) = x$. Therefore, the solution of the IVP is $y(x) = -1/2 + 3/2e^{x^2}$.

29. Here $f(x) = x$ and $g(x) = e^{2x}$. Therefore, the solution of the IVP is $y(x) = e^{2x}/3 + (2e - e^3/3)e^{-x}$.

30. Here $f(x) = \cos x$ and $g(x) = \cos x$. Therefore, the solution of the IVP is $y(x) = 1 - e^{-\sin x}$.

31. Start by rewriting the DE in the form $y' + \frac{2}{x} y = \frac{\sin x}{x}$. Then $f(x) = 2/x$ and $g(x) = (\sin x)/x$, so $F(x) = 2 \ln x$, $e^{F(x)} = x^2$, and $\int_{\pi/2}^x e^{F(t)} g(t) \, dt = \sin x - x \cos x - 1$. Therefore, the solution of the IVP is
$$y(x) = \frac{\pi^2/4 + \sin x - x \cos x - 1}{x^2}.$$

32. (a) $(z(x))' = (1 - n)(y(x))^{-n} y'(x)$. Therefore,
$$\frac{1}{1 - n} z'(x) + f(x)z(x) = \frac{1}{1 - n}(1 - n)(y(x))^{-n} y'(x) + f(x)(y(x))^{1-n}$$
$$= (y(x))^{-n} (y'(x) + f(x)y(x))$$
$$= (y(x))^{-n} (g(x)(y(x))^n) = g(x).$$

(b) The DE $P' - kCP = -kP^2$ is a Bernoulli equation. Thus, by part (a), $z(t) = P^{-1}$ is a solution of the first-order linear DE $z' + kCz = k$. Using part (b) of Exercise 26 with $f(t) = kC$ and $g(t) = k$, the solution of this DE is $z(t) = be^{-kCt} + 1/C$ and b is a constant. A bit of algebra now leads to the result:
$$P(t) = \frac{C}{bCe^{-kCt} + 1}.$$

(c) Let $z(x) = (y(x))^{-2}$. By part (a), z is a solution of the DE $z' + 2z = -2$. The solution of this DE is $z(x) = Ce^{-2x} - 1$. Therefore, a solution of the given DE is $y(x) = \pm \left(Ce^{-2x} - 1\right)^{-1/2}$.

33. (b) The DE is a Bernoulli equation with $n = -1$. The solution is $y(x) = x\sqrt{C + 2\ln x}$.

13.1 Polar coordinates and polar curves

1. $(\sqrt{2}, \pi/4)$; $(\sqrt{2}, -7\pi/4)$; $(-\sqrt{2}, -3\pi/4)$; $(-\sqrt{2}, 5\pi/4)$
2. $(\sqrt{2}, 3\pi/4)$; $(\sqrt{2}, -5\pi/4)$; $(-\sqrt{2}, 7\pi/4)$; $(-\sqrt{2}, -\pi/4)$
3. $(2, \pi/3)$; $(2, -5\pi/3)$; $(-2, 4\pi/3)$; $(-2, -2\pi/3)$
4. $(2, 5\pi/6)$; $(2, 7\pi/6)$; $(-2, \pi/6)$; $(-2, -\pi/6)$
5. $(\pi, 0)$; $(-\pi, \pi)$; $(\pi, 2\pi)$
6. $(\pi, \pi/2)$; $(\pi, -3\pi/2)$; $(-\pi, -\pi/2)$; $(-\pi, 3\pi/2)$
7. $(\sqrt{5}, 1.107)$; $(\sqrt{5}, -5.176)$; $(-\sqrt{5}, 4.249)$
8. $(\sqrt{5}, 2.034)$; $(\sqrt{5}, -4.249)$; $(-\sqrt{5}, 5.176)$
9. $(\sqrt{17}, 1.326)$; $(\sqrt{17}, -4.957)$; $(-\sqrt{17}, 4.467)$
10. $(\sqrt{10001}, 1.561)$; $(\sqrt{10001}, -4.722)$; $(-\sqrt{10001}, 4.702)$
11. $(0.596, 0.931)$; $(0.596, -5.352)$; $(-0.596, 4.072)$
12. $(0.596, 4.072)$; $(0.596, -2.211)$; $(-0.596, 0.931)$
13. $(\sqrt{2}, \sqrt{2})$
14. $(\sqrt{2}, \sqrt{2})$
15. $(\sqrt{3}/2, 1/2)$
16. $(42, 0)$
17. $(a, 0)$
18. $(-a, 0)$
19. $(0.540, 0.841)$
20. $(-0.540, -0.841)$
21. $(-0.832, 1.82)$
22. $(-0.832, -1.82)$
23. $(\sqrt{2}/2, \sqrt{2}/2)$
24. $(0.447, 0.894)$
25. Points of the form $(a, 0)$ where $a \in \mathbb{R}$ have this property.
26. (a)

θ	0	$\frac{\pi}{6}$	$\frac{\pi}{3}$	$\frac{\pi}{2}$	$\frac{2\pi}{3}$	$\frac{5\pi}{6}$	π	$\frac{7\pi}{6}$	$\frac{4\pi}{3}$	$\frac{3\pi}{2}$	$\frac{5\pi}{3}$	$\frac{11\pi}{6}$	2π
r	2	2.5	2.866	3	2.866	2.5	2	1.5	1.134	1	1.134	1.5	2

(c) The limaçon is symmetric with respect to the y-axis.

(d) The r-values are symmetric about $\theta = \pi/2$.

SECTION 13.1 POLAR COORDINATES AND POLAR CURVES

27. (a)

θ	0	$\frac{\pi}{6}$	$\frac{\pi}{3}$	$\frac{\pi}{2}$	$\frac{2\pi}{3}$	$\frac{5\pi}{6}$	π	$\frac{7\pi}{6}$	$\frac{4\pi}{3}$	$\frac{3\pi}{2}$	$\frac{5\pi}{3}$	$\frac{11\pi}{6}$	2π
r	2	1.866	1.5	1	0.5	0.134	0	0.134	0.5	1	1.5	1.866	2

The cardioid is symmetric with respect to the x-axis.

28. (a)

θ	0	$\frac{\pi}{6}$	$\frac{\pi}{3}$	$\frac{\pi}{2}$	$\frac{2\pi}{3}$	$\frac{5\pi}{6}$	π	$\frac{7\pi}{6}$	$\frac{4\pi}{3}$	$\frac{3\pi}{2}$	$\frac{5\pi}{3}$	$\frac{11\pi}{6}$	2π
r	-1	-0.73	0	1	2	2.73	3	2.73	2	1	0	-0.73	-1

(b) $r = 0$ when $\theta = \pi/3$ and $\theta = 5\pi/3$.

(c) $r < 0$ when $0 \leq \theta < \pi/3$ and when $5\pi/3 < \theta \leq 2\pi$.

41. (a) The polar points $(1, 0)$, $(1, 2\pi)$, and $(-1, \pi)$ all represent the Cartesian point $(1, 0)$. (Use $x = r\cos\theta$ and $y = r\sin\theta$.)

(b) The polar points $(-1, \pi/4)$, $(-1, 9\pi/4)$, and $(1, 5\pi/4)$ all represent the Cartesian point $(-\sqrt{2}/2, -\sqrt{2}/2)$.

(c) $(\sqrt{2}, \pi/4 + 2k\pi)$ and $(-\sqrt{2}, \pi/4 + (2k-1)\pi)$

42. $x = 2$

43. $x^2 + y^2 = 16$

44. $x = y$

45. $r = 2\sin\theta \implies r^2 = 2r\sin\theta \implies x^2 + y^2 = 2y$

46. $r = 3$

47. $r = 4\csc\theta$

48. $\tan\theta = 2$

49. $r = 2\cos\theta$

50. (a) $a > 1$

(b) When $a = 1$ the limaçon has a "dimple" rather than an inner loop.

(c) As $a \to 0$ the graph becomes a circle centered at the origin.

(d) The graphs are reflections of each other across the y-axis.

(e) As $a \to \infty$, the graph becomes circular.

13.2 Calculus in polar coordinates

1. The results are straightforward applications of the product rule: $(f \cdot g)' = f' \cdot g + f \cdot g'$.

2. $\dfrac{dy}{dx} = \dfrac{-\sin^2\theta + (1+\cos\theta)\cos\theta}{-\sin\theta\cos\theta - (1+\cos\theta)\sin\theta}$
   $= \dfrac{\sin^2\theta - \cos^2\theta - \cos\theta}{(2\cos\theta + 1)\sin\theta} = \dfrac{1 - 2\cos^2\theta - \cos\theta}{(2\cos\theta + 1)\sin\theta}$
   $= \dfrac{(1 - 2\cos\theta)(1 + \cos\theta)}{(2\cos\theta + 1)\sin\theta}$

3. (b) $\dfrac{dy}{dx} = \dfrac{\sin\theta + \theta\cos\theta}{\cos\theta - \theta\sin\theta}$. Thus, the spiral has a horizontal tangent line wherever $\theta = -\tan\theta$ and a vertical tangent line wherever $\theta = \cot\theta$.

   (c) The polar point $(1, 1)$ is the point $(\cos 1, \sin 1)$ in Cartesian coordinates. The slope of the tangent line at the polar point $(1, 1)$ is $m = (\sin 1 + \cos 1)/(\cos 1 - \sin 1) \approx -4.588$. Thus, the equation of the desired tangent line is $y = m(x - \cos 1) + \sin 1 \approx -4.59(x - 0.54) + 0.84$.

4. (b) $\dfrac{dy}{dx} = \dfrac{2\cos\theta\sin\theta + (1 + 2\sin\theta)\cos\theta}{2\cos^2\theta - (1 + 2\sin\theta)\sin\theta} = \dfrac{4\cos\theta\sin\theta + \cos\theta}{2\cos^2\theta - 2\sin^2\theta - \sin\theta}$. Thus, at the polar point $(1, 0)$ the limaçon has slope $1/2$. Thus, the tangent line is described by the equation $y = (x - 1)/2$.

   (c) The polar point $(0, 7\pi/6)$ corresponds to the Cartesian point $(0, 0)$. The slope of the tangent line at this point is $\sqrt{3}/3$ so an equation of the tangent line is $y = \sqrt{3}x/3$.

   (d) The polar point $(0, 11\pi/6)$ corresponds to the Cartesian point $(0, 0)$. The slope of the tangent line at this point is $-\sqrt{3}/3$ so an equation of the tangent line is $y = -\sqrt{3}x/3$.

   (e) Thus, the limaçon has horizontal tangent lines when $\cos\theta = 0$ or $\sin\theta = -1/4$—at the points $(3, \pi/2)$, $(-1, 3\pi/2)$, $(1/2, 3.3943)$, and $(1/2, 6.0305)$.

5. (b) $\dfrac{dy}{dx} = \dfrac{-a\sin^2\theta + \cos\theta + a\cos^2\theta}{-2a\sin\theta\cos\theta - \sin\theta}$. The limaçon has a vertical tangent whenever the denominator in this expression is zero (i.e., when $\theta = 0$ or $\cos\theta = -1/2a$). Thus, there will be three vertical tangent lines if and only if $|a| \leq 1/2$.

6. area $= \dfrac{1}{2}\displaystyle\int_0^\pi 1\, d\theta = \pi/2$

7. area $= 9\pi/2$

8. area $= \pi a^2/2$

9. area $= \beta/2$

10. area $= \displaystyle\int_0^{\pi/2} f(\theta)^2\, d\theta + \int_{11\pi/6}^{2\pi} f(\theta)^2\, d\theta = \dfrac{1}{2}\int_{-\pi/6}^{7\pi/6} f(\theta)^2\, d\theta = 2\pi + \dfrac{3\sqrt{3}}{2}$

11. area $= \dfrac{1}{2}\displaystyle\int_{7\pi/6}^{11\pi/6} f(\theta)^2\, d\theta = \pi - 3\sqrt{3}/2$

12. area $= 1/2$

13. area $= m/2$

14. (a) One leaf lies between $\theta = -\pi/2n$ and $\theta = \pi/2n$. The area of this leaf is $\dfrac{1}{2}\displaystyle\int_{-\pi/2n}^{\pi/2n} \cos^2(n\theta)\, d\theta = \pi/4n$.

    (b) The area of all $2n$ leaves is $\pi/2$ (i.e., one-half of the area of the circle $r = 1$).

SECTION 13.2 CALCULUS IN POLAR COORDINATES

15. (a) One leaf lies between $\theta = -\pi/2n$ and $\theta = \pi/2n$. The area of this leaf is $\dfrac{1}{2}\displaystyle\int_{-\pi/2n}^{\pi/2n} \cos^2(n\theta)\,d\theta = \pi/4n$.

    (b) The area of all n leaves is $\pi/4$ (i.e., one-fourth of the area of the circle $r = 1$).

16. area $= \displaystyle\int_0^{2\pi} \theta^2\,d\theta = 4\pi^3/3$

17. area $= \left(e^{4\pi} - 1\right)/4$

18. area $= \displaystyle\int_{2\pi}^{4\pi} \dfrac{\ln(t)^2}{2}\,dt \approx 15.664$. (The integral can be done in closed form, but is messy.)

19. area $= \dfrac{1}{2}\displaystyle\int_{-\pi/3}^{\pi/3} d\theta - \dfrac{1}{2}\displaystyle\int_{-\pi/3}^{\pi/3} \left(\tfrac{1}{2}\sec\theta\right)^2 d\theta = \dfrac{\pi}{3} - \dfrac{\sqrt{3}}{4}$

20. area $= \dfrac{1}{2}\displaystyle\int_{-\arccos a}^{\arccos a} d\theta - \dfrac{1}{2}\displaystyle\int_{-\arccos a}^{\arccos a} (a\sec\theta)^2\,d\theta = \arccos a - a^2\tan(\arccos a) = \arccos a - a\sqrt{1-a^2}$

21. $x = 2\cos t$, $y = 2\sin t$, $0 \le t \le 2\pi$. The graph is a circle of radius 2 with center at the origin.

22. $x = 2\cos t$, $y = 2\sin t$, $0 \le t \le \pi$. The graph is the top half of a circle of radius 2 with center at the origin.

23. $x = 1$, $y = \tan t$, $-\pi/4 \le t \le \pi/4$. The graph is the vertical line segment from $(1, -1)$ to $(1, 1)$.

24. $x = \cot t$, $y = 1$, $\pi/4 \le t \le 3\pi/4$. The graph is the horizontal line segment from $(1, 1)$ to $(-1, 1)$.

25. $x = t\cos t$, $y = t\sin t$, $0 \le t \le 2\pi$. The graph is one loop of a spiral.

26. $x = \cos^2 t$, $y = \cos t \sin t$, $0 \le t \le \pi$. The graph is a circle of radius $1/2$ with center at $(1/2, 0)$.

27. $x = \cos(2t)\cos t$, $y = \cos(2t)\sin t$, $0 \le t \le 2\pi$. The graph is the 4-leaf rose shown at the beginning of the section.

28. $x = \cos t + \cos^2 t$, $y = \sin t + \cos t \sin t$, $0 \le t \le 2\pi$. The graph is the cardioid shown at the beginning of the section.

29. $\left(\dfrac{dy}{d\theta}\right)^2 + \left(\dfrac{dx}{d\theta}\right)^2 = \left(f'(\theta)\sin\theta + f(\theta)\cos\theta\right)^2 + \left(f'(\theta)\cos\theta - f(\theta)\sin\theta\right)^2$ which implies that $dx/d\theta$

    $= \left(f'(\theta)^2 \sin^2\theta + 2f(\theta)f'(\theta)\sin\theta\cos\theta + f(\theta)^2 \cos^2\theta\right) +$
    $\qquad \left(f'(\theta)^2 \cos^2\theta - 2f(\theta)f'(\theta)\sin\theta\cos\theta + f(\theta)^2 \sin^2\theta\right)$
    $= f'(\theta)^2 + f(\theta)^2 > 0$

    and $dy/d\theta$ are not both simultaneously zero.

14.1 Three-dimensional space

1. (a) The surface is unrestricted in the x-direction.
   (b) The surface is unrestricted in the y-direction.

2. The graph in the xy-plane is a circle; in xyz-space it's a pipe centered on the z-axis.

3. All the graphs are cylinders, unrestricted in the x-direction.

   (a) Two planes, crossing diagonally along the x-axis.
   (b) The x-axis.
   (c) A parabolic "tent," unrestricted in the x-direction.
   (d) The empty set, because the equation $y^2 + z^2 = -1$ has no solutions.

4. (a) $x^2 + y^2 + z^2 = 4$
   (b) $(x-1)^2 + (y-1)^2 + (z-1)^2 = 1$
   (c) $x^2 + z^2 = 1$
   (d) $x^2 + y^2 = 4$
   (e) $z = \sin x$ and $z = \cos x$ are two possibilities

5. Yes, the graph is "cylindrical" in both the x-direction and the y-directions. The trace in the xz-plane (where $y = 0$) is the line $z = 3$. The trace in the yz-plane (where $x = 0$) is the line $z = 3$. The trace in the xy-plane (where $z = 0$) is empty.

6. Completing the square in y and z gives $x^2 + y^2 - 6y + z^2 - 4z = 0 \iff x^2 + (y-3)^2 + (z-2)^2 = 13$. Thus we have a sphere of radius $\sqrt{13}$, centered at $(0, 3, 2)$.

7. (a) Setting $x = 0$ in $x^2 + y^2 + z^2 = 1$ gives $y^2 + z^2 = 1$—the unit circle in the yz-plane.
   (b) Setting $y = 0$ in $x^2 + y^2 + z^2 = 1$ gives $x^2 + z^2 = 1$—the unit circle in the xz-plane.
   (c) Draw half circles in each of the three coordinate planes; join them to get part of the sphere.
   (d) Setting $z = 1/2$ in $x^2 + y^2 + z^2 = 1$ gives $x^2 + y^2 = 3/4$—the circle of radius $\sqrt{3}/2$ centered at $(0, 0)$ in the plane $z = 1/2$.
   (e) Setting $z = a$ in $x^2 + y^2 + z^2 = 1$ gives $x^2 + y^2 = 1 - a^2$—the circle of radius $\sqrt{1-a^2}$ centered at $(0, 0)$ in the plane $z = a$. If $a = 2$, there is no intersection.

9. All parts follow by manipulating the formula $d(P, Q) = \sqrt{(x-a)^2 + (y-b)^2 + (z-c)^2}$.

10. The desired 8 points are $(1, 1, 1)$, $(1, 1, -1)$, $(1, -1, 1)$, $(1, -1, -1)$, $(-1, 1, 1)$, $(-1, 1, -1)$, $(-1, -1, 1)$, $(-1, -1, -1)$.

11. (a) $3x - 1y = -5$
    (b) $mx - 1y = -b$
    (c) $0x + 1y = 3$
    (d) $1x + 0y = 1$

12. (a) If $A = B = 0$, then $Ax + By = C$ becomes $0 = C$, which is either trivial or impossible.
    (b) The line $Ax + By = C$ has slope $-A/B$, so lines with $B = 0$ have undefined slope.
    (c) Setting $x = 0$ in $Ax + By = C$ gives $By = C$, or $y = C/B$. Thus, lines with $B = 0$ have no y-intercept.
    (d) Setting $y = 0$ in $Ax + By = C$ gives $Ax = C$, or $x = C/A$. Thus, lines with $A = 0$ have no x-intercept.

13. (a) If $A = B = C = 0$, then the equation becomes $0 = D$, which is either trivial or impossible.

Section 14.1 Three-dimensional Space

(b) Setting $y = 0$ and $z = 0$ in $Ax + By + Cz = D$ gives $Ax = D$, so the x-intercept is $x = D/A$. Thus any plane with $A = 0$ and $D \neq 0$ (such as $y + z = 1$) has no x-intercept.

(c) Setting $x = 0$ and $y = 0$ in $Ax + By + Cz = D$ gives $Cz = D$, so the z-intercept is $z = D/C$. Thus any plane with $C = 0$ and $D \neq 0$ (such as $x + y = 1$) has no z-intercept.

14. (a) Setting $y = 0$ in $x + 2y + 3z = 3$ gives the trace $x + 3z = 3$; this line intercepts the x- and z-axes at $x = 3$ and $z = 1$, respectively.

    (b) Setting $x = 0$ in $x + 2y + 3z = 3$ gives the trace $2y + 3z = 3$; this line intercepts the y- and z-axes at $y = 3/2$ and $z = 1$, respectively.

    (c) Setting $x = 1$ in $x + 2y + 3z = 3$ gives the trace $1 + 2y + 3z = 3$, or $2y + 3z = 2$.

15. (a) Setting $y = 0$ and $z = 0$ gives $x = 1$—that's the x-intercept. Similarly, $y = 2$ and $z = 4$ are the y- and z-intercepts. Connecting these points in the first quadrant gives a picture of the plane.

    (b) The traces are $2y + z = 4$ in the yz-plane; $4x + z = 4$ in the xz-plane; $4x + 2y = 4$ in the xy-plane. These are the lines joining the points found in part (a).

16. (a) The line $1x + 0y = 3$ intercepts the x-axis but not the y-axis.

    (b) The plane $x = 1$ in xyz-space intercepts only the x-axis, at the point $x = 1$.

    (c) The plane $x + 2y = 1$ in xyz-space intercepts the x-axis at $x = 1$ and the y-axis at $y = 1/2$.

    (d) Any plane of the form $bz + cy = d$, where all coefficients are non-zero, hits the y-axis and the z-axis but not the x-axis.

17. The midpoint has coordinates $((x_1 + x_2)/2, (y_1 + y_2)/2, (z_1 + z_2)/2)$; one shows that these coordinates satisfy the equation of the given plane.

18. Various answers are possible.

19. The left and right systems have different orientation.

14.2 Functions of several variables

1. (a) $g(x, y) = x^2 + y^2$ has domain $\mathbb{R}^2$; the range is $[0, \infty)$
   (b) $h(x, y) = x^2 + y^2 + 3$ has domain $\mathbb{R}^2$; the range is $[3, \infty)$
   (c) $j(x, y) = 1/(x^2 + y^2)$ has domain all of $\mathbb{R}^2$ except for the origin; the range is $(0, \infty)$
   (d) $k(x, y) = x^2 - y^2$ has domain $\mathbb{R}^2$; the range is $(-\infty, \infty)$
   (e) The domain of $m(x, y) = \sqrt{1 - x^2 - y^2}$ is all points inside and on the unit circle in $\mathbb{R}^2$; the range is $[0, 1]$.

2. Always, $f(x, y) = y - x^2$ and $g(x, y) = x - y^2$.

   (a) Level curves are a family of parabolas.
   (b) Both are families of parabolas. The level curves of f open upward and are symmetric about the y-axis. The level curves of g open to the right and are symmetric about the x-axis.
   (c) The two surfaces are the same except for their orientation: f is a curved surface that arches over the y-axis while g arches over the x-axis.

3. (a) In the rectangle All level curves of f are circles centered at the origin, except the curve $z = 0$, which is just the origin.
   (b) All level curves of g are circles centered at the origin, except for except the curve $z = 2$, which is just the origin.
   (c) The curves themselves are identical, but their labels are different.
   (d) Both graphs are paraboloids, opening upward. The g-graph is one unit higher.

4. (a) All the level curves are straight lines with slope 2/3.
   (b) These level curves are also straight lines—the same lines as in the previous part.
   (c) The level lines are the same for both functions, but their labels are different. Specifically, the level line $z = c$ for f is the level line $z = -c$ for g.
   (d) All level curves of linear functions are straight lines.
   (e) Both graphs are planes. At each (x, y), the two planes have opposite z-coordinates.

6. (a) In weather language, $T(0, 0) = 15$ means that the temperature was 15 degrees Celsius at noon CST in Los Angeles on January 1, 1996.
   (b) Level curves of T, in weather language, are curves along which the temperature remains constant. As a rule, such "isotherms" run east-and-west.
   (c) $T(1400, 1100) = -15$
   (d) Near the coldest (or warmest) spot in the country, the level curves should be closed curves, shrinking down to the cold spot.

7. (a) The domain and range of f are $\mathbb{R}^2$ and $[0, \infty)$, respectively.
   (b) The graph of f is a cone in xyz-space.
   (c) The level curve of f through $(3, 4)$ is a circle of radius 5, centered at the origin.
   (d) All level curves of f are circles centered at the origin.

8. (a) $f(x, y) = |x - 1|$
   (b) The graph of f is a vee-shaped trough, parallel to the y-axis.
   (c) The level curve of f through $(3, 4)$ is the straight line $x = 3$.

SECTION 14.2 FUNCTIONS OF SEVERAL VARIABLES

   (d) All level curves of f are straight lines parallel to the y-axis.
9. (a) The level curve $z = c$ is a straight lines of the form $3x - 2y = c$. The lines are equally spaced, all with slope $2/3$.
   (b) The level line $z = 0$ has equation $3x - 2y = 0$.
   (c) The general formula is $L(x, y) = 3x - 2y$.
   (d) The graph of L is a plane; this shape is indeed consistent with the level curves plotted earlier.

14.3 Partial derivatives

1. (a) $f_x(x, y) = 2x$; $f_y(x, y) = -2y$

   (b) $f_x(x, y) = 2xy^2$; $f_y(x, y) = 2x^2 y$

   (c) $f_x(x, y) = \dfrac{2x}{y^2}$; $f_y(x, y) = \dfrac{-2x^2}{y^3}$

   (d) $f_x(x, y) = -y \sin(xy)$; $f_y(x, y) = -x \sin(xy)$

   (e) $f_x(x, y) = -\sin(x)\cos(y)$; $f_y(x, y) = -\cos(x)\sin(y)$

   (f) $f_x = \dfrac{-\sin(x)}{\cos(y)}$; $f_y = \dfrac{\sin(x)\sin(y)}{\cos^2(y)}$

   (g) $f_x(x, y, z) = y^2 z^3$; $f_y(x, y, z) = 2xyz^3$; $f_z(x, y, z) = 3xy^2 z^2$

   (h) $f_x(x, y, z) = -yz \sin(xyz)$; $f_y(x, y, z) = -xz \sin(xyz)$; $f_z(x, y, z) = -xy \sin(xyz)$

2. (a) Because $f(3) = 9$ and $f'(3) = 6$, it follows that the linear approximation function has equation $L(x) = 9 + 6(x - 3)$.

   (b) The graphs of f and L are close together near $x = 3$; in particular, the graph of f looks similar to a straight line.

   (c) The inequality $|f(x) - L(x)| < 0.01$ holds for x in the interval $2.9 < x < 3.1$. This can be found either graphically, by zooming, or algebraically. The latter approach can be done as follows. Notice first that

   $$|f(x) - L(x)| = |x^2 - (9 + 6(x - 3))| = |x^2 - 6x + 9| = |x - 3|^2.$$

   Thus

   $$|f(x) - L(x)| < 0.01 \iff |x - 3|^2 < 0.01 \iff |x - 3| < 0.1.$$

   This is equivalent to saying that $2.9 < x < 3.1$.

3. (a) Since $f(9) = 3$ and $f'(3) = 1/6$, it follows that the linear approximation function L has formula $L(x) = 3 + (x - 9)/6$.

   (b) Plotting f and L in the vicinity of $x = 9$ (e.g., on the interval $6 < x < 12$) shows that the two graphs are very close together.

   (c) Looking carefully at the graphs of f and L near $x = 9$ shows that the functions differ by less than 0.01 on the interval $8 < x < 10$. (The exercise can also be solved algebraically, but it's rather messy.)

4. (a) The fact that $f'(3) = 6$ means that the graph of f looks like a line of slope 6 near $x = 3$. (The graph bears this out.)

   (b) Both of the secant lines, from $x = 3$ to $x = 3.5$ and from $x = 3$ to $x = 3.1$, have slopes near 6.

   (c) The average rates of change $\Delta y/\Delta x$ of f over the intervals $[3, 3.5]$ and $[3, 3.1]$ are 6.5 and 6.1, respectively.

   (d) For $f(x) = x^2$, $\displaystyle\lim_{h \to 0} \dfrac{f(3+h) - f(3)}{h} = 6$. The answer is the derivative of f at $x = 3$.

5. (a) The fact that $f'(3) = 5$ means that the graph of f looks like a line of slope 5 near $x = 3$. (The graph bears this out.)

   (b) Both of the secant lines, from $x = 3$ to $x = 3.5$ and from $x = 3$ to $x = 3.1$, have slopes near 5.

   (c) The average rates of change $\Delta y/\Delta x$ of f over the intervals $[3, 3.5]$ and $[3, 3.1]$ are 5.5 and 5.1, respectively.

   (d) For $f(x) = x^2 - x$, $\displaystyle\lim_{h \to 0} \dfrac{f(3+h) - f(3)}{h} = 5$. The answer is the derivative of f at $x = 3$.

Section 14.3 Partial Derivatives

6. (In fact, the function $g(x, y) = y^2 + y - x^2$, so $g_y = 2y + 1$ and $g_x = -2x$. But these formulas aren't necessary for doing the exercise.)

    (a) From the table, $g_x(1, 1) \approx (g(1.01, 1) - g(1, 1))/0.01 = -2.01$. Similarly, $g_y(1, 1) \approx (g(1, 1.01) - g(1, 1))/0.01 = 3.01$.

    (b) The table reflects the fact that $g_x(0, 0) = 0$ in that the entries along the *row* through the $(0, 0)$-position are close to constant. The table reflects the fact that $g_y(0, 0) = 1$ in that the entries along the *column* through the $(0, 0)$-position vary at the same rate as y.

    (c) All parts are routine—the values of g, g_x, and g_y are all given by the table or in the previous part. (The point is that L is the linear approximation to g at $(0, 0)$.)

    (d) Using the formula $L(x, y) = y$, it's easy to fill in the table. The results are very similar to those in the corresponding part of the table of g-values.

    (e) To find the linear approximation function $M(x, y)$ with (i) $M(1, 1) = g(1, 1)$; (ii) $M_x(1, 1) = g_x(1, 1)$; (iii) $M_y(1, 1) = g_y(1, 1)$, we need to know or estimate the right-hand quantities. We know $g(1, 1) = 1$ from the table, and we estimated $g_x(1, 1) \approx -2$ and $g_y(1, 1) \approx 3$. This gives $M(x, y) = 1 + -2(x - 1) + 3(y - 1)$ as a reasonable estimate.

7. (a) The graph is a cylinder in the x-direction.

    (b) $f_x(x, y) = 0$; $f_y(x, y) = \cos y$. The fact that $f_x = 0$ reflects the fact that f is constant in x.

    (c) The linear approximation function L for f at the point $(0, 0)$ is $L(x, y) = y + 2$. Like f itself, L is independent of x.

8. (a) Mimicking the reasoning in Example 3 (page 403), gives $f_x(1.5, 1.5) \approx -1.5$ and $f_y(1.5, 1.5) \approx -4.5$.

    (b) The formulas $f_x = 2x - 3y$ and $f_y = -3x$ show that $f_x(1.5, 1.5) = -1.5$ and $f_y(1.5, 1.5) = -4.5$.

    (c) It follows from part (b), and the fact that $f(1.5) = 1.5$, that $L(x, y) = -1.5(x - 1.5) - 4.5(y - 1.5) + 1.5 = -1.5x - 4.5y + 10.5$.

9. (a) All the level curves are vertical lines.

    (b) From the diagram, $f_x(0, 0) \approx 1$ and $f_y(0, 0) = 0$.

    (c) From the diagram, $f_x(\pi/2, 0) \approx 0$ and $f_y(0, 0) = 0$.

    (d) All contour lines are vertical; this means that $f(x, y)$ is constant along vertical lines, which implies that $f_y(x, y) = 0$.

    (e) The fact that $f_x(x, y)$ is independent of y means that the rate of increase of f in the x-direction is the same for all y. The contour map of f reflects this fact in that contour lines are all vertical.

11. (a) All the level curves are straight lines; all have the same slope, and they're equally spaced.

    (b) The contour map shows that $f_x(0, 0) = 2$ and $f_y(0, 0) = 3$; one sees this by measuring how fast f increases in the x- and y-directions.

    (c) The contour map of f reflects the fact that both f_x and f_y are constant functions in the sense that all contour lines are straight and equally spaced.

    (d) The contour map shows that $f_x(x, y)$ is positive and $f_y(x, y)$ is negative in that moving to the right (in the x-direction) corresponds to increasing z, while moving upward (in the y-direction) corresponds to decreasing z.

13. (a) The partial derivatives of $f(x, y) = xy$ are $f_x(x, y) = y$ and $f_y(x, y) = x$. Thus, at the base point $(x_0, y_0) = (2, 1)$, $f(2, 1) = 2$, $f_x(2, 1) = 1$, and $f_y(2, 1) = 2$. Therefore $L(x, y) = 1(x - 2) + 2(y - 1) + 2 = x + 2y - 2$.

    (b) Level curves of $f(x, y) = xy$ are hyperbolas, of the form $xy = k$. Level curves of $L(x, y) = x + 2y - 2$ are straight lines, of the form $x + 2y - 2 = k$.

(c) The level curves of f and L look similar near $(2, 1)$. (E.g., all the level curves of L are lines, with slope $-1/2$. Similarly, the hyperbola $xy = 2$ (a level curve of f) has slope $-1/2$ at the point $(2, 1)$.

(d) In a small window around $(2, 1)$, the level curves of f and L appear almost identical—this is because L is the linear approximation to f at $(2, 1)$.

14. (a) The partial derivatives of $f(x, y) = x^2 - y^2$ are $f_x(x, y) = 2x$ and $f_y(x, y) = -2y$. Thus, at the base point $(x_0, y_0) = (2, 1)$, $f(2, 1) = 3$, $f_x(2, 1) = 4$, and $f_y(2, 1) = -2$. Therefore $L(x, y) = 4(x - 2) - 2(y - 1) + 3 = 4x - 2y - 3$.

(b) Level curves of $f(x, y) = x^2 - y^2$ are hyperbolas, of the form $x^2 - y^2 = k$. Level curves of $L(x, y) = 4x - 2y - 3$ are straight lines, of the form $4x - 2y - 3 = k$.

(c) The level curves of f and L look similar near $(2, 1)$. (E.g., all the level curves of L are lines, with slope 2. Similarly, the hyperbola $x^2 - y^2 = 3$ (a level curve of f) has slope 2 at the point $(2, 1)$.

(d) In a small window around $(2, 1)$, the level curves of f and L appear almost identical—this is because L is the linear approximation to f at $(2, 1)$.

15. (a) From the contour map of f we estimate $f_x(1, 2) = 2$ and $f_y(1, 2) = 4$.

(b) Symbolic differentiation of $f(x, y) = x^2 + y^2$ gives $f_x = 2x$ and $f_y = 2y$, so $f_x(1, 2) = 2$ and $f_y(1, 2) = 4$.

(c) It follows from the above and from the fact that $f(1, 2) = 5$ that $L(x, y) = 2(x - 1) + 4(y - 2) + 5 = 2x + 4y - 5$.

(d) Level curves $L(x, y) = k$ are all straight lines, with slope $-1/2$. Level curves $f(x, y) = k$ are all circles, centered at the origin. At the point $(1, 2)$, the level curves for f and L are tangent to each other.

16. (a) $f_x(x, y) = \cos x + y$; $f_y(x, y) = 2 + x$, so $f_x(0, 0) = 1$ and $f_y(0, 0) = 2$.

(b) The linear approximation to f at $(0, 0)$ is $L(x, y) = x + 2y$.

(c) All parts are routine, given the formulas. The two rows are have similar entries, especially at the left, because L approximates f closely near $(0, 0)$.

(d) Corresponding level curves of the two functions should appear similar, especially near $(0, 0)$.

17. (a) If $f(x, y) = x^2 + y^2$ and $(x_0, y_0) = (2, 1)$, then $L(x, y) = -5 + 4x + 2y$.

(b) If $f(x, y) = x^2 + y^2$ and $(x_0, y_0) = (0, 0)$, then $L(x, y) = 0$.

(c) If $f(x, y) = \sin(x) + \sin(y)$ and $(x_0, y_0) = (0, 0)$, then $L(x, y) = x + y$.

(d) If $f(x, y) = \sin(x) \sin(y)$ and $(x_0, y_0) = (0, 0)$, then $L(x, y) = 0$.

19. (a) The information given about partial derivatives implies that $L(x, y) = 6(x - 3) + 8(y - 4) + 25 = 6x + 8y - 25$.

(b) Use the formula above: $L(2.9, 3.1) = 6(-0.1) + 8(-0.1) + 25 = 23.6$; $L(3.1, 4.1) = 6(0.1) + 8(0.1) + 25 = 26.4$; $L(4, 5) = 39$.

(c) The function f can not be linear. If f were linear it would agree everywhere with L; but we saw above that $f(4, 5) \neq L(4, 5)$.

20. (a) The information given on partial derivatives implies that $L(x, y) = 0.6(x - 3) + 0.8(y - 4) + 5 = 0.6x + 0.8y$.

(b) Use the formula above: $L(2.9, 4.1) = 0.6(-0.1) + 0.8(0.1) + 5 = 5.02$; $L(4, 5) = 0.6(1) + 0.8(1) + 5 = 6.4$.

(c) The function g can not be linear. If g were linear it would agree everywhere with L—but we saw above that $g(4, 5) = \sqrt{41} \neq 6.4 = L(4, 5)$.

21. (a) The graph is "creased" along the x-axis. This suggests no y-derivative there.

Section 14.3 Partial Derivatives

(b) The limit
$$\lim_{h \to 0} \frac{f(h, 0) - f(0, 0)}{h} = \lim_{h \to 0} \frac{|h|}{h}$$
doesn't exist because
$$\lim_{h \to 0^+} \frac{|h|}{h} = 1 \quad \text{but} \quad \lim_{h \to 0^-} \frac{|h|}{h} = -1.$$

(c) Since $f(x, 0) = 0$, it follows that $f_x(0, 0) = 0$. On the graph, the surface is horizontal in the x-direction at the origin.

(d) Since $f(y, \pi/2) = 0$, it follows that $f_y(0, \pi/2) = 0$.

(e) Zooming in on the graph near $(0, \pi/2)$ shows that at this point the surface is "flat" in the y-direction.

(f) One possibility is $g(x, y) = |x| \cos y$.

22. (a) Graphs show a crease along the y-axis; this suggests that f_x may not exist there. f_y exists everywhere.

(b) Since $f(x, 0) = 0$, it follows that $f_x(x, 0) = 0$.

(c) Since $f(x, 1) = |x|$, it follows that $f_x(x, 1)$ doesn't exist.

14.4 Optimization and partial derivatives: a first look

1. (a) The ant's altitude remains constant.
   (b) The ant rises all the way; it's highest at (0.5, 1). At that point, its altitude is 0.5.

2. (a) The eggs go above the black circles.
   (b) For example, $(0, \pi/2)$ is a local maximum point, $(0, -\pi/2)$ is a local minimum point, and $(\pi/2, 0)$ is a saddle point.
   (c) Because $f_x = -\sin(x)\sin(y)$ and $f_y = \cos(x)\cos(y)$, there are four stationary points in $[-3, 3] \times [-3, 3]$: they're at $(0, \pi/2)$, $(0, -\pi/2)$, $(\pi/2, 0)$, and $(-\pi/2, 0)$.
   (d) The maximum value is 1; the minimum is -1.

3. (a) The function f has stationary points $(1, \pi/2)$ and $(1, -\pi/2)$. The first is a local minimum, the second a local maximum, apparently.
   (b) The partial derivatives are $f_x(x, y) = (2x - 2)\sin(y)$; $f_y(x, y) = (x^2 - 2x)\cos(y)$. It follows that the stationary points are at $(0, 0)$, $(2, 0)$, $(1, \pi/2)$ and $(1, -\pi/2)$.
   (c) The functions $f(-3, y) = 15\sin(y)$ has minimum value -15 at $y = -\pi/2$ and maximum value 15 at $y = \pi/2$.

4. (a) There's a local maximum at $(0, 0)$
   (b) There's a saddle at $(0, 0)$
   (c) There's a local minimum at $(0, 0)$
   (d) There's a saddle at $(1, 2)$.

5. The function $L(x, y) = 1 + 2x + 3y$ has no stationary points, because $L_x(x, y) = 2$ and $L_y(x, y) = 3$ for all (x, y).

6. (a) A plane can have a stationary points only if the plane is horizontal. In this case, every point is a stationary point.
   (b) $L_x = a$ and $L_y = b$, so there are stationary points only if $a = b = 0$. (c can have any value.) In this case, every point is stationary.

7. (a) All points on the y-axis (where $x = 0$) are stationary; they're local minimum points.
   (b) Since $f(x, y) = x^2$, $f_x = 2x$ and $f_y = 0$. Therefore, every point with $x = 0$ is stationary. (This agrees with part (a).)

8. (a) $g(x, y) = y^2$ works; so do many others
   (b) $h(x, y) = -(x - 1)^2$ works; so do many others
   (c) $k(x, y) = (x - 3)^2 + (y - 4)^2$ works; so do many others

14.5 Multiple integrals and approximating sums

1. Doing this by hand gives $\frac{1}{4}f\left(\frac{1}{8}\right) + \frac{1}{4}f\left(\frac{3}{8}\right) + \frac{1}{4}f\left(\frac{5}{8}\right) + \frac{1}{4}f\left(\frac{7}{8}\right)$. Using $f(x) = x^2$, and working this out as a fraction, gives 21/64.

   Maple's value for M_{100}, the midpoint sum with 100 subdivisions, is 0.333325.

2. The double sum has nine terms, each of the form $1/9 \sin(a)\sin(b)$, where (a, b) is the midpoint of one of the nine subrectangles. The total sum, evaluated numerically, is about 0.2133.

   Maple's value with $n = 10$, i.e., the midpoint sum with 100 subdivisions, is 0.211498.

3. The triple midpoint sum has nine terms, each of the form $8abc$, where (a, b, c) is the midpoint of one of the 8 subcubes. The total sum, evaluated numerically, is 512.

   Maple's value with $n = 2$, i.e., the triple midpoint sum with 8 subdivisions is 512. With $n = 4$, *Maple* still gives 512—this happens to be the exact answer.

4. (a) The level curves $z = 1, z = 2, z = 3$, etc., pass through the midpoints of the subdivisions. From this one concludes that the midpoint sum adds up to 64.

   (b) *Maple* agrees that the double midpoint sum is 64.

   (c) The level curves split each subrectangle in half; as a result, the midpoint sum commits no error in approximating the integral.

5. (a) Either the level curves or the formula can be used to estimate values of f at the midpoints; they lead to the estimate $I \approx M_{16} = 168$ (or something near that).

   (b) *Maple* calculates the double midpoint sum as 168.

   (c) Since the surface rises faster and faster as away from the origin, the midpoint sum somewhat underestimates the integral.

14.6 Calculating integrals by iteration

1. (a) If $R = [0, 1] \times [0, 1]$, then $\iint_R \sin(x) \sin(y) \, dA = (\cos(1))^2 - 2\cos(1) + 1 \approx 0.2113$.

    (b) If $R = [0, 1] \times [0, 1]$, then $\iint_R \sin(x + y) \, dA = -\sin(2) + 2\sin(1) \approx 0.7736$.

    (c) If $R = [0, 4] \times [0, 4]$, then $\iint_R (x^2 + y^2) \, dA = \dfrac{512}{3} \approx 170.6667$.

    (d) If $V = [0, 1] \times [0, 2] \times [0, 3]$, then $\iiint_R x \, dV = 3$.

    (e) If $V = [0, 1] \times [0, 2] \times [0, 3]$, then $\iiint_R y \, dV = 6$.

2. (a) $\iint_R (x + y) \, dA = \int_{x=0}^{x=1} \int_{y=x^2}^{y=x} (x + y) \, dy \, dx = \dfrac{3}{20} = 0.15$.

    (b) $\iint_R x \, dA = \int_{x=0}^{x=1} \int_{y=x^2}^{y=\sqrt{x}} x \, dy \, dx = \dfrac{3}{20} = 0.15$.

    (c) $\iint_R 1 \, dA = \int_{x=0}^{x=1} \int_{y=0}^{y=\sqrt{1-x^2}} 1 \, dy \, dx = \dfrac{\pi}{4} \approx 0.7854$.

3. (a) $\iint_R (x + y) \, dA = \int_{y=0}^{y=1} \int_{x=y}^{x=\sqrt{y}} (x + y) \, dx \, dy = \dfrac{3}{20} = 0.15$.

    (b) $\iint_R x \, dA = \int_{y=0}^{y=1} \int_{x=y^2}^{x=\sqrt{y}} x \, dx \, dy = \dfrac{3}{20} = 0.15$.

    (c) $\iint_R 1 \, dA = \int_{y=0}^{y=1} \int_{x=0}^{x=\sqrt{1-y^2}} 1 \, dx \, dy = \dfrac{\pi}{4} \approx 0.7854$.

4. (a) Integrating first in y, then in x, $\iint_R (x + y) \, dA = \int_{x=-1}^{x=1} \int_{y=x^2}^{y=1} (x + y) \, dy \, dx = 4/5$.

    (b) Integrating first in x, then in y, $\iint_R (x + y) \, dA = \int_{y=0}^{y=1} \int_{x=-\sqrt{y}}^{x=\sqrt{y}} (x + y) \, dy \, dx = 4/5$.

5. (a) Integrating first in y, then in x, $\iint_R x \, dA = \int_{x=0}^{x=1} \int_{y=0}^{y=e^x} x \, dy \, dx = 1$.

    (b) Integrating first in x, then in y, requires that the y-interval be broken up into two parts:

    $$\iint_R x \, dA = \int_{y=0}^{y=1} \int_{x=0}^{x=1} x \, dx \, dy + \int_{y=1}^{y=e} \int_{x=\ln y}^{x=1} x \, dx \, dy = 1/2 + 1/2 = 1.$$

6. (a) Single-variable calculus say that the area is the ordinary integral $\int_a^b f(x) \, dx$.

    (b) Writing the double integral in iterated form gives

    $$\iint_R 1 \, dA = \int_{x=a}^{x=b} \int_{y=0}^{y=f(x)} 1 \, dy \, dx = \int_{x=a}^{x=b} y\Big]_{y=0}^{y=f(x)} dx = \int_{x=a}^{x=b} f(x) \, dx;$$

    this is the same formula as in one-variable calculus.

14.7 Double integrals in polar coordinates

1. (a) The area of a wedge with outer radius r_2 and making angle $\Delta\theta$ at the origin is $\dfrac{r_2^2}{2}\Delta\theta$. A polar rectangle is the difference of two such wedges; the result follows with a little computation.

   (b) If we set $a = r$, $b = r + \Delta r$, and $\beta - \alpha = \Delta\theta$ in part (a), then the result follows.

2. The answer is 2/3 in all cases. As iterated integrals:

   (a) $I = \displaystyle\int_{x=-1}^{x=1} \left(\int_{y=0}^{y=\sqrt{1-x^2}} y\, dy \right) dx.$

   (b) $I = \displaystyle\int_{y=0}^{y=1} \left(\int_{x=-\sqrt{1-y^2}}^{x=\sqrt{1-y^2}} y\, dx \right) dy.$

   (c) $I = \displaystyle\int_{\theta=0}^{\theta=\pi} \left(\int_0^1 r^2 \sin\theta\, dr \right) d\theta.$

3. The cylinder shown in the picture has volume π; a cone of this type has 1/3 the volume of the cylinder. It follows that (as we found in the section) $I_2 = 2\pi/3$.

4. (a) The solid lies under a surface with parabolic cross sections.

   (b) $I_1 = 2\pi/3$, no matter how you slice it.

5. (a) The cardioid has area $3\pi/2$.

   (b) The triangle has area 1/2; note that it's bounded by the polar curves $\theta = 0$, $\theta = \pi/4$, and $r = \sec\theta$.

   (c) The circle in question has polar equation $r = \sin\theta$, for $0 \leq \theta \leq \pi$. The area inside is $\pi/4$.

6. (a) The integral reduces to $\displaystyle\int_0^\pi (1 + \sin\theta)\, d\theta = \pi + 2.$

   (b) The surface lies above the unit circle in the xy-plane; the surface is defined by the function $g(r, \theta) = 1 - r^2$. Thus, in polar form, the integral is $\displaystyle\int_0^{2\pi} \int_0^1 (1 - r^2) r\, dr\, d\theta = \frac{\pi}{2}.$

   (c) The surface lies above the unit circle in the xy-plane; the surface is defined by the function $g(r, \theta) = 1 - r$. Thus, in polar form, the integral is $\displaystyle\int_0^{2\pi} \int_0^1 (1 - r) r\, dr\, d\theta = \frac{\pi}{3}.$

3.8 Inverse Trigonometric Functions and Their Derivatives

1. $\arcsin(1) = \pi/2$
2. $\arcsin\left(\sqrt{3}/2\right) = \pi/3$
3. $\arccos(-1) = \pi$
4. $\arccos\left(-\sqrt{2}/2\right) = 3\pi/4$
5. $\arctan(1) = \pi/4$
6. $\arctan\left(\sqrt{3}\right) = \pi/3$
7. $\arcsin(0.8) \approx 0.93$
8. $\arccos(0.6) \approx 0.93$
9. $\arctan(0.4) \approx 0.38$
10. $\arctan(3) \approx 1.25$
11. (a) No. The gaps in the *table* reflect gaps in the functions' *domains*. For instance, arcsin(2) is *undefined* because there *is* no number x such that $\sin x = 2$.
    (b) For any $x \in [-1, 1]$, $\arcsin x + \arccos x = \pi/2 \approx 1.57$.
12. (a) The domain of the arccosine function is $[-1, 1]$. If $|x| \geq 1$, then $1/x \in [-1, 1]$. However, if $|x| < 1$, $1/x \notin [-1, 1]$. It follows that the domain of the arcsecant function is $x \in \mathbb{R}$ such that $|x| \geq 1$.
    (b) The range of the arcsecant function is $[0, \pi/2) \cup (\pi/2, \pi]$.
13. $\sin(\arccos x) = \sqrt{1 - x^2}$
14. $\tan(\arcsin x) = x/\sqrt{1 - x^2}$
15. $\sin(\arctan x) = x/\sqrt{1 + x^2}$
16. $\cos(\arctan x) = 1/\sqrt{1 + x^2}$
17. $f'(x) = 2/(1 + 4x^2)$
18. $f'(x) = 2x/(1 + x^4)$
19. $f'(x) = \frac{1}{2}\left((1 - x^2)\arcsin x\right)^{-1/2}$
20. $f'(x) = \frac{1}{2}\left(x(1 - x)\right)^{-1/2}$
21. $f'(x) = e^x/\sqrt{1 - e^{2x}}$
22. $f'(x) = e^{\arctan x}/\left(1 + x^2\right)$
23. $f'(x) = \left(x\left(1 + (\ln x)^2\right)\right)^{-1}$
24. $f'(x) = -2/\sqrt{1 - (2x + 3)^2}$
25. $f'(x) = 2/\sqrt{x^4 - 4x^2}$
26. $f'(x) = \dfrac{\arccos x + \arcsin x}{\sqrt{1 - x^2}(\arccos x)^2}$

SECTION 3.8 INVERSE TRIGONOMETRIC FUNCTIONS AND THEIR DERIVATIVES

27. $f'(x) = 2x \arctan(\sqrt{x}) + x^{3/2}/(2+2x)$

28. $f'(x) = \left[\sqrt{1-x^2}(2+\arcsin x)\right]^{-1}$

29. $F(x) = \arctan x$

30. $F(x) = \arcsin x$

31. $F(x) = \arctan(2x)$

32. $F(x) = \arctan(x/3)$

33. $F(x) = \arcsin(x/3)$

34. $F(x) = \frac{1}{2}\arcsin(2x)$

35. $F(x) = \arctan(e^x)$

36. $F(x) = \frac{1}{2}\arcsin(x^2)$

37. $F(x) = (\arctan x)^2$

38. $F(x) = \arctan(\sin x)$

39. $F(x) = e^{\arcsin x}$

40. $F(x) = \arctan(\ln x)$

41. $F(x) = \ln(\arctan x)$

42. $x = \pm\pi/3 \approx \pm 1.047, x = \pm 7\pi/3 \approx 7.33$.

43. $x \approx 1.107, x \approx 4.249, x \approx -2.034,$ and $x \approx -5.176$.

44. The tangent function has *vertical* asymptotes at $x = \pi/2 \pm k\pi$, where k is any positive integer. The arctangent function has *horizontal* asymptotes only at $y = \pm\pi/2$. They're the reflections of the two vertical asymptotes left after the tangent function is restricted.

45. (a) The range of the arcsine function is $[-\pi/2, \pi/2]$. Since 5 is not in this interval, no input to the arcsine function will yield 5.

    (b) The equation $\arcsin(\sin x) = x$ holds when $-\pi/2 \leq x \leq \pi/2$.

    (c) This is a straightforward application of the chain rule and the identity $\sqrt{1-\sin^2 x} = |\cos x|$.

    (d) *In the interval* $-\pi/2 \leq x \leq \pi/2$, the graph shown is the line $y = x$. For other values of x, the graph can be thought of as the result of "folding" the line $y = x$ back and forth, to stay within the vertical range (the interval $[-\pi/2, \pi/2]$) of the inverse sine function.

46. (a) domain: $|x| < 1$. The arcsine function has a vertical tangent line at $x = \pm 1$ so its derivative does not exist at these points.

    (b) domain: $-\infty < x < \infty$. The domain of the derivative of the arctangent function is the same as the domain of the arctangent function.

    (c) domain: $|x| > 1$. The domain of the derivative of the arcsecant function is *not* the same as the domain of the arcsecant function.

47. (b) $\lim_{x\to\infty} f(x) = \pi/2;\ \lim_{x\to-\infty} f(x) = -\pi/2;\ \lim_{x\to\infty} f'(x) = \lim_{x\to-\infty} f'(x) = 0$. These results imply the lines $y = -\pi/2$ and $y = \pi/2$ are horizontal asymptotes of the arctangent function.

(c) f is an odd function.

(d) f is increasing on $(-\infty, \infty)$ because f' is positive on this interval.

(e) f is concave up on $(-\infty, 0)$ and concave down on $(0, \infty)$. f has an inflection point at $x = 0$.

48. (b) $\lim_{x \to 1^-} f(x) = \pi/2$; $\lim_{x \to -1^+} f(x) = -\pi/2$; $\lim_{x \to 1^-} f'(x) = \lim_{x \to -1^+} f'(x) = \infty$. These results imply that the lines $x = -1$ and $x = 1$ are vertical asymptotes of the arcsine function.

    (c) f is an odd function.

    (d) f is increasing on $(-\infty, \infty)$ because f' is positive on this interval.

    (e) f is concave up on $(0, 1)$ and concave down on $(-1, 0)$. f has an inflection point at $x = 0$.

    (f) The graph of f' is decreasing on $(-1, 0)$ and increasing on $(0, 1)$.

49. (b) $\lim_{x \to 1^-} f(x) = 0$; $\lim_{x \to -1^+} f(x) = \pi$; $\lim_{x \to 1^-} f'(x) = \lim_{x \to -1^+} f'(x) = -\infty$. These results imply that the lines $x = -1$ and $x = 1$ are vertical asymptotes of the arccosine function.

    (c) f is neither even nor odd.

    (d) f is decreasing on $(-\infty, \infty)$ because f' is negative on this interval.

    (e) f is concave up on $(-1, 0)$ and concave down on $(0, 1)$. f has an inflection point at $x = 0$.

    (f) The graph of f' is increasing on $(-1, 0)$ and decreasing on $(0, 1)$.

50. (b) $\lim_{x \to 1^+} f(x) = 0$; $\lim_{x \to -1^-} f(x) = \pi$; $\lim_{x \to 1^+} f'(x) = -\infty$; $\lim_{x \to -1^-} f'(x) = \infty$. These results imply that the lines $x = -1$ and $x = 1$ are vertical asymptotes of the arcsecant function.

    (c) $\lim_{x \to \infty} f(x) = \lim_{x \to -\infty} f(x) = \pi/2$; $\lim_{x \to \infty} f'(x) = \lim_{x \to -\infty} f'(x) = 0$. These results imply that the line $y = \pi/2$ is a horizontal asymptote of the arcsecant function.

    (d) f is neither even nor odd.

    (e) f is increasing on $(-\infty, -1) \cup (1, \infty)$ because f' is positive on these intervals.

    (f) f is concave up on $(-\infty, -1)$ and concave down on $(1, \infty)$. f has no inflection points.

    (g) The graph of f' is increasing on $(-\infty, -1)$ and decreasing on $(1, \infty)$.

51. Let $f(x) = \arctan x - x/(1 + x^2)$ and $g(x) = x - \arctan x$. Now,

$$f'(x) = \frac{1}{1+x^2} - \frac{1+x^2 - x(2x)}{(1+x^2)^2} = \frac{1+x^2 - (1-x^2)}{(1+x^2)^2} = \frac{2x^2}{(1+x^2)^2} > 0$$

for all x, and

$$g'(x) = 1 - \frac{1}{1+x^2} = \frac{1+x^2}{1+x^2} - \frac{1}{1+x^2} = \frac{x^2}{1+x^2} > 0$$

for all x. Since $f'(x) > 0$ and $g'(x) > 0$ for all x, and since $f(0) = g(0) = 0$, $f(x) > 0$ and $g(x) > 0$ for all $x \geq 0$.

52. (b) Let $f(x) = \arctan x + \arctan(1/x)$. Now, $f'(x) = (1+x^2)^{-1} - \left[x^2\left(1 + \frac{1}{x^2}\right)\right]^{-1} = 0$ so f is a constant function. Since $f(1) = 2\arctan(1) = \pi/2$, $f(x) = \pi/2$ for all $x > 0$.

    (c) When $x < 0$, $f(x) = -\pi/2$.

53. Differentiate both sides of the identity given and verify that the derivatives are equal. This establishes that the two sides differ from each other by a constant. Evaluating both sides for a particular value of x (e.g., $x = 0$) shows that this constant is zero.

54. Differentiate both sides of the identity given and verify that the derivatives are equal. This establishes that the two sides differ from each other by a constant. Evaluating both sides for a particular value of x (e.g., $x = 0$) shows that this constant is zero.

55. Differentiate both sides of the identity given and verify that the derivatives are equal. This establishes that the two sides differ from each other by a constant. Evaluating both sides for a particular value of x (e.g., $x = 0$) shows that this constant is zero.

56. Differentiate both sides of the identity given and verify that the derivatives are equal. This establishes that the two sides differ from each other by a constant. Evaluating both sides for a particular value of x (e.g., $x = 0$) shows that this constant is zero.

57. Differentiate both sides of the identity given and verify that the derivatives are equal. This establishes that the two sides differ from each other by a constant. Evaluating both sides for a particular value of x (e.g., $x = 0$) shows that this constant is zero.

58. (a) Let $f(x) = 2\arcsin x$ and $g(x) = \arccos(1 - 2x^2)$. Since $f'(x) = g'(x)$, $f(x) = g(x) + C$. Since $f(0) = g(0)$, $C = 0$.

    (b) $-2\arcsin x = \arccos(1 - 2x^2)$ when $-1 \leq x \leq 0$.

59. (a) $f'(x) = 1/(1+x^2) = (\arctan x)' \implies f(x) = C + \arctan x$. Since $f(0) = \pi/4$, $f(x) = \pi/4 + \arctan x$ when $x < 1$.

    (b) $\lim_{x \to 1^-} f(x) = \pi/2$

    (c) When $x > 1$, $x = 1/y$ where $0 < y < 1$. Therefore, $f(x) = f(1/y) = \arctan\left(\frac{1+1/y}{1-1/y}\right) = \arctan\left(\frac{1+y}{y-1}\right) = -\arctan\left(\frac{1+y}{1-y}\right) = -f(y)$. Thus, $\lim_{x \to 1^+} f(x) = \lim_{y \to 1^-} -f(y) = -\pi/2$.

60. (a) Draw a right triangle which has a hypotenuse of length x and whose other two sides have length 1 and $\sqrt{x^2 - 1}$, respectively. When θ is chosen to be the angle adjacent to the side of length 1, one can see that $\cos\theta = 1/x$ (i.e., $\theta = \text{arcsec } x$) and $\tan\theta = \sqrt{x^2 - 1}$.

    (b) $\tan(\text{arcsec } x) = -\sqrt{x^2 - 1}$

61. (a) Let $x = \arctan u$ and $y = \arctan v$. Then, $\tan(\arctan u + \arctan v) = (u + v)/(1 - uv) \implies \arctan u + \arctan v = \arctan\left(\frac{u+v}{1-uv}\right)$.

    (b) The identity in part (a) is valid when $-\pi/2 < \arctan x + \arctan y < \pi/2$.

62. (a) $\arctan(1/2) + \arctan(1/3) = \arctan(1) = \pi/4$

    (b) $2\arctan(1/4) + \arctan(7/23) = \bigl(\arctan(1/4) + \arctan(1/4)\bigr) + \arctan(7/23) = \arctan(8/15) + \arctan(7/23) = \arctan(1) = \pi/4$

Chapter 4

Applications of the Derivative

4.1 Differential Equations and Their Solutions

1. $y' = t$

2. $y' = t \neq y$

3. $y' = -e^t \neq -y$

4. $y' = -e^{-t} = -y$

5. $y' = -42e^{-t} = -y$

6. $y' = kCe^{kt} = k(y - A)$

7. $y' = e^t + t \neq y + t$

8. $y' = t\exp(t^2) \neq ty$

9. $y' = t\exp(t^2/2) = ty$

10. No. $y' - 2y/t = 2t^3 + 3t - 2(t^4/2 + 3t^2/2 + 1/4)/t = t^3 - 1/(2t) \neq t^3$.

11. (a) The coffee reaches $100°$ F at $t = 2\ln(35/125)/\ln(95/125) \approx 9.28$ minutes.

    (b) The formula for $z(t)$ has the form $z(t) = 65 + Ae^{kt}$. The values $z(0) = 190$ and $z(2) = 180$ imply that $A = 125$ and $k = \ln\left(\sqrt{115/125}\right) \approx -0.04169$. Thus, $z(t) = 65 + 125e^{\ln\sqrt{115/125}\,t} \approx 65 + 125e^{-0.04169t}$.

    (c) Solving $z(t) = 100$ for t gives $t = 2\ln(7/25)/\ln(23/25) \approx 30.53$ minutes.

    (d)

    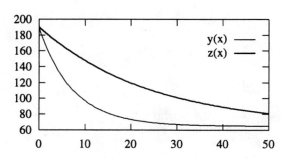

12. (a) The algebra is a bit messy, but it works out: $P'(t) = \dfrac{d}{dt}\left(\dfrac{1}{1+de^{-t}}\right) = \dfrac{de^{-t}}{(1+de^{-t})^2} = P(1-P)$.

    (b) Setting $d = 1$, so that $P(t) = \dfrac{1}{1+e^{-t}}$, solves the IVP $P' = P(1-P)$; $P(0) = 1/2$.

    (c) Setting $d = -1/2$, so that $P(t) = \dfrac{1}{1-e^{-t}}$, solves the IVP $P' = P(1-P)$; $P(0) = 1/2$. Solve the IVP $P' = P(1-P)$; $P(0) = 2$.

    (d) The two solution graphs both seem to (and do) approach $y = 1$ as t increases.

13. $P'(t) = -C\left(1+de^{-kCt}\right)^{-2} \cdot \left(-kCde^{-kCt}\right) = \dfrac{kC^2de^{-kCt}}{\left(1+de^{-kCt}\right)^2}$.

    $C - P = Cde^{-kCt}/\left(1+de^{-kCt}\right)$ so $kP(C-P) = P'$.

SECTION 4.1 DIFFERENTIAL EQUATIONS AND THEIR SOLUTIONS

14. (a) Let h be the height of the tank. When the tank is full, the water level is dropping at a rate proportional to $\sqrt{h}$ whereas when the tank is half full the water level is dropping at a rate proportional to $\sqrt{h/2}$.
    (b) $y' = \frac{1}{2}(C - kt)(-k) = -k\sqrt{y}, C - kt \geq 0$
    (c) If $0 \leq t \leq 60$, $y(t) = (6 - t/10)^2/4$; it's easy to see that $y(60) = 0$.
    (d) The curve is a parabola, with vertex at $(60, 0)$, and passing through $(0, 9)$.
    (e) $y'(t) = -k\sqrt{y(t)} = -0.1$ when $t = 40$ minutes.

15. (a) $v'(t) = 1$ means that the object accelerates upward at 1 meter per second per second.
    (b) $h'(t) = -3$ means that the object falls at a constant rate of 3 meters per second.
    (c) $v'(t) = -0.01v^2(t)$, means that the acceleration is proportional, but opposite in direction, to the square of the velocity.

16. (a) The function $v(t) = t$ solves the IVP $v'(t) = 1$; $v(0) = 0$. At $t = 10$, the (upward) velocity is therefore $v(10) = 10$ meters per second.
    (b) The function $h(t) = -3t + 100$ solves the IVP $h'(t) = -3$; $h(0) = 100$. Thus $h(10) = 70$; i.e., the object has height 70 meters at $t = 10$ seconds.
    (c) It's a straightforward calculation to see that if $v(t) = 100/(t+C)$, then $v'(t) = -100/(t+C)^2 = -0.01v^2$, as desired.
    (d) The previous part shows that that the function $v(t) = 100/(t+20)$ solves the IVP $v' = -0.01v^2$; $v(0) = 5$. Thus $v(30) = 2$ and $v(80) = 1$. In physical language, the answers mean that after 30 and 80 seconds, respectively, the upward velocities are 2 meters per second and 1 meter per second. (At $t = 0$, the upward velocity was 5 meters per second.)

17. (a) $v' = -5$ (or $v'(t) = -5$).
    (b) $p' = 15$ (or $p'(t) = 15$).
    (c) $v' = kv$; k is negative because friction works in the direction *opposite* to the velocity.
    (d) $p' = k\sqrt{p}$.

18. (a) $P(t)$ is an increasing function, so $P'(t)$ must never be negative. The factor $M - P(t)$ causes the rate of learning to decrease as the value of the performance function approaches the maximum. Thus, the rate of learning is large when $M - P(t)$ is large and approaches zero as $P(t)$ approaches M. The value of P never exceeds M because $P' = 0$ when $P = M$.
    (b) If $P(t) = M - Ae^{-kt}$, then $P'(t) = Ake^{-kt} = kAe^{-kt} = k(M - P)$, as desired.
    (c) From the previous part, solutions are of the form $P(t) = M - Ae^{-0.05t}$. Since $P(0) = 0.1M$, it follows that $A = 0.9M$. Therefore, $P(t) = M - 0.9Me^{-0.05t}$.
    We want t such that $P(t) = 0.9M$, so we solve the equation $M - 0.9Me^{-0.05t} = 0.9M$ for t. The solution is $t = 20\ln 9 \approx 44$ hours.

19. The rate at which the influenza spreads is described by the differential equation $P' = kP(3000 - P)$ where k is a constant.

20. (a) Since the outside temperature is $-10°$ C and the coffee is initially $90°$ C, the temperature $y(t)$ of the coffee at time t is $y(t) = -10 + 100e^{kt}$. For the foam cup, $k = -0.05$, so the temperature reaches $70°$ C at time $t = -\ln(80/100)/0.05 \approx 4.46$ minutes after leaving the store. For the cardboard cup, $k = -0.08$, so the temperature reaches $70°$ C after $t = -\ln(80/100)/0.08 \approx 2.79$ minutes.
    After $t = 5$ minutes, the coffee in the foam cup is $-10 + 100e^{-0.25} \approx 68°$ C while the coffee in the paper cup is $-10 + 100e^{-0.4} \approx 57°$ C.
    (b) In the store, the temperature of the coffee at time t is $y(t) = 25 + 65e^{kt}$. Therefore, the coffee reaches $70°$ C after $t = \ln(45/65)/k$ minutes. This implies that Boris's coffee reaches $70°$ C in approximately 7.4 minutes and Natasha's coffee reaches this temperature in approximately 4.6 minutes.
    After $t = 5$ minutes, Boris's coffee is approximately $76°$ C and Natasha's coffee is approximately $69°$ C.

(c) If the coffee is to be 70° C after 5 minutes outdoors, k must be chosen so that $70 = -10 + 100e^{5k}$. This implies that $k \leq \ln(80/100)/5 \approx -0.0446$.

4.2 More Differential Equations: Modeling Growth

1. $y(x) = 100e^{0.1x}$

2. $y(x) = e^{-0.0001x}$

3. $y(x) = 2^x$

4. $y(x) = y(0)e^{(x \ln 2)/10} = y(0)2^{x/10}$

5. Mimicking the solution procedure in the text, $k = 1/90$ and $d = 9$. Thus, $P(t) = 10/\left(1 + 9e^{-t/9}\right)$.

    (a) $P(10) = 10/(1 + 9e^{-10/9}) \approx 2.52$ thousand flies; $P(100) = 10/(1 + 9e^{-100/9}) \approx 9.999$ thousand flies.

    (b) $P(t) = 5000$ when $t = 9\ln(9) \approx 19.78$ days.
    $P(t) = 9000$ when $t = 9\ln(81) \approx 39.55$ days.
    $P(t) = 9900$ when $t = 9\ln(891) \approx 61.13$ days.

    (c) The population is growing fastest when $P(t) = 5000$. This occurs when $t = 9\ln(9) \approx 19.78$ days.

6. (a) The DE $P' = 0.08P$ means the same thing as 8% interest, compounded continuously.

    (b) The IVP $P'(t) = 0.1P(t) - 100$ expresses the given policy; the second term represents the yearly fee.

    (c) The function $P(t) = 2000e^{0.08t}$ solves solves the IVP $P' = 0.08P$, $P(0) = 2000$. The value after 10 years is $P(10) \approx 4451.08$ dollars.

    (d) The function $P(t) = 1000 + 1000e^{0.08t}$ solves solves the IVP $P' = 0.1P - 100$, $P(0) = 1000$. The value after 10 years is $P(10) \approx 3718.28$ dollars.

7. Let $B(t)$ be the weight of The Blob at time t hours after noon on Wednesday. Since The Blob grows at a rate proportional to its size, its growth is described by the differential equation $B' = kB$ and, therefore, $B(t) = Ae^{kt}$ for some numbers A and k. Since $B(0) = 1$ and $B(4) = 4$, $A = 1$ and $k = \frac{1}{4}\ln 4$. Thus, The Blob will weigh 3×10^{15} at time $t = 4\ln\left(3 \times 10^{15}\right)/\ln 4 \approx 102.83$ hours (or, equivalently, in about 4.28 days).

8. Let $M(t)$ be the amount of mold present at time t. Since the mold grows at a rate proportional to the amount present, the function M satisfies a differential equation of the form $M' = kM$. Thus, $M(t) = M(0)e^{kt}$ where $M(0) = 2$ grams is the amount of mold present at time $t = 0$ and k is a constant to be determined.

    Since there are 5 grams of mold after 2 days, $M(2) = 5 = 2e^{2k}$. Thus, $5/2 = e^{2k}$ so $k = 0.5\ln 2.5 \approx 0.45815$. It follows that $M(t) = 2(5/2)^{t/2}$. Therefore, at the end of 8 days, the amount of mold is $M(8) = 2(5/2)^4 = 78.125$ grams.

9. (a) Since the bottle becomes full at $t = 24$ hours and the volume of the culture doubles every hour, the bottle must be half-full at time $t = 23$ hours.

    (b) Let $V(t)$ be the volume of the culture at time t. Then, since the bottle is full at time $t = 24$, the bottle is less than 1% full when $V(t) < 0.01V(24)$ or $V(t)/V(24) < 0.01$.

    The volume of the culture grows at a rate proportional to the amount present, so by Theorem 1, page 290, $V(t) = V(0)e^{kt}$. Since $V(1) = 2V(0)$, $k = \ln 2$ so $V(t) = V(0)2^t$. It follows that the bottle is less than 1% full when $2^{t-24} < 0.01$; that is, when $t < 24 + \log_2(0.01) \approx 17.356$ hours. Thus, the bottle is less than 1% full approximately 72% of the time.

10. Since the fungus grows at a rate proportional to its current weight, the amount of fungus $F(t)$ at time t may be found by solving the differential equation $F' = kF$. The general solution of this (separable) differential equation is $F(t) = Ae^{kt}$.

    To determine values for the constants A and k we use the data given in the problem. Let $t = 0$ correspond to 5:30 p.m. Then, $F(0) = 10 = Ae^{k \cdot 0}$ so $A = 10$. Similarly, if t is measured in hours, 6:15 p.m. corresponds to $t = 3/4$. Because the fungus weighed 12 g at 6:15 p.m., $F(3/4) = 12 = 10e^{3k/4}$. We can solve this equation for k using logarithms; the result is $k = 4\ln 1.2/3 \approx 0.2431$.

Evidently Dr. Howell was killed about the time the weight of the fungus reached 13 g. Thus, we need to find the time t that satisfies the equation $13 = 10e^{0.2431t}$. Using logarithms again, we find that $t \approx 1.08$. Since 1.08 hours is about 65 minutes and $t = 0$ is 5:30 p.m., we (and Sara Abrams) conclude that Dr. Howell was killed at about 6:35 p.m.

11. Let $R(t)$ denote the amount of oil remaining in the well at time t. Then $R'(t) = kR(t)$ is the rate at which oil is being pumped from the well at time t and $R(t) = R(0)e^{kt}$. The conditions $R(0) = 10^6$ and $R(6) = 5 \times 10^5$ imply that $k = -\frac{1}{6}\ln 2$.

    (a) Since $R' = kR$, $R'(6) = \left(-\frac{1}{2}\ln 2\right)\left(6 \times 10^5\right) = -10^5 \ln 2 \approx -69315$ barrels per year. Thus, oil is being pumped out at a rate of approximately 69,315 barrels per year.

    (b) $R(t) = 5 \times 10^4$ when $t = -6\ln 0.05 / \ln 2 \approx 25.932$ years.

12. Let $C(t)$ be the amount of C^{14} present in the skeletal fragments at time t (in years). For convenience, let t_{now} be the time the analysis was conducted, and $t = 0$ correspond to the date of death. At the time the analysis was performed, $C(t_{now})/C(0) = 0.0625$. Because the rate of decay of C^{14} is proportional to the amount of C^{14} present, $C'(t) = kC(t)$ for some constant k. The solution of this differential equation is $C(t) = C(0)e^{kt}$.

    The value of k can be determined from the information given about the half-life of C^{14}. Since the half-life of C^{14} is 5728 years, $0.5 = e^{5728k}$ implies that $k = -(\ln 2)/5728$. Therefore, $C(t_{now})/C(0) = 0.0625 = e^{-(\ln 2/5728)t_{now}}$. From this it follows that $t_{now} = -\ln(0.0625) \cdot 5728/\ln 2 \approx 22{,}912$. Thus, the fragments date from about 22,912 years ago.

13. (a) At time $t = 0$, the concentration of salt in the mixture is 0.1 lbs/gal. No salt is added to the mixture, so the amount of salt leaving the tank is proportional to the concentration of salt in the tank. Thus, $S(t)$, the amount of salt in the tank at time t, is the solution of the differential equation $S'(t) = -\frac{5}{100}S(t)$ where t is measured in minutes.

    (b) Since $S(0) = 10$, the solution of the differential equation from part (a) is $S(t) = S(0)e^{-0.05t} = 10e^{-0.05t}$.

    (c) After one hour, the amount of salt left in the tank will be $S(60) = 10e^{-0.05 \cdot 60} = 10e^{-3} \approx 0.49787$ pounds.

4.3 Linear and Quadratic Approximation; Taylor Polynomials

1. If $f(x) = \dfrac{1}{1-x}$, $n = 3$, and $x_0 = 0$, then $P_3(x) = 1 + x + x^2 + x^3$.

2. If $f(x) = \sin x + \cos x$, $n = 4$, $x_0 = 0$, then $P_4(x) = 1 + x - \dfrac{x^2}{2} - \dfrac{x^3}{6} + \dfrac{x^4}{24}$.

3. If $f(x) = \ln x$, $n = 3$, $x_0 = 1$, then $P_3(x) = x - 1 - \dfrac{(x-1)^2}{2} + \dfrac{(x-1)^3}{3}$.

4. If $f(x) = \tan x$, $n = 2$, $x_0 = 0$, then $P_2(x) = x$.

5. If $f(x) = \sqrt{x}$, $n = 3$, $x_0 = 4$, then $P_3(x) = 1 + \dfrac{x}{4} - \dfrac{(x-4)^2}{64} + \dfrac{(x-4)^3}{512}$.

6. $f(x) = \sin x$ has $\ell(x) = x$, $q(x) = x$.

7. $f(x) = \cos x$ has $\ell(x) = 1$; $[-0.1415, 0.1415]$
   $q(x) = 1 - x^2/2$; $[-0.7028, 0.7028]$

8. $f(x) = \tan x$ has $\ell(x) = x$, $q(x) = x$.

9. $f(x) = e^x$ has $\ell(x) = 1 + x$, $q(x) = 1 + x + x^2/2$.

10. $f(x) = \arctan x$ has $\ell(x) = x$, $q(x) = x$.

11. $f(x) = \arcsin x$ has $\ell(x) = x$, $q(x) = x$.

12. (a) The tangent line at $x = 3$ passes through (3, 5) and (0, 10), so it has slope $-5/3$; thus $g'(3) = -5/3$.
    (b) The linear approximation to g at $x = 3$ is $l(x) = 5 - 5/3(x - 3)$, so $l(2.95) = 5 - 5/3(-1/20) \approx 61/12$.

13. (b) $q(x) = (a - 64b + 4096c) + (b - 128c)x + cx^2$.
    (c) $q'(x) = (b - 128c) + 2cx$. Notice that this agrees with what one gets by differentiating the expression in the previous part.

14. (a) The graph of an odd function is symmetric about the origin. All graphs shown have that property.
    (b) The *even-order* Maclaurin polynomials P_2, P_4, P_6, and P_8 are the same as the odd-order Maclaurin polynomials P_1, P_3, P_5, and P_7. (This happens because the sine function is odd. So, therefore, are all of its Maclaurin polynomials.)

15. (a) Easy calculations show that $P_6(x) = 1 - \dfrac{x^2}{2} + \dfrac{x^4}{24} - \dfrac{x^6}{720}$. The lower-order polynomials can now be read off.
    (b) Both g and all the Maclaurin polynomials are even. The graphs "reflect" this by being symmetric about the y-axis.
    (c) The derivatives $P_1'(x)$, $P_3'(x)$, $P_5'(x)$, $P_7'(x)$ of the *sine* polynomials are, respectively, the Maclaurin polynomials P_0, P_2, P_4, and P_6 for $g(x) = \cos x$. This is no great surprise, since $f'(x) = g(x)$.

16. (a) We'll use the linear approximation $l(x)$ at $x = 1$. Since $f(1) = 0$ and $f'(1) = \sin 1 \approx 0.84147$, $l(x) = 0 + \sin 1(x - 1)$; $l(0.5) = 0 + \sin 1(-0.5) \approx 0.42074$.
    (b) Whether the estimate above is too big or too small depends on the concavity of f between $x = 0.5$ and $x = 1$. Notice that $f''(x) = 2x \cos(x^2)$; thus $f''(x) > 0$ for x in $[-0.5, 1]$; so f is concave *up*, and so the linear approximation *underestimates* f.
    (c) $f''(1) = 2\cos 1 \approx 1.08060$; therefore the quadratic approximation at $x = 1$ has the form $q(x) = f(1) + f'(1)(x - 1) + \dfrac{f''(1)}{2}(x - 1)^2 = 0 + \sin 1(x - 1) + \cos 1(x - 1)^2$. Therefore $q(0.5) \approx -0.28566$.

17. (a) We know that $g(5) = 2$; the graph shows that $g'(5) = 1$. Thus the linear approximation to g at $x = 5$ is $\ell(x) = 2 + 1(x - 5)$; it gives the (crude) estimate $g(0) \approx \ell(0) = -3$.

    (b) We know that $g(5) = 2$; the graph shows that $g'(5) = 1$. The slope of the g' graph at $x = 5$ is about -2, so $g''(5) \approx -2$. Thus the quadratic approximation to g at $x = 5$ is $q(x) = 2 + (x - 5) - (x - 5)^2$; it gives the (crude) estimate $g(8) \approx q(8) = -4$.

18. The linear approximation to h at $x = 2$ is $\ell(x) = 3 - 2(x - 2)$. Therefore $\ell(3) = 1$. How close is this estimate to $h(3)$?

    To answer, we use the error bound inequality for linear approximations. Since $-2 \leq h''(x) \leq 1$, we have $|f''(x)| \leq 2$, i.e., $K = 2$. With this K and $x = 3$, the formula says that $|f(3) - \ell(3)| \leq K(3 - 2)^2/2 = 1$, i.e., $|f(3) - 1| \leq 1$. This means that $0 \leq f(3) \leq 2$, as desired.

19. (a) $\ell_p(t) = 25t$; $\ell_p(1) = 25$; $\ell_p(-1) = -25$

    (b) $q_p(t) = 25t + t^2$; $q_p(1) = 26$; $q_p(-1) = -24$

    (c) $\ell_v(t) = 25 + 2t$; $\ell_v(1) = 27$

    (d) $\pm 3/2$ meters per second

20. (a) $\ell_p(t) = 25t$ meters; $\ell_p(1) = 25$ meters; $\ell_p(-1) = -25$ meters

    (b) $q_p(t) = 25t - t^2$ meters; $q_p(1) = 24$ meters; $q_p(-1) = -26$ meters

    (c) $\ell_v(t) = 25 - 2t$ meters per second; $\ell_v(1) = 23$ meters per second

    (d) $\pm 3/2$ meters per second

21. (a) $\ell(t) = 100$ meters; $\ell(1) = 100$ meters

    (b) $q(t) = 100 - 4.9t^2$ meters; $q(1) = 95.1$ meters

22. (a) $q(x) = 1 - x^2/2$

    (b) $|f(x) - q(x)| \leq 1/6$

    (c) The actual approximation error is less than that allowed by the error bound formula.

23. (a) $q(x) = (x - 1) - \frac{1}{2}(x - 1)^2$

    (b) $f'''(x) = 2/x^3$ so $16/27 \leq f'''(x) \leq 16$ if $1/2 \leq x \leq 3/2$. Therefore,

    $$|f(x) - q(x)| \leq \frac{16}{6}(x - 1)^3 \leq \frac{16}{6}\left(\frac{1}{2}\right)^3 = \frac{1}{3}.$$

    (c) The actual approximation error is less than that allowed by the error bound formula.

24. (a) To estimate $\sqrt{103}$, use the linear approximation to $f(x) = \sqrt{x}$ at $x = 100$. Since $f'(x) = 1/(2\sqrt{x})$, we have $f'(100) = 1/20$. Therefore the linear approximation at $x = 100$ is

    $$\ell(x) = f(100) + f'(100)(x - 100) = 10 + \frac{x - 100}{20}.$$

    Thus $\ell(103) = 10 + 3/20 = 10.15$. A calculator gives $\sqrt{103} \approx 10.14889$; the difference is $10.15 - 10.14889 \approx 0.00111$.

    (b) Let $f(x) = \sqrt[3]{x}$; consider the linear approximation at $x = 27$. Arguing as in the previous exercise gives $\ell(x) = 3 + \frac{1}{27}(x - 27)$. Thus $f(29) \approx \ell(29) = 3 + 2/27 \approx 3.07407$. A calculator gives $\sqrt[3]{29} \approx 3.07232$; the difference is $3.07407 - 3.07232 \approx 0.00176$.

    (c) Let $f(x) = \tan x$; consider the linear approximation at $x = \pi/6$. Then $f'(\pi/6) = \sec^2(\pi/6) = 4/3$. Arguing as above gives $\ell(x) = 1/\sqrt{3} + \frac{4}{3}(x - \pi/6)$. Notice that 31 degrees ≈ 0.54105 radians $\approx \pi/6 + 0.01745$ radians. Thus $f(0.54105) \approx \ell(0.54105) = 1/\sqrt{3} + \frac{4}{3} \cdot 0.01745 \approx 0.60062$. A calculator gives $\tan 0.54105 \approx 0.60086$; the difference is $0.60086 - 0.60062 \approx 0.00024$.

SECTION 4.3 LINEAR AND QUADRATIC APPROXIMATION; TAYLOR POLYNOMIALS 187

(d) Let $f(x) = x^{10}$; consider the linear approximation at $x = 1$. Then $f'(1) = 10$. Arguing as above gives $\ell(x) = 1 + 10(x - 1)$. Thus $f(0.8) \approx \ell(0.8) = 1 + 10(-0.2) = -1$.

A calculator gives $0.8^{10} \approx 0.10737$. The difference between the estimate and the exact value is large, this time: $0.10737 - (-1) = 1.10737$. (The estimate is poor because $f(x)$ is poorly approximated by its tangent line near $x = 1$.)

25. (a) If $f(x) = \sqrt{x}$, then $f'(x) = x^{-1/2}/2$; $f''(x) = -x^{-3/2}/4$. On the interval $[100, 103]$, $|f''(x)| \leq 1/(4 \cdot 100^{3/2}) \approx 0.00025$. Hence $K = 0.00025$ works, and the theoretical error bound is $\frac{K}{2}(103 - 100)^2 = 0.000125 \cdot 9 = 0.001125$. The actual error in $\ell(x)$, computed above, was 0.00111.

(b) If $f(x) = x^{1/3}$, then $f'(x) = x^{-2/3}/3$; $f''(x) = -2/(9x^{5/3})$. On the interval $[27, 29]$, $|f''(x)| \leq 2/(9 \cdot 27^{5/3}) \approx 0.001$. Thus $K = 0.001$ works, and the theoretical error bound is $\frac{K}{2}(29 - 27)^2 = 0.0005 \cdot 4 = 0.002$. The actual error in $\ell(x)$, computed above, was 0.00176, a bit less than the bound above permits.

(c) If $f(x) = \tan x$, then $f'(x) = \sec^2 x$; $f''(x) = 2\sec^2 x \tan x$. On the interval $[30\pi/180, 31\pi/180]$, $|f''(x)| \leq 1.64$ (plot f'' to see why). Thus $K = 1.64$ works, and the theoretical error bound is $\frac{K}{2}(\pi/180)^2 = \approx 0.00025$. The actual error in $\ell(x)$, computed above, was 0.00024, a little less than the bound above permits.

(d) If $f(x) = x^{10}$, then $f''(x) = 90x^8$. On the interval $[0.8, 1]$, $|f''(x)| \leq 90$. Thus $K = 90$ works, and the theoretical error bound is $\frac{K}{2}(0.8 - 1)^2 = 45(0.2)^2 = 1.8$. The actual error in $\ell(x)$, computed above, was 1.10737, less than the bound above permits.

26. (a) $f(x) = f'(x) = f''(x) = e^x$, so $f(0) = f'(0) = f''(0) = 1$. Therefore, $\ell(x) = 1 + x$, $q(x) = 1 + x + x^2/2$, $E_1(x) = e^x - \ell(x) = e^x - 1 - x$, and $E_2(x) = e^x - q(x) = e^x - 1 - x - x^2/2$.

(c) $|E_2(x)| \leq |E_1(x)|$ if $-1 \leq x \leq 1$

(d) The graph of E_1 resembles a quadratic curve and the graph of E_2 resembles a cubic curve.

(e) No, they just look that way. The first nonzero term in the Taylor series for $E_1(x)$ is of degree 2 so its graph appears quadratic when x is near zero. Similarly, the first nonzero term in the Taylor series for $E_2(x)$ is of degree 3 so its graph appears cubic when x is near zero.

27. (a) $f(x) = \ln x$, $f'(x) = 1/x$, and $f''(x) = -1/x^2$ so $f(1) = 0$, $f'(1) = 1$, and $f''(1) = -1$. Therefore, $\ell(x) = x - 1$, $q(x) = (x - 1) - (x - 1)^2/2$, $E_1(x) = \ln x - (x - 1) = \ln x - x + 1$, and $E_2(x) = \ln x - (x - 1) + (x - 1)^2/2 = \ln x + 3/2 - 2x + x^2/2$.

(c) $|E_2(x)| \leq |E_1(x)|$ if $0 \leq x \leq 2$

(d) The graph of E_1 resembles a quadratic curve and the graph of E_2 resembles a cubic curve.

(e) No, they just look that way. If $x_0 = 1$, the first nonzero term in the Taylor series for $E_1(x)$ is of degree 2 so its graph appears quadratic when x is near one. Similarly, if $x_0 = 1$, the first nonzero term in the Taylor series for $E_2(x)$ is of degree 3 so its graph appears cubic when x is near one.

28. (a) $f(x) = x^5$, $f'(x) = 5x^4$, and $f''(x) = 20x^3$ so $f(1) = 1$, $f'(1) = 5$, and $f''(1) = 20$. Therefore, $\ell(x) = 1 + 5(x - 1) = 5x - 4$, $q(x) = 1 + 5(x - 1) + 10(x - 1)^2 = 6 - 15x + 10x^2$, $E_1(x) = x^5 - \ell(x) = x^5 - 5x - 4$, $E_2(x) = x^5 - q(x) = x^5 - 10x^2 + 15x - 6$.

(c) $|E_2(x)| \leq |E_1(x)|$ if $0.28 \leq x \leq 2$

(d) The graph of E_1 resembles a quadratic curve and the graph of E_2 resembles a cubic curve.

(e) No, they just look that way. If $x_0 = 1$, the first nonzero term in the Taylor series for $E_1(x)$ is of degree 2 so its graph appears quadratic when x is near one. Similarly, if $x_0 = 1$, the first nonzero term in the Taylor series for $E_2(x)$ is of degree 3 so its graph appears cubic when x is near one.

4.9 Parametric equations, parametric curves

1. The curve is the upper unit semi-circle plotted from $(-1, 0)$ to $(0, 1)$ to $(1, 0)$.

2. The curve is the lower unit semi-circle plotted from $(-1, 0)$ to $(0, -1)$ to $(1, 0)$.

3. The curve is the right unit semi-circle plotted from $(0, -1)$ to $(1, 0)$ to $(0, 1)$.

4. The curve is the left unit semi-circle plotted from $(0, -1)$ to $(-1, 0)$ to $(0, 1)$.

5. The curve is the unit circle plotted clockwise from $(0, -1)$ to $(0, 1)$ to $(0, -1)$.

6. (a) $(2, 1)$, $(5, 3)$, $y = \frac{2}{3}(x - 2) + 1$
   (b) $(5, 3)$, $(2, 1)$, $y = \frac{2}{3}(x - 2) + 1$
   (c) $(0, b)$, $(1, m + b)$, $y = mx + b$
   (d) (a, c), $(a + b, c + d)$, $y = \frac{d}{b}(x - a) + c$
   (e) (x_0, y_0), (x_1, y_1), $y = \frac{y_1 - y_0}{x_1 - x_0}(x - x_0) + y_0$

7. In each case the idea is to calculate $\sqrt{f'(t)^2 + g'(t)^2}$; if the result is constant, then the curve has constant speed. Among the given choices only the last—$x = \sin(\pi t)$, $y = \cos(\pi t)$—has constant speed.

8. Just calculate: $\sqrt{f'(t)^2 + g'(t)^2} = \sqrt{a^2 + c^2}$, a constant.

9. (a) The spacing of bullets suggests that P moves quickly at $t = 3$, $t = 4$, $t = 9$, and $t = 10$, and slowly at $t = 0$, $t = 1$, $t = 6$, and $t = 7$.
   (b) The distance along the curve from $t = 2.5$ to $t = 3.5$ seems to be about 3 units. Thus P appears to travel about 3 units per second at $t = 3$.
   (c) Use the curve to estimate the speed of P at $t = 6$. The distance along the curve from $t = 5.5$ to $t = 6.5$ seems to be about 1 unit. Thus P appears to travel about 1 unit per second at $t = 6$.

10. The result is the sine curve $y = \sin x$ from $x = -8$ to $x = 8$. This happens because
$$x = t^3; \quad y = \sin t^3 \implies y = \sin x,$$
and
$$-2 \le t \le 2 \implies -8 \le t^3 = x \le 8.$$

11. (a) The result is the circle of radius 2, centered at $(2, 1)$.
    (b) Here's the calculation: Since $x = a + r \cos t$ and $y = b + r \sin t$,
    $$(x - a)^2 + (y - b)^2 = r^2(\cos t)^2 + r^2(\sin t)^2 = r^2.$$
    (c) Setting $x = 2 + \sqrt{13} \cos t$, $y = 3 + \sqrt{13} \sin t$, and $0 \le t \le 2\pi$, gives the circle of radius $\sqrt{13}$, centered at $(2, 3)$.
    (d) No proper "curve" results: for all t, (x, y) stays put at $(2, 3)$.

12. (a) The curve is an ellipse, centered at the origin, 4 units wide and 2 units tall. The quantities $2a$ and $2b$ measure the ellipse's width and height.
    (b) The same curve results if $0 \le t \le 4\pi$, but it's traversed twice, because the coordinate functions are 2π-periodic.

(c) Letting $x = 5\cos t$, $y = 3\cos t$, $0 \leq t \leq 2\pi$ works.

(d) Letting $x = 3\cos t$, $y = 5\cos t$, $0 \leq t \leq 2\pi$ works.

(e) Replacing $x = a\cos t$ and $y = b\sin t$ in the equation $\dfrac{x^2}{a^2} + \dfrac{y^2}{b^2} = 1$ shows that the equation holds as advertised.

(f) The "ellipse" looks like a circle if $a = b$. Its xy equation is $x^2 + y^2 = a^2$.

(g) An ellipse with $a = 1000$ and $b = 1$ looks like a very thin pancake—2000 units wide and 2 units tall.

(h) An ellipse with $b = 1000$ and $a = 1$ looks like a very thin pancake standing on edge—2000 units tall and 2 units wide.

13. (a) The origin corresponds to $t = 0$; $P(0.1) \approx (0.48, 0.56)$; $P(\pi/2) = (1, 0)$. Thus P starts at the origin and starts off in a northeasterly direction.

(b) Both x and y are 0 if and only if both $5t$ and $6t$ are integer multiples of π. This occurs only for $t = 0$, $t = \pi$, and $t = 2\pi$.

(c) Using the t-interval $0 \leq t \leq 4\pi$ would produce exactly the same curve, but it would be traversed twice.

14. (a) The curve is the line segment connecting $(0, 4)$ and $(2, 7)$.

(b) $y = \frac{3}{2}x + 4$

(c) The slope is $3/2$. The slope doesn't depend on t because the curve is a line segment—it has constant slope.

15. (a) The curve starts at $(at_0 + b, ct_0 + d)$ and ends at $(at_1 + b, ct_1 + d)$.

(b) $y = \dfrac{c}{a}(x - b) + d$

(c) $x = \dfrac{a}{c}(y - d) + b$

(d) If $a = c = 0$, the parametric curve is just the point (b, d).

16. (a) The curve C is in this case simply the point $(0, 0)$—the curve "goes" nowhere. Thus C doesn't really deserve to have a slope.

(b) The curve defined for $-1 \leq t \leq 1$ by $x = t^3$ and $y = t^3$ is really the line segment $y = x$ from $x = -1$ to $x = 1$—this certainly deserves to have slope 0 at $t = 0$. In this case, however, the theorem doesn't apply, because $f'(0) = 0$.

(c) The curve defined for $-1 \leq t \leq 1$ by $x = t$ and $y = t$ is also the line $y = x$ from $x = -1$ to $x = 1$ (as in the previous part). This time the theorem does apply: the curve C has slope 1 at $t = 0$.

17. (a) The model would be more realistic if it took wind resistance into account. To do so, one would need some mathematical information about wind resistance.

(b) Imitate the argument given for $f(t)$. Notice, too, that if $g(t) = 7 - 16t^2$, then $g'' = -32$, $g(0) = 7$, and $g'(0) = 0$, just as claimed.

(c) By definition, $s(t) = \sqrt{f'(t)^2 + g'(t)^2} = \sqrt{150^2 + (-32t)^2} = \sqrt{22500 - 1024t}$. Plotting this function over the interval $0 \leq t \leq 0.661$ (when the ball hits the ground) gives almost a horizontal line—the velocity changes very little over the short time interval.

18. The pitch is *not* a strike. $x(t) = 60.5$ when $t = 0.605$; $y(0.605) = 1.14$.

The ball crosses the plate at ≈ 101.86 ft/sec.

19. (a) If $x = f(t) = s_0 t$ and $y = g(t) = 7 - 16t^2$ it's easy to check directly that $f''(t) = 0$, $f'(0) = s_0$, $f(0) = 0$, $g''(t) = -32$, $g'(0) = 0$, and $g(0) = 7$. These are the necessary conditions.

(b) The ball reaches home plate when $f(t) = s_0 t = 60.5$, i.e., at $t = 60.5/s_0$ seconds.

(c) The trajectory is parabolic for any $s_0 > 0$. (If $s_0 = 0$, the ball drops straight down.) This can be seen by eliminating t. Since $x = s_0 t$, $t = x/s_0$, so $y = 7 - 16t^2 = 7 - 16x^2/s_0^2$. This is the equation of a parabola in the xy-plane.

20. (a) $x(t) = 60.5$ at time $t = 20\left(e^{363/2000} - 1\right)/9 \approx 0.44225$. Thus, the air-dragged ball takes about 0.0389 seconds longer to reach the plate.

    (b) $y(t) \approx 3.8706$ feet at the time when $x(t) = 60.5$

    (c) The ball's speed at time t is $s(t) = \sqrt{(x'(t))^2 + (y'(t))^2}$. Thus, when $x = 60.5$, the ball's speed is approximately 125.9 ft/sec.

    (d) $y = 0$ at time $t = \sqrt{7}/4 \approx 0.66144$. At this time $x \approx 86.851$ feet.

21. Now, $x = 200\ln(3t/4 + 1)$.

    (a) $x(t) = 60.5$ at time $t = 4\left(e^{121/400} - 1\right)/3 \approx 0.47098$. Thus, the air-dragged ball takes about 0.0677 seconds longer to reach the plate.

    (b) $y(t) \approx 3.4508$ feet at the time when $x(t) = 60.5$

    (c) When $x = 60.5$, the ball's speed is approximately 111.87 ft/sec.

    (d) When $y = 0$, $x \approx 80.569$ feet.